Design and Optimization of Mechanical Engineering Products

K. Kumar
Birla Institute of Technology, India

J. Paulo Davim
University of Aveiro, Portugal

A volume in the Advances in Mechatronics and Mechanical Engineering (AMME) Book Series

IGI Global
DISSEMINATOR OF KNOWLEDGE

Published in the United States of America by
IGI Global
Engineering Science Reference (an imprint of IGI Global)
701 E. Chocolate Avenue
Hershey PA, USA 17033
Tel: 717-533-8845
Fax: 717-533-8661
E-mail: cust@igi-global.com
Web site: http://www.igi-global.com

Library of Congress Cataloging-in-Publication Data

Names: Kumar, K. (Kaushik), 1968- editor. | Davim, J. Paulo, editor.
Title: Design and optimization of mechanical engineering products / K. Kumar and J. Paulo Davim, editors.
Description: Hershey, PA : Engineering Science Reference, [2018] | Includes bibliographical references.
Identifiers: LCCN 2017016456| ISBN 9781522534013 (hardcover) | ISBN 9781522534020 (ebook)
Subjects: LCSH: Machine design.
Classification: LCC TJ230 .D427 2018 | DDC 620/.0042--dc23 LC record available at https://lccn.loc.gov/2017016456

British Cataloguing in Publication Data
A Cataloguing in Publication record for this book is available from the British Library.

Advances in Mechatronics and Mechanical Engineering (AMME) Book Series

J. Paulo Davim
University of Aveiro, Portugal

ISSN:2328-8205
EISSN:2328-823X

Mission

With its aid in the creation of smartphones, cars, medical imaging devices, and manufacturing tools, the mechatronics engineering field is in high demand. Mechatronics aims to combine the principles of mechanical, computer, and electrical engineering together to bridge the gap of communication between the different disciplines.

The **Advances in Mechatronics and Mechanical Engineering (AMME) Book Series** provides innovative research and practical developments in the field of mechatronics and mechanical engineering. This series covers a wide variety of application areas in electrical engineering, mechanical engineering, computer and software engineering; essential for academics, practitioners, researchers, and industry leaders.

Coverage

- Control Methodologies
- Intelligent Navigation
- Biologically Inspired Robotics
- Manufacturing Methodologies
- Sustainable and green manufacturing
- Control Systems Modelling and Analysis
- Bioengineering Materials
- Computer-Based Manufacturing
- Intelligent Sensing
- Tribology and surface engineering

IGI Global is currently accepting manuscripts for publication within this series. To submit a proposal for a volume in this series, please contact our Acquisition Editors at Acquisitions@igi-global.com or visit: http://www.igi-global.com/publish/.

The Advances in Mechatronics and Mechanical Engineering (AMME) Book Series (ISSN 2328-8205) is published by IGI Global, 701 E. Chocolate Avenue, Hershey, PA 17033-1240, USA, www.igi-global.com. This series is composed of titles available for purchase individually; each title is edited to be contextually exclusive from any other title within the series. For pricing and ordering information please visit http://www.igi-global.com/book-series/advances-mechatronics-mechanical-engineering/73808. Postmaster: Send all address changes to above address.

Titles in this Series

For a list of additional titles in this series, please visit: www.igi-global.com/book-series

Numerical and Analytical Solutions for Solving Nonlinear Equations in Heat Transfer
Davood Domiri Ganji (Babol Noshirvani University of Technology, Iran) and Roghayeh Abbasi Talarposhti (University of Mazandaran, Iran)
Engineering Science Reference • copyright 2018 • 275pp • H/C (ISBN: 9781522527138) • US $195.00 (our price)

Nanofluid Technologies and Thermal Convection Techniques
Ramesh Chand (Government Arya Degree College Nurpur (HP), India)
Engineering Science Reference • copyright 2017 • 229pp • H/C (ISBN: 9781683180067) • US $185.00 (our price)

Mathematical Concepts and Applications in Mechanical Engineering and Mechatronics
Mangey Ram (Graphic Era University, India) and J. Paulo Davim (University of Aveiro, Portugal)
Engineering Science Reference • copyright 2017 • 495pp • H/C (ISBN: 9781522516392) • US $215.00 (our price)

Handbook of Research on Advances and Applications in Refrigeration Systems and Technologies
Pedro Dinis Gaspar (University of Beira Interior, Portugal) and Pedro Dinho da Silva (University of Beira Interior, Portugal)
Engineering Science Reference • copyright 2015 • 924pp • H/C (ISBN: 9781466683983) • US $545.00 (our price)

Engineering Creative Design in Robotics and Mechatronics
Maki K. Habib (The American University in Cairo, Egypt) and J. Paulo Davim (University of Aveiro, Portugal)
Engineering Science Reference • copyright 2013 • 366pp • H/C (ISBN: 9781466642256) • US $195.00 (our price)

Computational Methods for Optimizing Manufacturing Technology Models and Techniques
J. Paulo Davim (University of Aveiro, Portugal)
Engineering Science Reference • copyright 2012 • 395pp • H/C (ISBN: 9781466601284) • US $195.00 (our price)

Multi-Agent Systems for Traffic and Transportation Engineering
Ana Bazzan (Instituto de Informática, Universidade Federal do Rio Grande do Sul, Brazil) and Franziska Klügl (Örebro University, Sweden)
Information Science Reference • copyright 2009 • 446pp • H/C (ISBN: 9781605662268) • US $195.00 (our price)

IGI Global
DISSEMINATOR OF KNOWLEDGE

701 East Chocolate Avenue, Hershey, PA 17033, USA
Tel: 717-533-8845 x100 • Fax: 717-533-8661
E-Mail: cust@igi-global.com • www.igi-global.com

Table of Contents

Detailed Table of Contents

Globally, new products are launched daily. They provide answers to common or specialised problems, enrich lifestyles, provide alternatives to old solutions, amuse us etc. Companies that develop, design, manufacture, market and sell these products seek commercial compensation in the short, medium or long term "success", however you measure it. Subsequently, New Product Development is a major issue for most companies as they seek to reduce time to market, reduce the development cycle, access new technologies and develop more and better products and services. New products that can successfully compete in local, national and global markets are a key concern for the majority of companies, so successful NPD is fundamental to both stimulating and supporting economic growth. It is a subject, which has received and continues to receive much attention, particularly in seeking to improve its effectiveness and efficiency. This chapter reviews the NPD process and considers the variables associated with the different approaches, which may be needed when developing a new product.

Hand layup method is normally used for manufacturing composite parts with large and complex geometry where as Resin Transfer Moulding (RTM) process is a better substitute, but is sparingly used due to lack of proper manufacturing technology. Developments of RTM manufacturing process always requires a proper mould design. In addition, difficulty in the tooling design and mould fabrication cost increases with increase in size and complexity of the component. The scale down strategy of full scale product avoids bigger size mould requirements, prototype bulk production for product testing and quality check at the starting phase of product development. Moreover, the mould scale down strategy can be used to validate the process and the product with less capital input. In this work, a methodology to develop a RTM technology for manufacturing a complex and large composite structure utilizing the process simulation were demonstrated using high speed cab front as selected component.

Mixed flow pumps have got a wide range of application such as that in irrigation, flood control, dewatering and power station cooling systems. A number of design templates have been in use to design the mixed flow pump impeller blades including free vortex theory and mean stream line theory. Here blade of the impeller of the mixed flow pump has been designed using Mean stream line theory and the free vortex theory and stresses arising due to static loading conditions has been obtained for both the cases. Comparison analysis for the equivalent stresses developed due to structural loading using various materials for the blade was undertaken. It was found that the stress for the blade designed using mean stream line theory is always on the higher side in comparison to that designed using free vortex theory. Further, out of the four materials of construction, titanium and copper were found to be suitable for blade designed using free vortex theory and the mean stream line theory respectively.

Seismic base isolation is one of the most widely implemented and accepted seismic protection systems and is a relatively recent and evolving technology. The most common isolation system used is Laminated Lead Rubber Bearings (LLRB). They combine the function of isolation and energy dissipation in a single compact unit, giving structural support, horizontal flexibility, damping, and a re-centering force in a single unit. The force deformation behavior of LLRB is modeled as bilinear system with viscous damping. In this paper, a comprehensive design of LLRB is presented. Accurate evaluation of the structural properties and precise modeling of isolation devices are of utmost importance in predicting the response of the structure during the earthquakes. Base isolation has now been used in numerous buildings in countries like Italy, Japan, New Zealand, and USA. Base isolation is also useful for retrofitting of important buildings (like hospitals and historic buildings).

Reverse engineering has become the one of the most relevant concepts in modern design doctrines. Advances in technology demand shorter lead time in the overall product development stage, especially in the automobile sector. Hence as a study in reverse engineering, the author has reverse engineered a

Friction-stir processing (FSP) is a property enhancement technique which, not only removes the defects of initial casting process, but also improves the microstructure of the metals and metal matrix composites (MMCs). The process is based on frictional heating which results in a considerable dynamic plastic deformation within the metals. FSP can be specifically applied to develop fine-grained microstructures throughout the thickness of metal surface, to impart super plasticity and ensure homogeneous distribution of reinforced particles, if any. This chapter is a dedicated effort to consolidate the latest developments contributed by different researchers in last few years. The work covers various components and parameters, selected and used, for FSP to obtain specific desired results. Also, it includes past researches to exhibit various changes in mechanical properties with a keen focus on morphological study (by scanning electron microscopy) of these MMCs. In the last, a brief discussion on application and future scope of FSP processed MMC materials, is presented.

Nowadays, the research and engineering attention stimulated towards development of environmentally gentle materials to satisfy the energy needs of the society through renewable resources. The growing cognizance in the production and consumption of renewable energy for civilization necessities has directed towards the growth in the wind energy utilization. The wind energy is a leading renewable and sustainable energy resource and key answer to the global energy problem. The rotor blades of wind turbines are its integral parts and traditional materials used for blade manufacturing are carbon or glass fibre reinforced polymer composites owing to their low density and high strength to stiffness ratio. But the non renewability and adverse environmental effect during their processing and disposal forced the researchers to look out for some biodegradable and light weight natural plant fibres for reinforcement in polymeric resin to produce required polymer composites. In the present work, application of bio filler based epoxy composite is proposed to be used as wind turbine rotor blades.

The research work aims at investigating the performance of functionally graded annular fin. The work involves computation of efficiency and effectiveness of such fins and compares the fin performances for different geometry and grading parameters. It is also assumed that the temperature gradient only exists in one direction i.e. along the radius. A general second order governing differential equation has been derived for all the profiles and grading. The thickness of fin is considered to be a power function of radial co-ordinate. The functional grading of thermal conductivity is assumed to be a power function of radial co-ordinate and also consists of parameters, namely grading parameters, varying which different grading combinations can be carried out. The efficiency and effectiveness of the annular fin of different geometry and grading combinations have been calculated and plotted. The performance analysis reveals the dependence of thermal behaviour of annular fins on geometry and grading parameter.

Section 3
Product and Process Optimization

Chapter 11
Optimizing Material Removal Rate Using Artificial Neural Network for Micro-EDM..................... 209
Ananya Upadhyay, Birla Institute of Technology Mesra, India
Ved Prakash, Central Mechanical Engineering Research Institute (CSIR), India
Vinay Sharma, Birla Institute of Technology Mesra, India

Machining can be classified into conventional and unconventional processes. Unconventional Machining Process attracts researchers as it has many processes whose physics is still not that clear and they are highly in market-demand. To predict and understand the physics behind these processes soft computing is being used. Soft computing is an approach of computing which is based on the way a human brain learns and get trained to deal with different situations. Scope of this chapter is limited to one of the soft computing optimizing techniques that is artificial neural network (ANN) and to one of the unconventional machining processes, electrical discharge machining process. This chapter discusses about micromachining on Electric Discharge Machining, its working principle and problems associated with it. Solution to those problems is suggested with the addition of powder in dielectric fluid. The optimization of Material Removal Rate (MRR) is done with the help of ANN toolbox in MATLAB.

Chapter 12
Multi-Criteria Decision Making for Optimization of Product Development Under Green Manufacturing Environment.. 234
Sumit Bhowmik, National Institute of Technology, India
Jagadish, National Institute of Technology, India

At present, increasing environmental awareness and stringent environmental regulations of the competent authority, the use of green manufacturing (GM) technique is an important topic for the present industries. To overcome the environmental issues, the manufacturing industries use various advanced manufacturing processes for optimal product development. The advance manufacturing process generates large amounts of toxic substances results in various environmental issues during the optimal product manufacturing. Minimization of environmental issues and the amount of waste generated are strongly depends on its process and response parameters. Thus, optimization of process parameters for GM is essential and is a multi-criteria decision making (MCDM) optimization problems. This chapter provides an overview of applications of some MCDM methods for optimization followed by detailed fundamental aspects of optimization issues in green manufacturing. The work proposed an integrated method consisting of AHP coupled with MOORA and validated through an experimental case study.

The present chapter aims to determine optimal tribo-testing condition for minimum coefficient of friction and wear depth of electroless Ni-P, Ni-P-W and Ni-P-Cu coatings under lubrication using grey relational analysis. Electroless Ni-P, Ni-P-W and Ni-P-Cu coatings are deposited on AISI 1040 steel substrates. They are heat treated at suitable temperatures to improve their hardness. Coating characterization is done using scanning electron microscope, energy dispersive X-Ray analysis and X-Ray diffraction techniques. Typical nodulated surface morphology is observed in the scanning electron micrographs of all the three coatings. Phase transformation on heat treating the deposits is captured through the use of X-Ray diffraction technique. Vicker's microhardness of the coatings in their as-deposited and heat treated condition is determined. Ni-P-W coatings are seen to exhibit the highest microhardness. Friction and wear tests under lubricated condition are carried out following Taguchi's experimental design principle. Finally, the predominating wear mechanism of the coatings is discussed.

Inconel 625 is a nickel-chromium based super-alloy which is mostly used in high end applications due to its excellent chemical and fabrication properties combined with high strength and outstanding corrosion resistance. Wire EDM is most commonly the used process for machining this material and gives high accuracy and precision in the machined parts. In this paper, an attempt is made to undertake an experimental investigation to optimize the input variables like pulse on time, pulse off time, peak current, and servo voltage in Wire EDM for achieving simultaneous optimization of cutting rate, surface roughness, dimensional deviation, and wire wear ratio. Taguchi's L9 orthogonal array with one replication has been used as an experimental design and then analysis of variance (ANOVA) has been applied to determine the significant parameters and their contributions on response variables. Taguchi method has been combined with grey relational analysis for multi-characteristic optimization. Finally, results are validated with the help of confirmation experiments.

Preface

Product is something that may be offered to a marketplace for attention, acquisition, use, or consumption which may satisfy a wish or want. In business and engineering, development is the complete method of delivery of a brand new or changed product to promote to rework a market chance into a product on the market purchasable. Products are often classified as tangible or intangible. A tangible product could be an object which will be perceived by touch or vision like a machine, instrument, equipment, gadget, vehicle etc. whereas an intangible product could be a product which will solely be perceived indirectly like an insurance, software etc. A decent understanding of client desires and needs, the competitive atmosphere and also the nature of the market represent the highest needed factors for the success of a launched product for an increased existence.

Cost, time, and quality are the most important variables that drive the client wants. Aimed toward these three variables, firms continuous develop / improve practices and techniques to satisfy the client necessities and reciprocally increase their market share and profit. There are several uncertainties and challenges throughout the method which are the main concerns for the management and usage of best practices eliminates hurdles or barriers. Most organizations realize that all products have a limited lifespan, and so products need to be developed and modified continuously by the company to stay in business. Designing and optimizing a new product is typically a huge part of any manufacturing process and in long run extends product life span.

The main objective of the book is product development and target audience are all Academics Students, Researchers and Industry Practitioners, Engineers, Research Scientists/ Academicians involved in design of Mechanical Engineering products.

The chapters in the book has been categorized in three sections, namely Section 1: Design and Development of New Component/Product and Usage of Alternative Design Methodology for Betterment; Section 2: Development of New or Alternative Material for an Existing Component/Product; and Section 3: Product and Process Optimization.

Section 1 contains Chapter 1 to Chapter 6, whereas Section 2 has Chapter 7 to Chapter 10, and Section 3 with Chapters 11 to 14

Section 1 starts with Chapter 1 which introduces the readers to different approaches to New Product Development and the various variables associated to them. It is a fact that new products are launched daily across the world providing answers to common or specialized problems, enrich lifestyles, provide alternatives to old solutions, amuse us etc. New Product Development is a major issue for most companies as they seek to reduce time to market, reduce the development cycle, access new technologies and develop more and better products and services. Companies that develop, design, manufacture, market and sell these products seek commercial compensation in the short, medium or long term “success”. Subsequently, new products that can successfully compete in local, national and global markets are a key concern for the majority of companies, so this chapter reviews the New Product Development process and considers the variables associated with the different approaches, which may be needed when developing a new product.

Chapter 2 indicates development of a proper Resin Transfer Moulding process for a proper mould design. It is very much required for composite parts that are large and complex geometry and would be difficult to manufacture by hand layup method. The RTM process is a better substitute to it, but is not used readily due to the lack of proper manufacturing technology. In addition, the difficulty in the tooling design and mould fabrication cost increases with increase in size and complexity of the component. The scale down strategy of full scale product avoids bigger size mould requirements, prototype bulk production for product testing and quality check at the starting phase of product development. Moreover, the mould scale down strategy can be used to validate the process and the product with less capital input. In this chapter, a methodology to develop a RTM technology for manufacturing of a complex and large composite structure (high speed cab front) utilizing the process simulation were demonstrated.

Chapter 3 highlights the usage of two different methodologies namely Free vortex theory and mean stream line theory for designing of mixed flow pumps used in a wide range of application such as that in irrigation, flood control, dewatering and power station cooling systems. In this blades of the impeller of the mixed flow pump has been designed using Mean stream line theory and the free vortex theory and stresses arising due to static loading conditions has been obtained for both the cases. Comparison analysis for the equivalent stresses developed due to structural loading using various materials for the blade was undertaken. It was found that the stress for the blade designed using mean stream line theory is always on the higher side in comparison to that designed using free vortex theory.

In Chapter 4 Base Isolation System for Buildings is designed. Base isolation has now been implemented in numerous buildings in countries like Italy, Japan, New Zealand, and USA. It has also proven to be effective for retrofitting of important buildings (like hospitals and historic buildings). Seismic base isolation is one of the most widely implemented and accepted seismic protection systems and is a relatively recent and evolving technology. The most common isolation system used is Laminated Lead Rubber Bearings (LLRB). They combine the function of isolation and energy dissipation in a single compact unit, giving structural support, horizontal flexibility, damping, and a re-centering force in a single unit. The force deformation behavior of LLRB is modeled as bilinear system with viscous damping. In this chapter, a comprehensive design of LLRB is presented. Accurate evaluation of the structural properties and precise modeling of isolation devices are of utmost importance in predicting the response of the structure during the earthquakes.

Chapter 5 elaborates application of reverse engineering technique for designing a twin tube shock absorber which can be used in an automobile where presently mono tube shock absorbers are being used. The whole work has been done on a virtual platform form generating 3D CAD model of the damper to CFD analysis of the same. The data obtained was used as the datum for the design modifications and performance enhancement of the part. Subsequently similar analyses were conducted on the modified assemblies and most optimal one was highlighted.

In Chapter 6 a Louvered Cylindrical Mixing Pipe was designed with an aim of Maximum Air Entrainment. In order to achieve the same, Conservation equations of mass, momentum and energy have been solved numerically along with a two equation based k-ε turbulent model to determine the air entrainment into a mixing pipe. Two different louvered pipe configurations (one with entrance face closed to atmosphere while the other was opened) were used to entrain air into the mixing pipe and by changing various design parameters, the resulting effects were studied. For both the configurations, the opening area and the number of nozzles (assumed exhaust pipe) had significant effect in bringing down the exhaust temperature especially the configuration with its entrance face opened to atmosphere and temperature as low as 309 K (approx.) was successfully achieved through the analysis. This ends the section which was dedicated to designing and development of new Components or products and also usage of alternative design methodology for betterment.

Section 2, dedicated to development of new or alternative material for an existing component or product, starts with Chapter 7. The chapter proposes composite based material as a replacement for braces of orthotic calipers where presently aluminium alloys are being used. Components were fabricated using both thermoset and thermoplast matrix material with compatible reinforcement. The components were tested for some of the mechanical properties both by experimentation and by simulation. The findings were quite encouraging as it not only had a higher mechanical strength but also had a much lower weight to volume ratio, making it more adaptable for the locomotion affected patients.

Chapter 8 describes Friction-stir processing as a property enhancement technique which, not only removes the defects of initial casting process, but also improves the microstructure of the metals and metal matrix composites. The technique can be specifically applied to develop fine-grained microstructures throughout the thickness of metal surface, to impart super plasticity and ensure homogeneous distribution of reinforced particles, if any. This chapter is a dedicated effort to consolidate the latest developments contributed by different researchers in last few years. The work covers various components and parameters, selected and used to obtain specific desired results. Also, it includes past researches to exhibit various changes in mechanical properties with a keen focus on morphological study of these MMCs.

In Chapter 9 looking at the environmental issues, energy generation using renewable resources has been targeted. Again in order to move further the blades of the rotor for the wind turbines have been developed using biodegradable and light weight natural plant fibres for reinforcement in polymeric resin to produce required polymer composites.

Chapter 10 investigates the performance of annular fin with functionally graded composite materials. The chapter involves computation of efficiency and effectiveness of such fins and compares the fin performances for different geometry and grading parameters assuming unidirectional temperature gradient i.e. along the radius. A general second order governing differential equation has been derived for all the profiles and grading considering thickness of fin and functional grading of thermal conductivity to be a power function of radial co-ordinate. The performance analysis reveals the dependence of thermal behaviour of annular fins on geometry and grading parameter. This ends the section dedicated to development of new or alternative material for an existing component or product.

The last section, i.e., Section 3, dedicated to product and process optimization, commences with Chapter 11. The chapter introduces application of a soft computing technique i.e. artificial neural network for performance optimization of Micro Electro Discharge Machining. Micro Electro Discharge Machining is a non-conventional or nontraditional machining technique currently having high market demand due to accuracy, ability towards complicated machining as well as machining of extremely hard materials like Tungsten Carbide. This chapter discusses about micromachining on Electric Discharge Machining, its working principle and problems associated with it. Solution to those problems is suggested with the addition of powder in dielectric fluid. The optimization of Material Removal Rate was done with the help of ANN toolbox in MATLAB software.

Today all forums and organizations are very much concerned about Global Warming. The next chapter in this section, Chapter 12 introduces multi-criteria decision making optimization tool towards green manufacturing. Presently industries, especially manufacturing industries, are implementing using various advanced manufacturing processes for optimal product development. The advance manufacturing process generates large amounts of toxic substances results in various environmental issues during the optimal product manufacturing. Minimization of environmental issues and the amount of waste generated are strongly depends on its process and response parameters. Thus, optimization of process parameters for Green Manufacturing is essential and is a problem. This chapter provides an overview of applications of some multi-criteria decision making methods for optimization followed by detailed fundamental aspects of optimization issues in green manufacturing. The work proposed an integrated method consisting of AHP coupled with MOORA and validated through an experimental case study.

Most of the moving product usually undergoes wear and tear. The situation becomes very critical in case of coated products. Chapter 13 aims to determine optimal tribo-testing condition for minimum coefficient of friction and wear depth of electroless Ni-P, Ni-P-W and Ni-P-Cu coatings under lubrication using Taguchi coupled multi objective optimization tool namely, grey relational analysis. Electroless Ni-P, Ni-P-W and Ni-P-Cu coatings are deposited on AISI 1040 steel substrates. Coating characterization was ensured by studying results of scanning electron microscope monographs, energy dispersive X-Ray analysis and X-Ray diffraction techniques. Friction and wear tests under lubricated condition were carried out following Taguchi's experimental design principle. Finally, the predominating wear mechanisms of the coatings were discussed and the optimal parameters were suggested.

The last chapter of the book and the section Chapter 14 optimizes the input variables like pulse on time, pulse off time, peak current, and servo voltage in Wire Electro Discharge Machine for achieving simultaneous optimization of cutting rate, surface roughness, dimensional deviation, and wire wear ratio for machining of Inconel 625. Inconel 625 is a nickel-chromium based super-alloy which is mostly used in high end applications due to its excellent chemical and fabrication properties combined with high strength and outstanding corrosion resistance. Here using Taguchi's L9 orthogonal array with one replication experimental investigation was undertaken. Taguchi method has been combined with grey relational analysis for multi-characteristic optimization and analysis of variance (ANOVA) has been applied to determine the significant parameters and their contributions on response variables. Finally, results were validated with the help of confirmation experiments.

First and foremost, the editors would like to thank God. In the process of putting this book together it was realized how true this gift of writing is for anyone. You have given the power to believe in passion, hard work and pursue dreams. This could never have been done without the faith in You, the Almighty. The Editors would also like to thank all the Chapter Contributors, the Reviewers, the Editorial Advisory Board Members, Book Development Editor and the team of Publisher IGI Global for their availability for work on this editorial project.

Kaushik Kumar
Birla Institute of Technology, India

J. Paulo Davim
University of Aveiro, Portugal

Section 1

Design and Development of New Component/Product and Usage of Alternative Design Methodology for Betterment

Chapter 1
A Conceptual Analysis on the Different Approaches to New Product Development and the Variables Associated With Modelling the Process

Jonathan D. Owens
University of Salford, UK

Andrew M. Atherton
Lancaster University, UK

ABSTRACT

Globally, new products are launched daily. They provide answers to common or specialised problems, enrich lifestyles, provide alternatives to old solutions, amuse us etc. Companies that develop, design, manufacture, market and sell these products seek commercial compensation in the short, medium or long term "success", however you measure it. Subsequently, New Product Development is a major issue for most companies as they seek to reduce time to market, reduce the development cycle, access new technologies and develop more and better products and services. New products that can successfully compete in local, national and global markets are a key concern for the majority of companies, so successful NPD is fundamental to both stimulating and supporting economic growth. It is a subject, which has received and continues to receive much attention, particularly in seeking to improve its effectiveness and efficiency. This chapter reviews the NPD process and considers the variables associated with the different approaches, which may be needed when developing a new product.

DOI: 10.4018/978-1-5225-3401-3.ch001

INTRODUCTION

In the hotly competitive global environment of the twenty-first century New Product Development (NPD) is being seen as a major consideration for most organisations (Cooper, 2011, 2008; Dillon *et al*, 2005; Song and Benedetto, 2008; Smith, 2011; Schmidt *et al*, 2009). Across the global business arena, it is seen as one of the key areas for focus, as companies seek to reduce time-to-market, access new technologies in order to develop more and better new products and services. As the development of such new products and services which can compete in local, national and global markets has thus become a key concern for most companies, so successful NPD is now more than ever being seen as a fundamental business tool to stimulating and supporting economic growth (Owens and Atherton, 2015). It is therefore a subject that has received and continues to receive much attention, particularly in seeking to improve effectiveness and efficiency. Therefore, this paper reviews the New Product Development (NPD) process and considers the variables associated with the different approaches that could be considered when developing a new product.

DEVELOPMENTAL AND IMPORTANCE OF NEW PRODUCTS

Various researchers (Beard and Easingwood, 1996; Cooper and Edgett, 2003; Akgun *et al*, 2005; MacCormack *et al*, 2012) have identified NPD as a critical element of business strategy for most companies. Even companies that do not wish to grow will still be at risk from competitors, new technologies and the changing needs of customers if NPD is not undertaken (Cooper, 2011). The industrial revolution changed the social-economic structure of Britain and the world. Prior to this, products were usually made and sold within a region, customer needs were easily communicated between the parties involved (as everything was virtually hand-made and made-to-order) and flexibility and uniqueness were almost certain. Investment in machinery and large factories prompted the development of new towns and cities. No area in Britain reflects this more than Lancashire with its cotton mills, where the customer and vendor operated geographically and relied upon brief descriptions and third parties to produce a marketable product. Mass production reduced the standard cost of products and opened the market to more customers. Henry Ford was the first great global exponent of this in 1913 with the production of the Ford Model T. Other automotive manufacturers adopted mass production techniques and although the motorcar was still a luxury item then, it became more accessible to a wider market. Technological advances, customer preferences and economic developments have changed the automotive industry into a highly competitive market. Research and development departments are constantly looking for variants, modifications and improvements that will attract a larger portion of the market (Li, 1999; Petrie, 2008). As a result, companies that do not adapt to and satisfy customers' needs generally fail (Cooper and Kleinschmidt, 2007; Gerwin and Ferris, 2004). The automotive industry is not unique in having to constantly change; indeed, most industry sectors are under increasing pressure to modify and improve their product ranges (Van Echtelt *et al*, 2008). Competition dictates that a larger share of, or the maintenance of a current position in the market place, will only be achieved by offering a more attractive product (Jobber, 2009; Wagner and Hoegl, 2006).

The Sourcing of New Product Ideas

The idea for a new product can originate from many sources (Cooper, 2011); for example, customers, employees, current products, competitors etc. The problem occurs in determining which products to develop and which to avoid. The risk of failure should not discourage the introduction of new products, however, though careful product screening should be utilised throughout when instigating development (Balasubramanian and Baumgardner, 2004; Holstrom *et al*, 2006; Johnsen, 2011). Cooper (2008) suggests the first stage of screening should avoid two types of error, these being:

1. **NO-GO Error:** Dismissal of a good idea.
2. **GO Error:** The development of a poor idea.

For example, research by Havener and Thorpe (1994) indicates that an assessment of a product should be thorough and must take into account much more than just the financial gain. They discuss the cases of both the fax machine and Post-it® notes, which would have been rejected if only the cost analysis had been undertaken.

The New Product Development Process

Theory on the introduction of new products concentrates on the development process from innovation to introduction to market (Hutlink and Hart, 1998; Goffin and Micheli, 2010). Concurrent engineering, design methods, technology applications and a range of other techniques that ensure the new product meets the customer requirement. However, new products can still fail in manufacturing terms as they are pushed 'over the wall' (Owens and Davies, 2000) with all the correct documentation, but little or no explanation of what it is intended to be used for. Subsequently, a reoccurring theme within in the literature supports the development of a structured and clear road map of the NPD process can help reduce this failure and the understanding of product newness categories. Figure 1 presents a product typology of 'newness' categories. Specifically, product design and development is the interest of this chapter.

'Good design' can be achieved when the product not only looks good, but it also does the job well (Kalyanaram and Krishnan, 1997; Jang *et al*, 2009; Hsu, 2006). Therefore, "designing" is differentiated, because it is a tool which can be applied during NPD to help turn an invention into a successful product, or to extend the usefulness of an existing innovation (Goffin and Micheli, 2010; Holstrom *et al*, 2006; *Osterlund*, 1997). Subsequently, NPD is a most appropriate term for this research, because it relies upon "design" activities carried out to deliver a product, which may, or may not, be an "innovation".

In order to undertake NPD, it would be prudent to have "*a formal blueprint, roadmap, template or thought process for driving a new product project team from the idea stage through to market launch and beyond*" - a formal NPD process (Cooper, 2005, 2011; Kleinschmit *et al*, 2007). However, as with many other things in the business world, a definitive process that provides continual success has not been forthcoming. This section reviews some of the different models that have been put forward to describe the process and further examines activities, methods and techniques, which have direct relevance to the area of developing quality products.

Figure 1. The classification of new products
Cooper, 2005.

New to the World Products.
These new products are the first of their kind and create and entirely new market. Examples of products in this category are the Sony Walkman and 3M's Post-It notes. This category represents up to ten percent of all new products.

New to the Company.
These are products new to a company, but not to the marketplace, enabling a company to enter an established market for the first time. For example, IBM was not the first to launch an office version laser printer, Hewlett Packard (HP) were, therefore it was not an innovation, but it did however represent a new product line for IBM. Approximately twenty percent of all new products correspond with this category.

Additions to existing Product Lines.
These are new items to a company and fit within an existing product line the company manufactures. They may also represent a fairly new product to the market place. For example, the introduction of HP smaller and considerably less expensive version of its laser printer, that was suitable for home computers. The printer was a new item to the LaserJet line. This product type represents approximately twenty-six percent of all new product launches.

Improvements in revisions to existing products.
These are essentially replacements of existing products. They offer improved performance or greater perceived value over the previous product. Similarly, this product type can represent up to twenty six percent of all new product launches.

Repositioning.
These are essentially new applications for existing products, and involve retargeting an old product to a new market segment for a different application. For example aspirin was considered a standard reliever for headache and fever symptoms, but due to safer, more effective and cheaper products it (aspirin) was superseded. However, new research demonstrated this product had other benefits, and resulted with the aspirin being marketed as prevention against blood clots, strokes and heart attacks. This product type can account for up to seven percent of all new product launches.

Cost Reductions.
These are new products designed to replace existing products in the line, and can yield similar benefits and performance. From a marketing perspective they are not new products but from a design and production perspective, they could represent significant change to the company. This product type can account for up to eleven percent of all new product launches.

Modelling the New Product Development Process

There are numerous examples of previous research (Bhaskaran and Krishnan, 2009; Hsu, 2006; Ogama and Pillar, 2006; Perks *et al*, 2005) where they have attempted to monitor the progress of the new product both prescriptively and descriptively. In addition, some researchers (Danese and Filippini; 2010; Schmidt *et al*, 2009) who are focussing on NPD process models often form dissimilar opinions as to their structure. That said there is agreement amongst researchers in that a generic NPD model cannot be created. However, an agreed commonality is from established research by Saren (1984) are the following five categories:

1. Departmental-stage models.
2. Activity-stage models.
3. Decision-stage models.
4. Conversion process models
5. Response models.

Figure 2 provides a summary of these different models by type. Discussion points on the usefulness of each model for research work and their practical use in the management of NPD are also illustrated.

Saren (1984) suggests that dividing methods into groups provides a useful point for an examination of how each model might purposefully be used in research; for example, who is involved in the innovation; at what point and in what order specified tasks are undertaken; upon what basis decisions are made; how inputs to a process become outputs; or the reactions to specified stimuli. However, he concluded that although each individual model is valid, in that it indicates something of the characteristics of the process, more work needs to be done on the holistic process of innovation in companies; this is something that is reiterated in more research carried out by Cooper and Kleinschmidt (2007).

More recent research by Noke and Hughes (2010) uses the nineteen sixties phased development model as a starting point for a comparison with some of the NPD process model ideas, which have been progressing from the late nineteen eighties in to the twenty first century. They (Noke and Hughes, 2010) also claim that modern stage gate methodologies, product and cycle-time excellence; and total design as the main examples of advancement in processes aimed at improving product success. Since there is a number of different ways to model the NPD process, and each way has associated with it its own specific strengths and weaknesses, this paper will discuss those that are recurrent in the most recent literature. Subsequently, the remainder of this paper will consider the origins, uses and limitations of five NPD modelling techniques, which are common and progressive:

1. Stage gate models.
2. Multiple convergent process.
3. Product and cycle-time excellence.
4. Total design.
5. Third generation NPD process.

THE STAGE GATE MODEL

The stage gate model breaks the NPD process into discrete identifiable stages, with each stage being followed by decision gates as illustrated in Figure 3. The number of stages can be increased to harmonise with both company and product development requirements. Each stage is designed to gather information required in order to progress the new product onto the next stage of development. In addition, each stage consists of a set of parallel activities undertaken by personnel from different fictional area within the business, but working together as a team. Subsequently, to manage risk via a stage-gate process, the parallel activities within a stage should be designed to gather important information as to drive both technical and businesses uncertainties associated with the new product, thereby facilitating built in feedback loops (Zhao *et al*, 1999; Atilgan-Inan *et al*, 2010). In recent years some researchers (Dayan and Basarir, 2010;

Figure 2. A summary of some NPD process models
Saren, 1984.

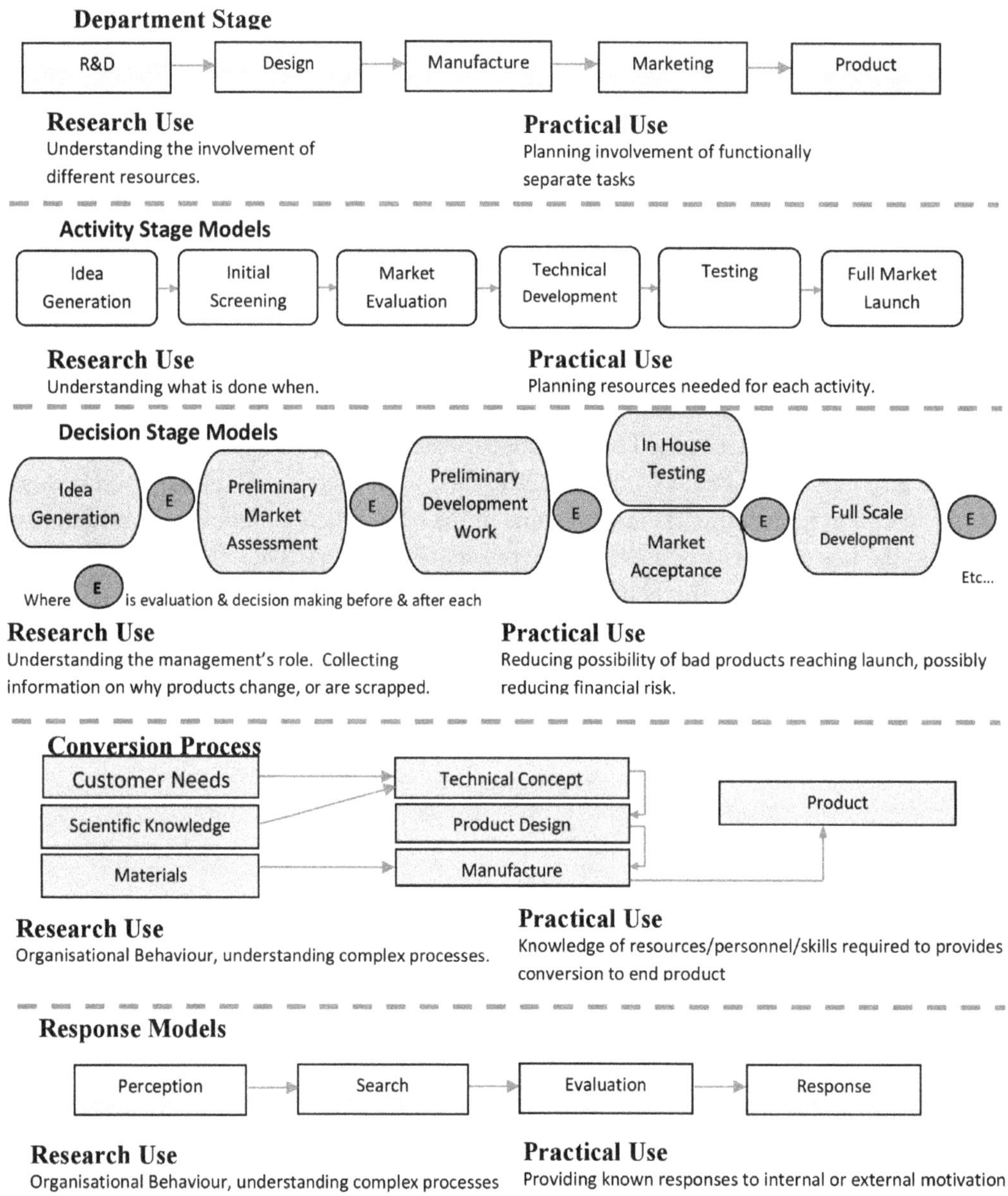

Smith, 2011; Troy *et al*, 2008; Fu; 2010) have suggested that successful product development is aided by following a stage gate decision process because it encourages activities to be undertaken by a core team of representatives from all functional departments. The stage gate models may help the reader to understand the management of the process and may also help prevent losses being made by revealing most errors early on, and before market launch, the products which will fail in an industrial situation by reducing failure risk in the comprehensive review implemented at the gate of each stage (Petrie, 2008).

Figure 3. Overview of a stage-gate NPD process
Cooper, 2008; 2011.

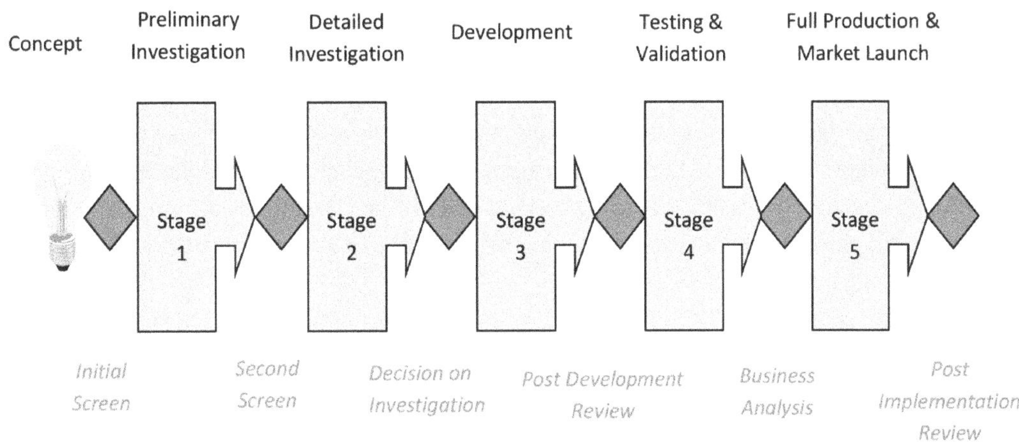

However, the model does not lead us to a means of ensuring that the product will meet the need of the user(s), or indeed the final customer. This type of system does have the potential to include "go/ no-go" decisions, based on whether the product is being designed to a high enough quality and one that will satisfy the user(s) and ultimate customer. On the other hand, without elucidation from someone providing a customer needs compliance emphasis, there is not necessarily any strong incentive to use this as a basis for what the product "must meet" or "should meet" when product management decision are being made at each of the gates.

Although, this model can be seen widely in practical use today in some industrial organisations (Cooper, 2008; Owens, 2007, 2009; Smith, 2011) there are some general problems which occur when following a stage gate model which are indicated by Cooper (1999; 2005; 2008). He adds later (Cooper, 2011) that even though the idea has been taken up in the last two decades with positive effect, stage gate process models are still not really usable because they are too time consuming, often have too many ways of waiting time, are too bureaucratic and have no provision for focus. Also, one author (Himmelfarb, 1992, p.10) provides a fairly severe and emotive set of comments by claiming that it "*creates products that are hard to make, that cost too much, that require too many expensive design changes, and that may or may not meet marketplace needs. It encourages isolation of functional areas and, worst of all, it' very slow.*" It should be noted that these observations do not bode well for the extra and important inclusion of both usability and customer needs compliance in this particular process model.

THE MULTIPLE CONVERGENT MODEL

A key limitation of the stage-gate model and linear NPD process in general is the interaction between the people involved within the process (Griffin, 1997; Shenhar, 2001; Hart *et al,* 2003; Petrie, 2008). One such method that helps address this interaction difficulty is the Multiple Convergent Model. It looks at differing approach to integrate people, both directly and explicitly, into the NPD process, thereby taking into consideration the documented limitations in other NPD process models (Brown, 2008).

The model takes into account the lessons learned that suggest that success comes from having quality inputs which are valid from multidisciplinary areas (Troy *et al*, 2008). There is much importance placed on the use of networks and the production of a model that breaks down multidisciplinary boundaries (Newey and Verreynne, 2011). The models views NPD as tasks with areas moving towards a common conclusion, but are required to come together at a number of different natural and integrative points for evaluation, as illustrated in Figure 4. In this way the multiple convergent model is similar to the stage gate models. However, where it differs is that the convergent model has multiple points, which it recognises as important to an iterative process. According to Hart (1995) the advantages of taking the process as a series of converging points for evaluation, followed by diversion into functional activities are:

- It accommodates iteration.
- It allows for iterative communication and evaluation within the functional groups.
- The framework can accommodate third parties easily.
- Methods for real integration of work from functional groups can be provided in the convergent points.

Despite the model being driven by converging points, the main disadvantage in practical use appears to be, ironically, that it may be too divergent. It converges for cross-functional decisions, but then separates out into each of the different functions to carry out the tasks. Hart *et al* (2003) describes a key element as the amount of information sharing that is modelled, however, horizontal communications between functional areas, are only modelled as happening during the evaluation or collation points and not during other activities. With so many points of convergence during the process, this model does not appear (Smith, 2011) to be conducive with efficiency and it seems to require a large amount of management effort to keep the process on track.

THE PRODUCT AND CYCLE-TIME EXCELLENCE MODEL

As a supplementary to the multiple convergent model there are alternative modelling approaches whereby the prime focus is to reduce the time-to-market, subsequently, concentrating on how to influence the economic outcome of the design process. The product and cycle-time excellence model McGrath *et al* (1992) is an example of this. It is similar to the stage-gate approach (Petrie, 2008), except that during the phase reviews the decisions are not made by a NPD team involved with developing the new product, but by a small number of senior managers who are the 'Product Approval Committee' (Brown, 2008). Under product and cycle time excellence, the process is seen as a funnel taking in lots of ideas, following the completion of five phases, producing new products, as illustrated in Figure 5. In an attempt to reduce the time it takes to develop a new product, the productivity model breaks down each of the five phases in the process into fifteen or twenty steps and then each of these steps into ten to thirty tasks. Database records can be kept on the timing for each of the tasks and thus the total development time can be judged for each new product (Suomala and Jokioinen, 2003).

The productivity and cycle time concepts also pay attention to the management of the process in more holistic ways. Of the seven major elements for this model, four (Jenkins *et al*, 1997) are directly overseeing the whole of the product development process in the company:

Figure 4. The early stage of the multiple convergent model
Hart and Baker, 1996.

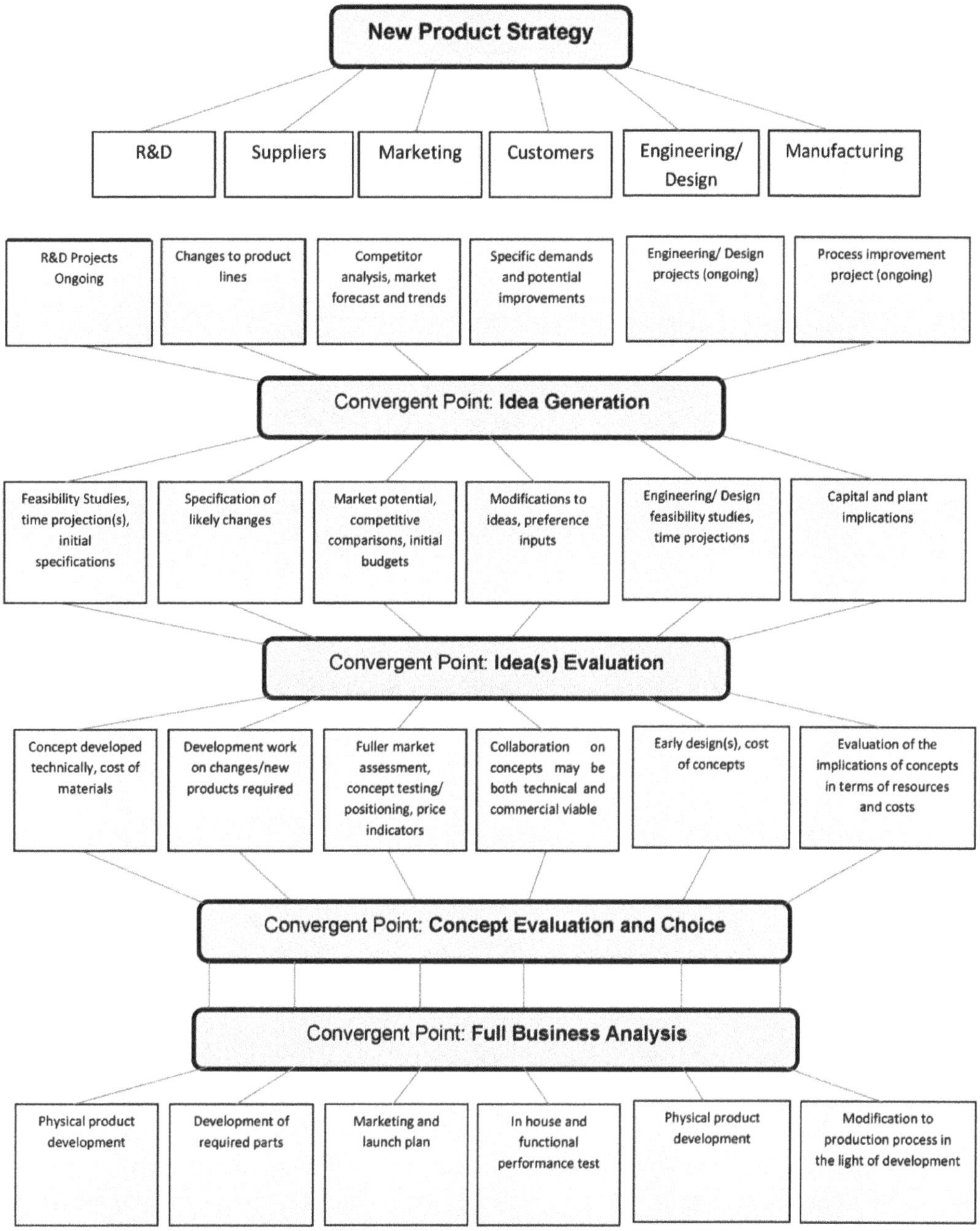

1. The provision of core teams during development
2. The use of a product strategy
3. The review and correct implementation of technology management
4. The endorsement of cross project management

Figure 5. The phase review process within the product and cycle time excellence model
Jenkins et al, 1997.

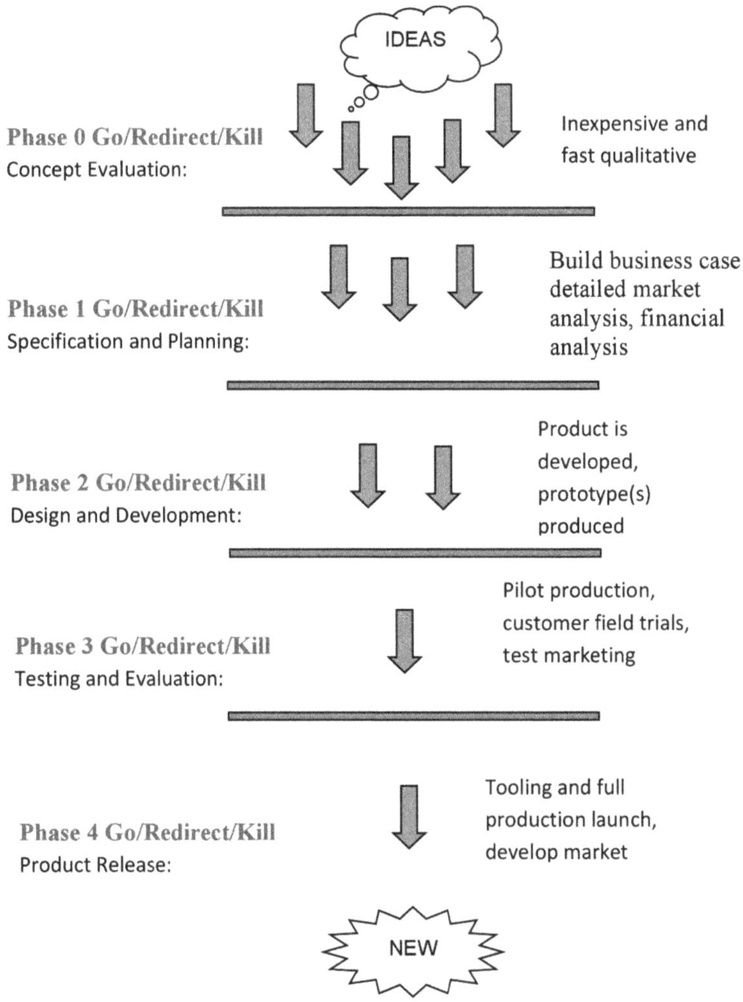

The authors of the product and cycle time excellence model also advocate the use of design techniques and automatic development tools that will help focus and streamline the development of the product (McGrath *et al,* 1992).

The product and cycle time excellence model is more than just a definition of the development process. It is aimed at efficiently managing the development of new products such that the product is produced on time and within budget, whilst using the optimum balance of skills and methods at the right point during the projects' progression (Akgun *et al*, 2005; Troy *et al*, 2008). However, these types of models, which are driven by productivity and cycle time reduction, rely upon putting senior management in an overriding position of authority and also upon splitting down the design process to a level so low that it can be timed. As well as the obvious philosophical discussions about specifying exactly the creative nature of design that these issues incite, both of these ideas seem regressive and are reminiscent of the work-study principles, based on Taylorism, which had its 'glory days' at the beginning of the twentieth century.

TOTAL DESIGN

Total design is seen as a broadly based business activity in which specialists collaborate in the investigation of a market, the selection of a projects, the conception and manufacture of a product, and in the provision of various kind of user support. (Pugh and Moreley, 1988, p1)

The models discussed previously in this paper revolve around breaking down the process into manageable portions by seeing the development process as a series of problems to be solved (Brown, 2008; Kleinschmidt *et al*, 2007; MacCormack, 2012). Various authors (Hart *et al*, 2003; Baldwin and Clark, 2000; Griffin, 1997; Moultrie *et al*, 2006; O'Dwyer and Ledwith, 2009; Nicholas *et al*, 2011) have criticised the way of focussing on parts of the problem and solving them one-by-one because they have found that they are often used without paying sufficient attention to the aspects of assimilating all of the problems together. A slightly different outlook to view design and development as a converging spiral, the spiral from of the design process attempts to emulate real-life, in that the design process is evolving (Oakley, 1990). The spiral moves from a formulation stage, to an evolution stage, through to a stage where transfer takes place and is followed by a reaction stage that returns the development to the formulation stage, as illustrated in Figure 6. This is different to other series or stage gate models as it relies upon interactive and overlapping stages that evolve.

The spiral form was a depiction used for Acar's (1966) triple-helix model of the product development process, which can be cross-sectioned at any point to reveal the interaction between specification, conceptual design and embodiment design. In the total design model championed by Pugh and his colleagues, the spiral is taken into more depth (Pugh, 1991; Pugh and Moreley, 1988; Hollins and Pugh, 1990).

The development of the total design model and subsequent publishing of Pugh have stimulated much discussion within engineering design circles (Brown, 2008; Goffin and Micheli, 2010; Hsu, 2006; Jenkins

Figure 6. A spiral model of the design process
Oakley, 1990.

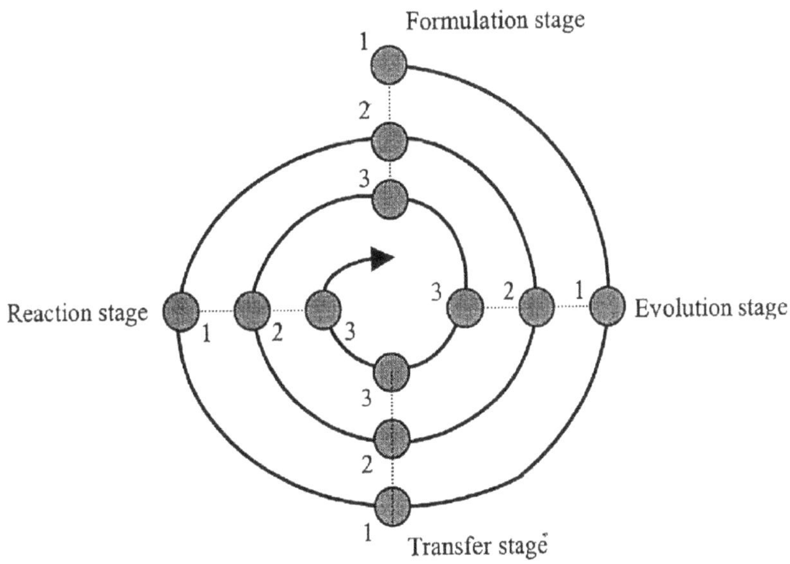

et al, 1997). Also, the total design philosophy is taught as a useful and useable model of best practices on a number of UK higher education institutes' engineering courses, especially because it emphasises the use of many different discipline independent tools and methods (Goffin and Micheli, 2010).

The total design model outlines six nominal spirals, which attempt to capture the main undertakings during the design process, all within an iterative environment. These 'design cores' are presented by Pugh (1991) as:

- Investigation of market/user needs and demands
- The development of the product design specification
- Conceptual design
- Detailed (technical) design
- Manufacture
- Selling (marketing)

Figure 7 illustrates how Pugh visualises the whole package of design activities, within a "...framework of planning and organisation...and how they fit into a business structure" (Pugh, 1991, p8).

The total design and its embellishment with detailed information on how to approach each of the "design cores" goes a long way to assist engineering designers practically undertake product design systemically. Pugh and his colleagues have devoted much literature (Pugh and Moreley, 1988; Pugh, 1991; Hollins and Pugh, 1990) to explaining methods and tools that can be used in conjunction with the total design philosophy. The model does acknowledge and capture many of the complexities of NPD and attempts to attract a cross disciplinary audience. It also explicitly acknowledges the place of design within the company's structure and long-term strategy (Jang *et* al, 2009; Sethi, 2000). However, much of the work is essentially a model and text for engineers, and gears itself more towards explaining business requirements to a technical audience, rather than explaining technical issues to a business audience.

The market and user needs "design core" does not clarify well enough for an inexperienced company or researcher to fully comprehend the importance of meeting customer needs to the success of the product (Carbonell and Rodriguez, 2006). However, having said that, there are numerous issues that have arisen from studying this approach, such as the strong emphasis on the Product Design Specification (PDS) and the recognition of informal paths of communication within the design team (Brown, 2008).

REQUIREMENTS CAPTURE PROCESS MODEL

The requirement capture model developed by Cooper *et al* (1999) captures the understanding of both the individual and the team who are involved in the development of the new product, and how these understandings relate to the customer requirements. This process is illustrated in Figure 8. It considers the internal and external variables of the new product comprehension by the parties involved in the NPD process, and how these may influence the data interpretation. Subsequently, the parties involved review the outputs from three levels:

1. Acquisition of data
2. Transformation of data
3. Generation of requirements

Figure 7. The total design activity model
Pugh, 1991, p. 11.

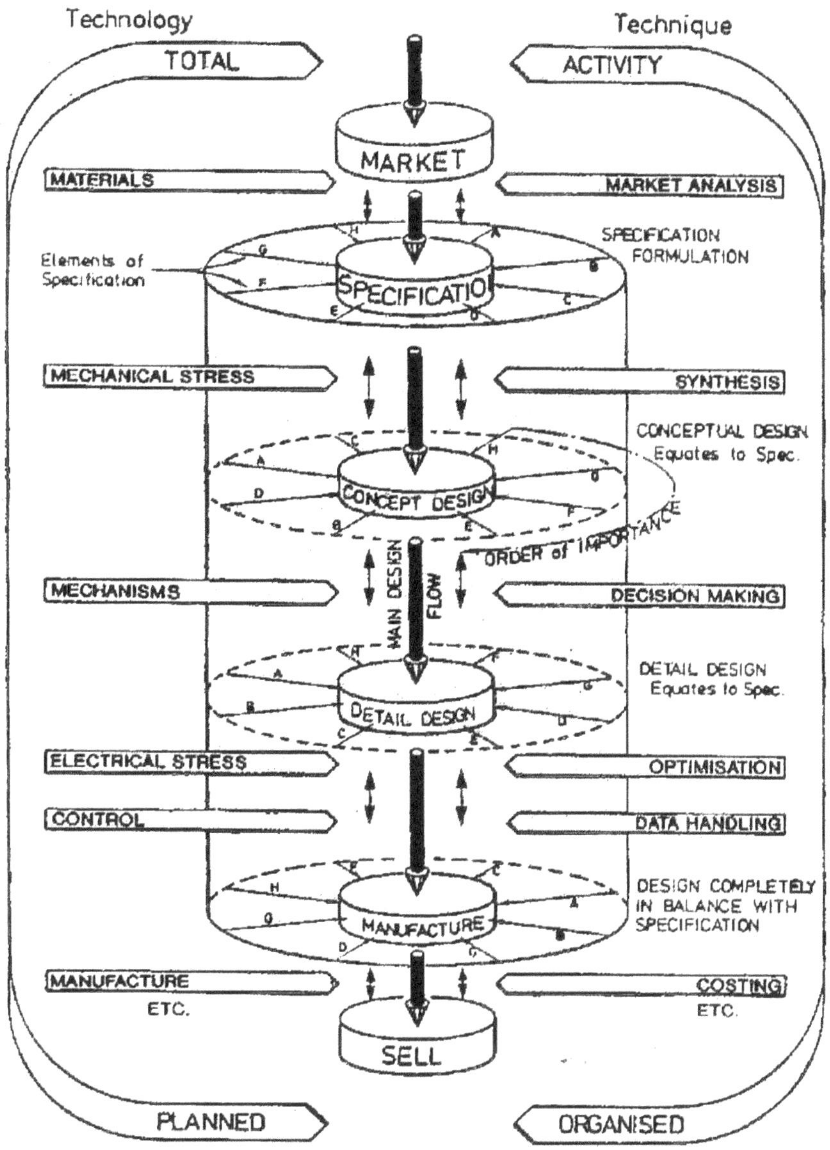

It focuses on where there is a collaborative comprehension of customer requirements and needs, in order to produce an agreed outcome and set of requirements. Although established, Cooper *et al* (1998) research is still a useful benchmark in that it provides a focus on dealing with information provided by the customer, in order to establish clear definition by the NPD team as to what the customer requires. Some key issues that are addressed in the model are:

- Different views and understandings (perceptions) of the same data are included
- Activities and events change the understating (perception) of a customer requirement(s)

Figure 8. Theoretical requirements capture process model
Cooper et al, 1998, p. 510.

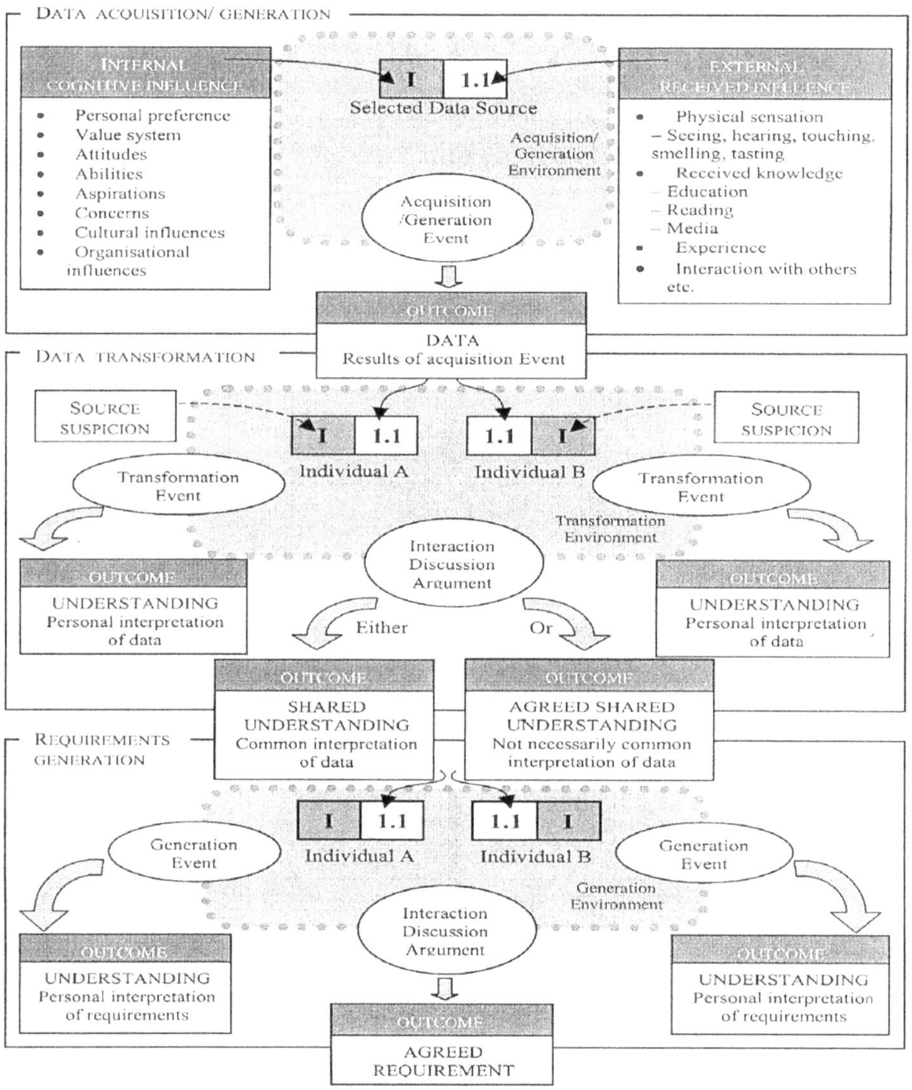

- Data acquisition and transformation events are required to gain an agreement on the definition of a product requirement(s)

Third Generation New Product Processes

The phase development models of the nineteen-sixties are referred to as the first generation of defined product processes (Dillon *et al*, 2005). The phase review process advocates sequential development stages, each carried out by different functional groups that complete their phase then pass on the results to the next phase and function (Ahmed, 1998; Barozak *et al*, 2009). First generation development processes are often referred to as "*over the wall*" (illustrated in Figure 9) processes because development is handed onto the next group, when the last is finished.

Figure 9. "Over the wall" concept
Owens and Davies, 2000.

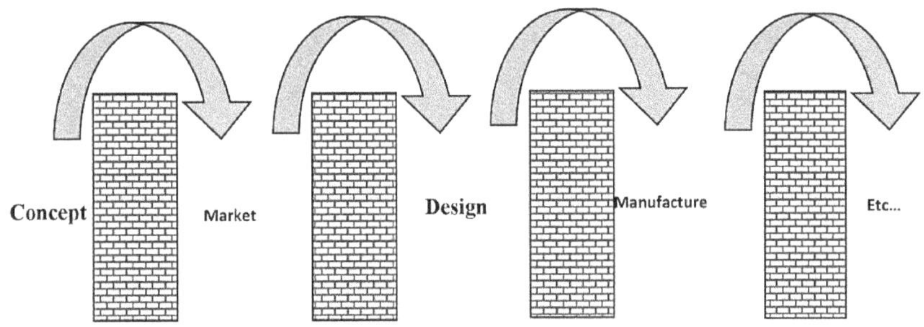

The second generation of product development processes are the processes of today – which are mainly based upon stage gate type models involving a cross discipline structure of one type or another (Troy *et al*, 2008). According to Cooper (2008) the third generation are the future way in which products should be produced. He suggests these third generation processes are relatively ill defined because they are still in development and should be developed and grown around the specific company. They are driven by the need to efficiently create new products and get them to market as quickly as possible, but with a much greater tolerance for calculating risk taking, which is a conflicting view to that of the previously discussed models driven by cycle time (Carbonell and Rodriguez, 2006; Chan, 2010).

Cooper *et al* (2002) suggests there is a distinct need to redress the balance from a restrictive linear process that only moves the product development forward when a decision is made on the outcome of a preceding activity, as illustrated in Figure 10. The ideas put forward by these third generation models tie in with the management practices of concurrent and simultaneous engineering.

This model would, inevitably require integration through software, hardware and 'human ware' or team facilitation (Ogama and Pillar, 2006; Reid and Brentani, 2004). It will allegedly work from a premise that attempts to maintain discipline, but allows a balance of action thoroughness and the need to move quickly (Cooper and Edgett, 2003). To answer problems that may arise from this basis of reasoning, four fundamental 'F's' have been defined (Cooper, 2011):

1. **Fluidity:** The model is fluid and adaptable, with overlapping and fluid stages for greater speed.
2. **Fuzzy Gates:** The model features conditional go decisions (rather than absolute ones), which are dependent on the situation.

Figure 10. Third generation process
Cooper, 2011.

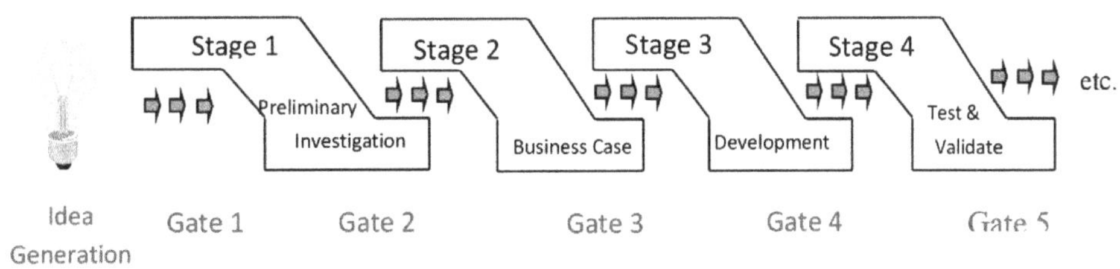

3. **Focused:** The model builds in prioritisation methods that look at the entire portfolio of projects (rather than one at a time) and focuses resources on the "best bets".
4. **Flexible:** The model is not a rigid stage gate process; each project is unique and has its own routing through the process.

The implications from the use of such a model is that everything becomes so much more difficult to define in absolute terms, making devising and understanding the product development process a more daunting task (Zollo *et al*, 2002). As a project progresses, the decisions made will be more complex and sophisticated and may be hard to place in context if the stages overlap too readily (Gerwin and Ferris, 2004). Thus, falling into an ad-hoc free-for-all system of product development seems a distinct possibility. Later Cooper (2005) also made some of these observations and suggests that this model will only work within a framework based on the second-generation stage gate models. He does not advocate a withdrawal from stages and gates, instead he realises that to make these systems really work product development should allow much more flexibility. A potential way of achieving this could be a move towards reducing the authoritative role of senior management and pushing the decision-making role of the NPD project team members and leaders (Jorgensen and Messner, 2009; Rodriguez *et al*, 2008).

SUMMARY: THE USEFULNESS OF A NPD PROCESS MODELLING APPROACH

The previous commentaries show the immense profusion and disparity as to the different approached to modelling the NPD process. This is by no means a definitive discussion, but rather a reflection to direct the reader to the vast multiplicity that exists in this field. The need to observe and comprehend the differing modelling approaches, and how one might portray the NPD process is that of practicality. For example, if one can somehow capture what a business does, and can realise the path the product development process takes, then perhaps it may be possible to have an improved and useful comprehension of how to enhance and improve the NPD process, then it may be possible to secure tangible benefits. Past research has suggested that positive consequences result from the existence of formal NPD processes. Also, research (Cooper, 2011; Moultrie *et al*, 2006; Petrie, 2008; Sethi, 2000) has suggested that the lack of understanding and implementation of product development processes in industry can account for poor product development performance.

Since many product development authors and practitioners have reported these positive results, it is no wonder that they are driven to try and capture the essence of good product development practices and processes. Therefore, in an effort to make the task of modelling the process more manageable, different authors have tried to summarise their complexities by generalising and minimising differences between companies and products (Noke and Radnor, 2004; Johnsen, 2011). However, because of this, the models are often only a representation of the process and are regularly produced by individual researchers as tools to investigate specific phenomena that occur during product development. Also, the reality of producing working models is the consideration of differences, which occur between what the literature describes and/or prescribes and what is actually done in reality because the nuances within each company are so difficult to encapsulate (Cooper *et al,* 1998; Jorgensen and Messner, 2009).

Figure 11. Model of the need assessment process
Holt et al, 1984.

Need Identification	A problem or a user need is perceived, often in a vague form. This is usually the initiation of the product innovation process.
Need Evaluation	Based on available information, the perceived need is analysed and evaluated; for example in connection with preparation of the proposal.
Need Clarification	This involves a systematic study of the user needs involved. It may be undertaken in connection with a feasibility study in the last part of the idea generation stage.
Need Specification	Based on assessed needs and their relative strength, relevant need requirements are specified.
Need Up-Dating	As the project moves ahead, the needs specified are up-dated at intervals in connection with development of the technology and planning of the marketing and manufacturing operations.

Given the plethora of product development models available and reviewed here, it would be reasonable to assume that there would be one, which specifically follows a customer's need through product development. However, although many pay more than just 'lip-service' to customer needs. None have been found that depict the whole of the process for NPD, with explicit emphasis on customer needs compliance. The closest is a descriptive list suggested by Holt *et al* (1984). Their list of stages during which different user needs issues are addressed is illustrated in Figure 11. This is a useful list, and does highlight different periods of need recognition, assessment and appraisal. Yet, it does not get to grips with the essence of product development interaction, process, iteration and communications required. These are issues that are being investigated in this research to be able to understand at least some of the facets of customer needs compliance during the product development process.

CONSIDERATIONS FOR FURTHER RESEARCH

In summary, it is perhaps apparent various researchers have provided a lot of different methods for depicting the NPD process. Although, one important finding is so far none of the methods have been specifically developed for the following customers' needs through the NPD process from concept to launch. However, the representation of the linkages within NPD practice that these models demonstrate is perhaps a useful starting point for further examination, as long as they are taken within their context and understanding of their limitations. Therefore, in order to advance this research, a review of the literature incorporating management and information provision for customer needs compliance issues, in order to ascertain of there is a correlation with the earlier comment.

REFERENCES

Acar, B. S. (1966). A new model for the design process. *Proceedings of the Institute of Mechanical Engineers, E*(210), 135-138.

Ahmed, P. X. (1998). Culture and climate for innovation. *European Journal of Innovation Management*, *1*(1), 30–43. doi:10.1108/14601069810199131

Akgun, A. E., Byrne, J., Keskin, H., & Lynn, G. S. (2005). Knowledge networks in NPD projects: A transactive memory perspective. *Information & Management*, *42*, 1105–1120. doi:10.1016/j.im.2005.01.001

Atilgan-Inan, E., Buyukkupcu, A., & Akinci, S. (2010). A content analysis of factors affecting NPD process. *Business and Economics Research Journal*, *1*(2), 87–100.

Baldwin, C. Y., & Clark, K. B. (2000). *Design Rules: The Power of Modularity*. Cambridge, MA: MIT Press.

Barczak, G., Griffin, A., & Khan, K. B. (2009). Trends and drivers of success in NPD practices: Results of the 2003 PDMA best practice study. *Journal of Product Innovation Management*, *26*(1), 3–23. doi:10.1111/j.1540-5885.2009.00331.x

Beard, C and Easingwood, C (1996). New Product Launch: Marketing Action and launch tactics for high technology products. *Journal of Industrial Marketing Management, 25*, 87-103.

Bhaskaran, S. R., & Krishnan, V. (2009). Effort, Revenue and Cost Sharing Mechanisms for Collaborative NPD. *Management Science*, *55*(7), 1152–1169. doi:10.1287/mnsc.1090.1010

Booz, Allen, and Hamilton. (1991). *New Product Management*. Booz, Allen and Hamilton Inc.

Brown, T. (2008). Design Thinking. *Harvard Business Review*, *86*(6), 84–92. PMID:18605031

Carbonell, P., & Rodriguez, A. I. (2006). The impact of market characteristics and innovation speed on perceptions of positional advantage and new product performance. *International Journal of Research in Marketing*, *23*(1), 1–12. doi:10.1016/j.ijresmar.2006.01.002

Chan, H. C. (2010). Linkage community based innovation and speed to market: The mediating role of the NPD process. *The International Journal of Organisational Innovation*, *2*(4), 49–60.

Cooper, R., Wooton, B., & Bruce, M. (1998). Requirements Capture: Theory and Practice. *Technovation*, *18*(8/9), 497–511. doi:10.1016/S0166-4972(98)00033-9

Cooper, R. G. (1999). From Experience: The invisible success factors in product innovation. *Journal of Product Innovation Management*, *16*(4), 333–351. doi:10.1016/S0737-6782(99)00005-3

Cooper, R. G. (2005). *Product Leadership: Pathways to Profitable Innovation*. Basic Books.

Cooper, R. G. (2008). The Stage-Gate® idea-to-launch process update, what's new and NexGen systems. *Journal of Product Innovation Management*, *25*(3), 213–232. doi:10.1111/j.1540-5885.2008.00296.x

Cooper, R. G. (2011). *Winning at New Products: Creating Value through Innovation* (4th ed.). Basic Books.

Cooper, R. G., & Edgett, S. J. (2003). Overcoming the crunch in resources for new product development. *Research Technology Management, 46*(3), 48-59.

Cooper, R. G., Edgett, S. J., & Kleinschmidt, E. J. (2002). Optimising the stage-gate process: What best practice companies do – II. *Research Technology Management, 45*(6), 43-51.

Cooper, R. G., & Kleinschmidt, E. J. (2007). Winning businesses in product development: Critical Success Factors. *Research Technology Management*, *50*(3), 52–66.

Danese, P., & Filippini, R. (2010). Modularity and the impact on new product development time performance: Investigating the moderating effects of supplier involvement and interfunctional integration. *International Journal of Operations & Production Management*, *30*(11), 1191–1209. doi:10.1108/01443571011087387

Dayan, M., & Basarir, A. (2010). Antecedents and consequences of team reflexivity in NPD projects. *Journal of Business and Industrial Marketing*, *25*(1), 18–29. doi:10.1108/08858621011009128

Dillon, T. A., Lee, R. K., & Matheson, D. (2005). Value Innovation: Passport to wealth creation. *Research Technology Management, 48*(2), 22-36.

Fu, Y. K. (2010). New Product success among SME's: An empirical study in Taiwan. *Journal of International Management Studies*, *5*(1), 147153.

Gerwin, D., & Ferris, S. J. (2004). Organising New Product Development Projects in Strategic Alliances. *Organisation Science, 15*(1), 22-38.

Goffin, K., & Micheli, P. (2010). Maximising the value of Industrial Design in NPD. *Research Technology Management*, *51*, 29–37.

Griffin, A. (1997). PDMA research on new product development practices. *Journal of Product Innovation Management*, *14*(6), 429–458. doi:10.1016/S0737-6782(97)00061-1

Harborne, P., & Johne, A. (2003). Creating a project climate for successful product innovation. *European Journal of Innovation Management*, *6*(2), 118–133. doi:10.1108/14601060310475273

Hart, S. J. (1989). Design orientation and market success. In The marketing initiative – Economic and Social Research Council (ESRC) studies in British Marketing. Prentice Hall. doi:10.1016/0142-694X(89)90043-4

Hart, S. J. (1995). Where we've been and where we're going in new product development research. In Product development – Meeting the challenge of the design-marketing interface. John Wiley and Sons.

Hart, S. J., & Baker, M. J. (1996). The multiple convergent processing model of new product development. In New Product Development – A reader. The Dryden Press.

Hart, S. J., Hutlink, E., Tzokas, N., & Commandeur, H. (2003). Industrial Companies' Evaluation Criteria in new Product Development gates. *Journal of Product Innovation Management*, *20*(1), 22–36. doi:10.1111/1540-5885.201003

Havener, C., & Thorpe, M. (1994). Customers can tell you what they want. *Management Review*, *14*(12), 42–45.

Himmelfarb, P. A. (1992). *Survival of the fittest – New product development during the 90's*. Englewood Cliffs, NJ: Prentice-Hall International.

Hollins, W., & Pugh, S. (1990). *Successful product design*. Butterworth.

Holstrom, J., Korhonen, H., Laiho, A., & Hartial, H. (2006). Managing product introductions across the supply chain: Findings from a development project. *Supply Chain Management, 11*(2), 121–130. doi:10.1108/13598540610652519

Holt, K., Geschka, H., & Peterlongo, G. (1984). *Need assessment – A key to user orientated product innovation*. John Wiley and Sons Ltd.

Hsu, Y. (2006). Comparative study of product design strategy and related design issues. *Journal of Engineering Design, 17*(4), 357–370. doi:10.1080/09544820600664655

Hultink, E. J., & Hart, S. (1998). The world's path to the better mousetrap: Myth or reality? An empirical investigation into the launch strategies of high and low advantage new products. *European Journal of Innovation Management, 1*(3), 106–121. doi:10.1108/14601069810230199

Jang, S., Yoon, Y., Lee, I., & Kim, J. (2009). Design Orientated NPD. *Research Technology Management*, 36–46.

Jenkins, S., Forbes, S., Durrani, T. S., & Banerjee, S. K. (1997). Managing the product development process [Part I: An assessment]. *International Journal of Technology Management, 13*(4), 359–378. doi:10.1504/IJTM.1997.001670

Jobber, D. (2009). *Principles and Practice of Marketing* (6th ed.). McGraw-Hill Education.

Johnsen, T. E. (2011). Supply network delegation and intervention strategies during supplier involvement in new product development. *International Journal of Operations & Production Management, 31*(6), 686–708. doi:10.1108/01443571111131999

Jorgensen, B., & Messner, M. (2009). Management control in NPD: The dynamics of managing flexibility and efficiency. *Journal of Management Accounting Research, 21*, 99–124. doi:10.2308/jmar.2009.21.1.99

Kahn, K. B., Barczak, G., & Moss, R. (2006). Establishing an NPD best practice framework. *Journal of Product Innovation Management, 23*(2), 106–116. doi:10.1111/j.1540-5885.2006.00186.x

Kaipia, R., & Holmstrom, J. (2007). Selecting the right planning approach for a product. *Supply Chain Management, 12*(1), 3–13. doi:10.1108/13598540710724347

Kalyanaram, G., & Krishnan, V. (1997). Deliberate product definition: Customising the product definition process. *JMR, Journal of Marketing Research, 34*(2), 276–286. doi:10.2307/3151864

Kleinschmidt, E. J., de Brentani, U., & Salomo, S. (2007). Performance of global new product development programs: A resource based view. *Journal of Product Innovation Management, 24*(5), 419–441. doi:10.1111/j.1540-5885.2007.00261.x

Li, T. (1999). The impact of the marketing-R&D interface on new product export performance: A contingency analysis. *Journal of International Marketing, 7*(1), 10–34.

MacCormack, A., Crandall, W., Henderson, P., & Toft, P. (2012). Do you need a NPD strategy? Aligning process with context. *Research Technology Management*, (Jan-Feb): 34–43. doi:10.5437/08956308X5501014

McGrath, M. E., Anthony, M. T., & Shipiro, A. R. (1992). Product development: Success through cycle time excellence. In *PRTM*. Butterworth-Heinemann Publishing.

Moultrie, J., Clarkson, P. J., & Probert, D. (2006). A tool to evaluate design performance in SME's. *International Journal of Productivity and Performance Management*, *55*(3 / 4), 184–216. doi:10.1108/17410400610653192

Moultrie, J., Clarkson, P. J., & Probert, D. (2007). Development of a design audit tool for SME's. *Journal of Product Innovation Management*, *24*(4), 335–368. doi:10.1111/j.1540-5885.2007.00255.x

Newey, L., & Verreynne, M.-L. (2011). Multilevel absorptive capacity and interorganisational new product development: A process study. *Journal of Management & Organization*, *17*(01), 39–55. doi:10.1017/S183336720000170X

Nicholas, J., Ledwith, A., & Perks, H. (2011). NPD best practice in SME and large organisations: Theory versus practice. *European Journal of Innovation Management*, *14*(2), 227–251. doi:10.1108/14601061111124902

Noke, H., & Hughes, M. (2010). Climbing the value chain: Strategies to create NPD capability in mature SME's. *International Journal of Operations & Production Management*, *30*(2), 132–154. doi:10.1108/01443571011018680

Noke, H., & Radnor, Z. J. (2004). Navigating innovation: A diagnostic tool supporting the process. *Journal of Manufacturing Technology Management*, *15*(2), 172–185. doi:10.1108/09576060410513751

Norman, P. A., & Peterson, T. (1999). Quality improvement through innovation. TQM and Innovation. In *Proceedings of the fourth international conference on ISO 9000 and TQM*. HKBU Press.

O'Dwyer, M., & Ledwith, A. (2009). Determinant of New Product performance in Small Firms. *International Journal of Entrepreneurial Behaviour & Research*, *15*(2), 124–136. doi:10.1108/13552550910944548

Oakley, M. (1990). Design and design management. In Design management- A handbook of issues and methods. Basil Blackwell Ltd.

Ogama, S., & Pillar, F. (2006). Reducing the risk of NPD. *MIT Sloan Management Review*, *47*(2), 65–71.

Osterlund, J. (1997). Competence management by informatics in R&D: The corporate level. *IEEE Transactions on Engineering Management*, *44*(2), 135–145. doi:10.1109/17.584922

Owens, J. D. (2007). Why do some UK SME's still find the implementation of a New Product Development process problematical? An Exploratory Investigation. *Management Decision Journal*, *45*(2), 235–251. doi:10.1108/00251740710727269

Owens, J. D. (2009). Modelling the NPD process: The value of a Product Development Process Model Approach, as a means for Business Survival in the 21st Century. In D. Jemielniak & J. Kociatkiewics (Eds.), Handbook of Research on Knowledge Intensive Organisations (pp. 208–227). Academic Press. doi:10.4018/978-1-60566-176-6.ch013

Owens, J. D., & Atherton, A. (2015). Can the development of a conceptual model for UK engineering companies help understand what the customer wants from new products? *Journal of Multidisciplinary Engineering Science and Technology*, *2*(11), 3155–3165.

Owens, J. D., & Davies, J. (2000). The Importance of a New Product Development (NPD) Process: Getting Started. *First European Conference on Knowledge Management*. Bled School of Management.

Perks, H., Cooper, R., & Jones, C. (2005). Characterising the role of design in NPD: An empirically derived taxonomy. *Journal of Product Innovation Management*, *22*(2), 111–127. doi:10.1111/j.0737-6782.2005.00109.x

Petrie, A. (2008). Developing products with a holistic process. *Design Management Review*, *19*(3), 68–73. doi:10.1111/j.1948-7169.2008.tb00131.x

Pugh, S. (1991). *Total design – Integrated methods for successful product engineering*. Addison-Wesley Publishing.

Pugh, S., & Moreley, I. E. (1988). *Total Design: Towards a theory of total design and Total Design: Some questions and some answers. Design division*. University of Strathclyde.

Reid, S. E., & de Brentani, U. (2004). The fuzzy front end of the new product development for discontinuous innovations: A theoretical model. *Journal of Product Innovation Management*, *21*(3), 170–198. doi:10.1111/j.0737-6782.2004.00068.x

Rodriguez, N. G., Perez, M. J. S., & Gutierrez, J A T. (2008). Can good organisational climate compensate for a lack of top management commitment to NPD? *Journal of Business Research*, *61*(2), 118–131. doi:10.1016/j.jbusres.2007.06.011

Saren, M A (1984). A classification and review of models of the intra-firm innovation process. *R&D Magazine, 14*(1), 11-24.

Schmidt, J., Sarangee, K., & Montoya, M. (2009). Exploring NPD project review practices. *Journal of Product Innovation Management*, *26*(5), 520–535. doi:10.1111/j.1540-5885.2009.00678.x

Sethi, R. (2000). New product Quality and Product Development Teams. *Journal of Marketing*, *64*(2), 1–14. doi:10.1509/jmkg.64.2.1.17999

Shenhar, A. J. (2001). One size does not fit all projects: Exploring classical contingency domains. *Management Science*, *47*(3), 394–414. doi:10.1287/mnsc.47.3.394.9772

Smith, A. D. (2011). Competitive approaches to NPD: A comparison of successful organisations in an unstable economic environment. *Team Performance Management*, *17*(3/4), 124–145. doi:10.1108/13527591111143682

Song, M., & Benedetto, A. D. (2008). Supplier's involvement and success of radical NPD in new ventures. *Journal of Operations Management*, *26*(1), 1–22. doi:10.1016/j.jom.2007.06.001

Suomala, P., & Jokionen, I. (2003). The pattern of success in product development: A case study. *European Journal of Innovation Management*, *6*(4), 213–227. doi:10.1108/14601060310500931

Taylor, A. (2010). The next generation: Technology adoption and integration through internal competition in NPD. *Organization Science*, *21*(1), 23–41. doi:10.1287/orsc.1080.0399

Troy, L. C., Hirunyawipada, T., & Paswan, A. K. (2008). Cross-functional and new product success: An empirical investigation of the findings. *Journal of Marketing*, *72*(Nov), 132–146. doi:10.1509/jmkg.72.6.132

Van Echtelt, F. E. A., Wynstra, F., Van Weele, A. J., & Duyesters, G. (2008). Managing supplier Involvement in NPD: A multiple case study. *Journal of Product Innovation Management*, *25*, 180–201. doi:10.1111/j.1540-5885.2008.00293.x

Wagner, S. M., & Hoegl, M. (2006). Involving suppliers in product development: Insights from R&D directors and project managers. *Industrial Marketing Management, 35*, 936-943.

Wright, I. C., & Swain, E. (1995). New product engineering: The quality paradox. *Journal of Engineering Design*, *6*(1), 49–56. doi:10.1080/09544829508907902

Zhao, Z., Yap, C. M., & Ang, B. W. (1999). A framework for facilitating quality tools' applications in new product development. *TQM and Innovation proceedings of the fourth international conference on ISO 9000 and TQM*. HKBU Press.

Zollo, M., Reuer, J. J., & Singh, H. (2002). Interorganisational routines and performance in strategic alliances. *Organization Science*, *13*(6), 701–713. doi:10.1287/orsc.13.6.701.503

Chapter 2
Development of Liquid Composite Moulded Thermoset Composite Automotive Parts Using Process Simulations:
Liquid Composite Moulding Process Simulations

Raghu Raja Pandiyan Kuppusamy
Birla Institute of Technology (BIT), India

ABSTRACT

Hand layup method is normally used for manufacturing composite parts with large and complex geometry where as Resin Transfer Moulding (RTM) process is a better substitute, but is sparingly used due to lack of proper manufacturing technology. Developments of RTM manufacturing process always requires a proper mould design. In addition, difficulty in the tooling design and mould fabrication cost increases with increase in size and complexity of the component. The scale down strategy of full scale product avoids bigger size mould requirements, prototype bulk production for product testing and quality check at the starting phase of product development. Moreover, the mould scale down strategy can be used to validate the process and the product with less capital input. In this work, a methodology to develop a RTM technology for manufacturing a complex and large composite structure utilizing the process simulation were demonstrated using high speed cab front as selected component.

INTRODUCTION

Resin Transfer Moulding (RTM) offers a number of advantages over Hand Lay Up (HLU) for fabrication of polymer matrix composite components such as higher reinforcement volume percentage, better distribution of resin and reinforcement, good surface finish and uniform thickness, high production volume due to lower cycle time, and less styrene emission. However still, HLU is being practiced by most of

DOI: 10.4018/978-1-5225-3401-3.ch002

the moulders for fabrication of complex and large components due to the very high cost of developing a proper mould design and determining most effective process parameters to produce the component with minimum mould filling time without the formation of dry spot. A proper mould design requires not only the geometry of the mould but also the effective injection strategy, pressure and temperature distributions. In addition, the difficulty in the mould design and also higher fabrication cost with increase in size and complexity of the component and chances of formation of dry spot is extremely high if the proper injection strategy is not used. Currently, RTM process is developed by trial and error methods, leading to high development cost and sub-optimum solution. Process simulation can be effectively used to determine the effective injection strategy including the position of injection gates and vents, and the most effective values of parameters for minimum mould filling time without formation of dry spot. The effective injection strategy, and the resulting pressure and temperature distribution can be utilized to design and fabricate the mould. A proper scaling down methodology based on process simulation can be used to manufacture a scaled down prototype for product testing and validating the process simulation of the full scale product. Thus the development of the RTM technology based on process simulation and scaled down prototyping will reduce the development cost and will enable the industry to use RTM technology instead of HLU for large and complex composite structures. While researchers have worked on different aspects on RTM, no effort has gone into develop RTM procedure with a scaled down component and use the data for scaling up to production level

LITERATURE REVIEW

A review of current approaches in modeling and simulation of the RTM process was presented by Shojaei et al (2003). It had been reported that the mould of shell like geometry having thickness much smaller than the length and width, was often considered to be two-dimensional. Therefore, the resin flow in a three-dimensional thin cavity had been modeled as a two-dimensional problem. Finite element method (FEM) was often preferred by the researchers to mathematically model the resin flow for the ease of application to all kind of geometries and boundary conditions can be modeled easily (Wang and Lee (1989), Fracchia et al. (1989), Bruschke and Advani (1990), Calhoun et al. (1996), Lim and Lee (2000), Phelan (1997)). The commercially available FEM packages such as ANSYS, LUSAS, NASTRAN, ABAQUS etc., cannot directly support RTM process simulation, whereas these can support mould filling simulations using homogenization technique with limitations of heat transfer and cure kinetics analysis. Moreover, these packages do not have separate process simulation module and this led research groups to develop their own codes such as LIMS (Advani (1994), Bruschke and Advani (1994)), LIMS3D (Varma and Advani (1994)), RTMSIM (Lin and Hahn (1996)) etc., to achieve their requirements. To find ever faster routes to RTM process simulation, commercially available specialized software packages such as PAM-RTM, RTM-Work that simulates various isothermal, non-isothermal, injection-compression, vacuum associated and auto clave moulding processes had been launched into the market. There had been a few work published on RTM process simulation for applications such as automobile parts (Verrey et al.(2006), Demaria et al. (2007))) using PAM-RTM software.

Objective and Scope

The objectives of this research work are to develop resin transfer moulding technology for large and complex composite structures by process simulation and to develop a scaling down methodology for manufacturing a scaled down prototype. RTM technology includes a proper mould design and determination of the most effective values of the process parameters for minimum mould filling time without a dry spot. Scaling down methodology will allow to manufacture scaled down prototype to validate the RTM Technology and to test the product for full scale performance.

Research Road Map

To meet the above objectives, the work flow of this research was structured with the selection of component for a specific application, characterization of raw material properties, process simulation to develop RTM technology using characterized raw material data, development of scaling down methodology and validation of the RTM technology using actual experiments. A schematic of this research road map is shown in Figure 1.

Figure 1. Road map of proposed research work

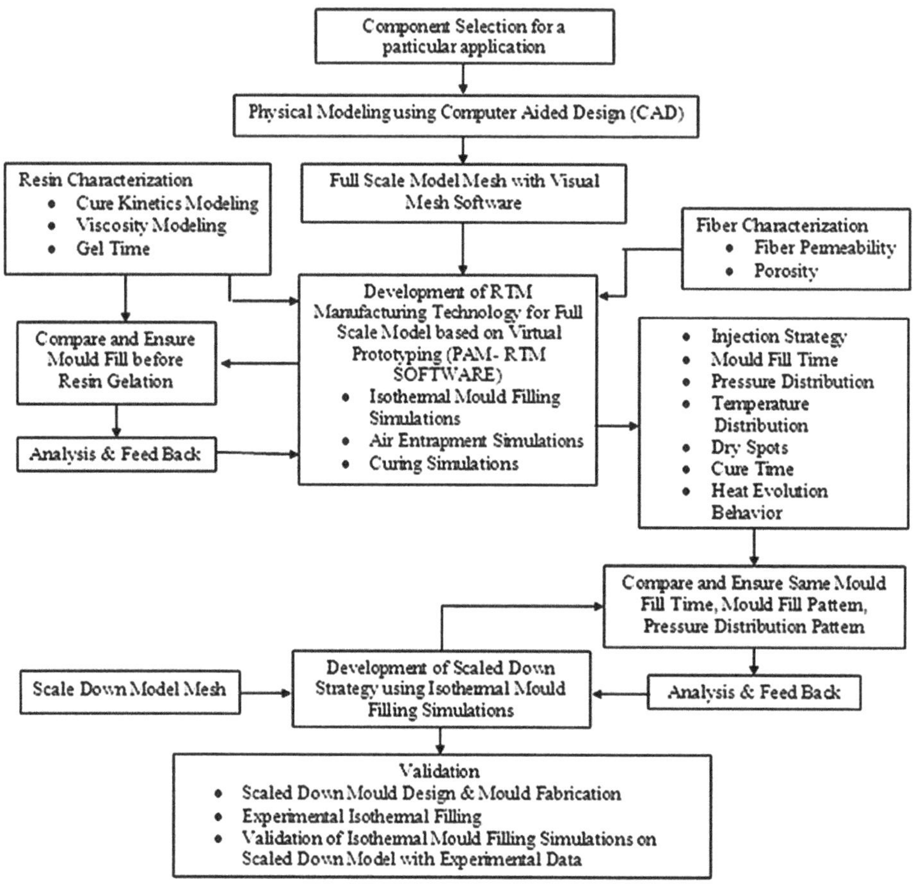

Component Selection

The cab front used by Indian Railways, which is large and complex in geometry and currently being manufactured by HLU, was chosen as the object of this study. The design of the cab front is shown in Figure 2. The two-dimensional CAD model of the cab front was meshed with Visual Mesh software. The geometry was meshed with 2D triangular elements with an element size of 0.06 m which is shown in Figure 3. Further reduction in element size increases the number of nodes and elements which in turn increases the load on computation without significant improvement in the simulation results.

Raw Material Characterization

Mould filling and subsequent curing are the significant processing stages considered in RTM process simulation. The mould filling involves the flow of viscous resin through the anisotropic reinforcement medium which depends on the permeability of the reinforcement medium and the resin viscosity. Resin viscosity depends on the extent of cure. During mould filling, the resin cure and hence the viscosity is changing with time. Hence the knowledge of permeability of the reinforcement medium, the resin cure kinetics and the interdependence of resin viscosity and cure kinetics (rheokinetics) are required for realistic simulation of RTM process. The gel time is another crucial parameter for the mould filling as the component geometry needs to be filled before the resin gels. The mould filling should be completed within 40% of the gel time to avoid the excessive increase in viscosity. The gel time and the curing be-

Figure 2. Schematic of the cab front

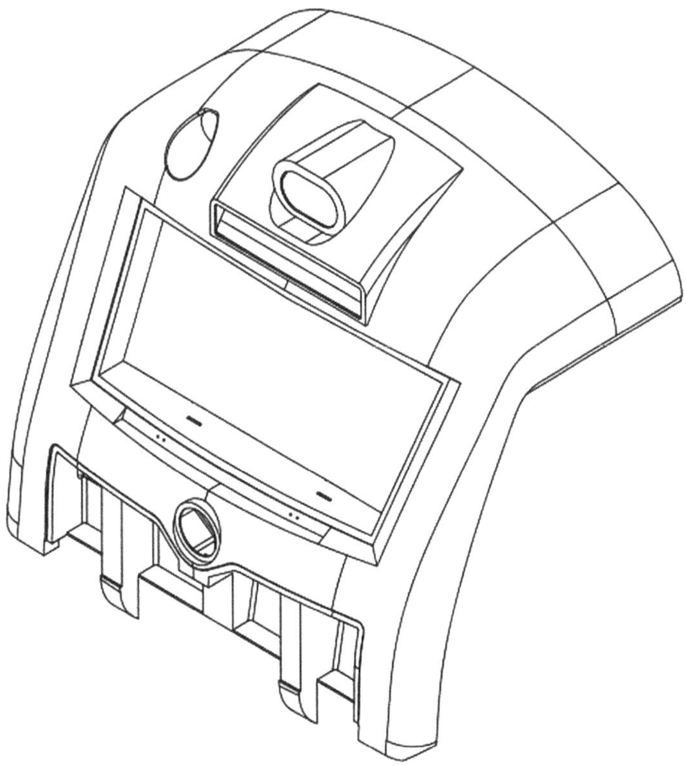

Figure 3. Discretized cab front

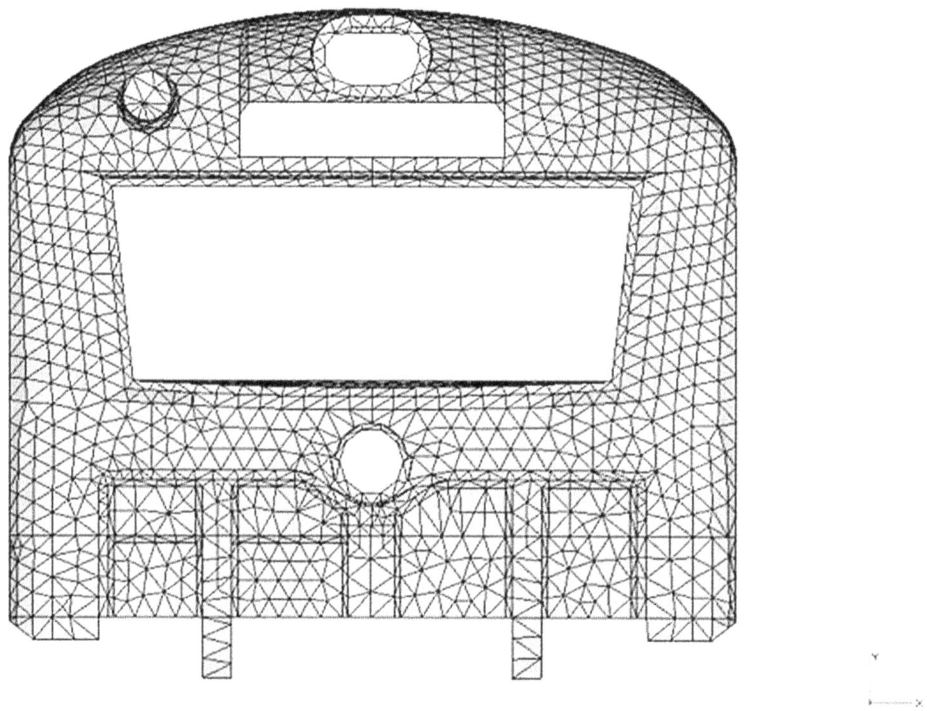

havior of the resin are the function of catalyst and accelerator concentrations. Optimization of accelerator and catalyst concentrations is required to obtain desired gel time. In the present work, cure kinetics, rheokinetics and gel time of the resin and the permeability of the reinforcement mat were characterized.

Materials

A commercially available, medium reactive unsaturated polyester resin specially formulated for RTM process was the resin matrix considered in this study. The medium reactive unsaturated polyester (UP) resin was tailored with a mixture of isophthalic acid, maleic anhydride and propylene glycol to an acid value of 15±2 mg KOH/g. Methyl ethyl ketone peroxide (MEKP) was used as an initiator/catalyst and a cobalt salt was employed as promoter/accelerator for the free radical polymerization of curing resin at ambient temperatures. Woven roven mat (WRM) of E-class glass fiber was used as the reinforcement material in this study.

Resin Gel Time

The gel time and peak-exotherm measurements for various initiator and promoter volume percentages were performed according to ASTM D2471. In the present work, accelerator and catalyst volumes were varied from 0.5 to 2.5 ml per 100 ml of resin. An empirical relationship between gel time (G) with accelerator (A) and catalyst (C) volume percentages was developed using non-linear regression analysis, as given in Equation 1.

$$G = 1694 * A^{-1.785} * C^{-1.682} \tag{1}$$

Resin Cure Kinetics

The extent of reaction during the curing process was measured experimentally using Differential Scanning Calorimetry (DSC) Q100 V 8.1 model of TA instruments operated under isothermal conditions and the heat flow vs. time was measured until the resin cure reaction was complete. To select suitable temperatures for the required isothermal experiments, a dynamic DSC run at a heating rate 10°C/min was first performed. From the dynamic scan, it was found that the onset of curing reaction was above 60°C and the maximum heat flow of curing reaction occurred around 105°C. Within this range, four temperatures of 70°C, 80°C, 90°C and 100°C were chosen for conducting isothermal curing experiments. The rate equation based on n^{th} order cure kinetics as given in Equation 2 was developed to quantify the rate of degree of cure ($\frac{d\alpha}{dt}$) as a function of degree of cure (α) and the temperature (T). The model parameters were obtained by a non-linear multiple regression method.

$$\frac{d\alpha}{dt} = 2.7 \times 10^{5} * \exp\left[\frac{-4982}{T}\right] * (1-\alpha)^{1.495}, \tag{2}$$

where time is in minutes.

Rheokinetics of Resin

Viscosity measurements were performed with a cup and bob viscometer, model Visco 88 Rheometer (Make: Bohlin Instruments). Tests were conducted at constant shear rate of 60 sec^{-1} under four isothermal conditions at 25°C, 40°C, 50°C and 60°C. The viscosity was measured as a function of time at constant temperature and shear rate till the onset of resin gelation. The measured Viscosity vs. Time data and the Cure vs. Time data obtained from Equation 2 were superimposed to develop the following relationship to describe the rheokinetics of the resin.

$$\eta(T,\alpha) = 5.8 * 10^{-7} * \exp\left[\frac{3649}{T}\right] - \left(8.3 * 10^{-3} * \alpha\right) \tag{3}$$

Reinforcement Permeability

Two methods have been used for the measurement of the permeability of the reinforcement which is described below.

Permeability Based on Flow Tracking by Conductivity Method

A constant flow 2D radial injection experimental set up was designed which tracks the flow front position and the pressure developed as a function of time during the flow of the fluid through the reinforcement medium. The set up uses radially arranged flow sensors to track the flow front and a pressure transducer to measure the change in pressure with time due to constant flow. The permeability was calculated using Darcy's equation. Distilled water, 70%(v/v) honey-water solution and 70%(v/v) glycerin-water solution were used as test fluids and chopped strand mat (CSM) and woven roven mat were the preforms used for the permeability measurements. The preform permeability was obtained between 10^{-8} and 10^{-6} m^2 with the different test fluids. Experimental results revealed that the test fluids have strong influence on the preform permeability.

Permeability Based on Flow Tracking by Visualization Technique

The permeability obtained using above method indicated that the permeability is dependent on the interactions of the test fluid and the reinforcement. To determine the effective permeability, which includes the interaction of the resin and reinforcement, during the resin mould fill time, the actual RTM mould filling experiments were conducted using the unsaturated polyester resin and the desired reinforcement mat. The mould filling simulation was carried out using PAM-RTM 2008 and a sensitivity analysis was performed using different values of permeability to match the experimental flow pattern with simulated flow pattern. Actual raw material parameters such as porosity, resin viscosity and density as given in Table 1, were used for the simulation. The results of the sensitivity analysis showed that for the permeabilities $K_x = 2x10^{-10}$ m^2 and $K_y = 5x10^{-10}$ m^2 matches the experimental and simulated flow patterns and the details of comparisons between experimental and simulated flow patterns are given in Figures 4 & 5 and Table 2.

Development of RTM Technology Using Process Simulations

The development of RTM technology for the selected composite component (cab front) included the prediction of (i) optimal positions of injection ports and air vents with minimum mould fill time and without dry spots, (ii) most effective process parameters such as injection pressure for the manufacturing process, (iii) distribution of pressure and temperature throughout the mould, (iv) amount of air in the mould and (v) curing time and the effect of temperature on resin cure. The RTM process simulations were performed for this purpose which were based on the finite element method. A commercial RTM simulation package, PAM RTM-2008, was used for the simulation. A number of simulation iterations were intuitively performed using different injection schemes varying the number and locations of injection ports and air gates. Both constant pressure injection and constant flow injection strategies were employed. For all the simulation trials, 10 mm diameter port was used. Preliminary simulations indicate that using constant flow injection method with more than 0.6 litres/minute flow produces excessive pressure drops of the order of 20 atm across the mould geometry which indicates that the flow rate above 0.6 litres/minute is not feasible. For constant pressure injection strategy, the injection pressure of 5 atm has been found to be a better choice considering the cost of mould and normal practice in industry where it operates below 5 atm for low pressure operation. Results indicates that the injection strategy with four injection points on the front face and four vents at the corners of the cab front as shown in Figure

Figure 4. Experimental flow front progression

Figure 5. Simulated flow front progression

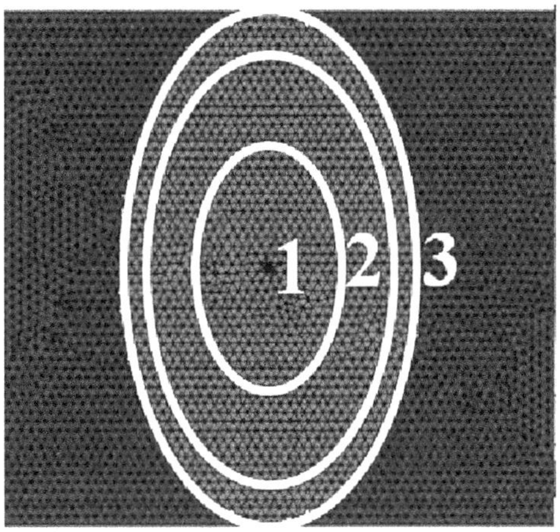

Table 1. Properties of raw materials

Property	Value
Resin viscosity	0.26 Pa.s
Resin Density	1100 Kg/m^3
Mat Porosity	0.5
Fiber density	2540 Kg/m^3

Mould cavity thickness = 1.5 mm
Injection pressure = 40 Psi ($2.75*10^5$ Pa)

Table 2. Comparisons on experimental and simulated flow fronts at specific intervals

Flow Fronts	Filling Time (Sec)	Experimental Flow Front Position (cm)		Simulated Flow Front Position (cm)	
		X	Y	X	Y
1	60	7.35	11.6	8	11.8
2	180	13	22	13	21
3	225	16.4	25	16	25

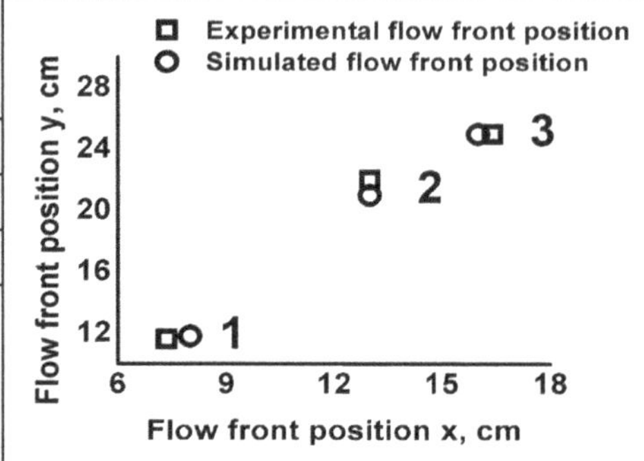

6 with constant pressure of 5 atmospheres found to be very effective, delivering an optimum mould fill time of 32 minutes without dry spots. The proposed injection strategy reduces the mould fill time to 32 minutes from several hours at constant pressure injection of 5 atm due to the better placement of injection ports and air vents. To determine the most effective values of process parameters such as injection pressure, viscosity and permeability, mould fill time was correlated as a function of these parameters, which are given in Equations 4, 5 & 6. Amount of the entrapped air was detected by imposing a vent pressure of 1 atmosphere at all four vents. The simulation results showed that micro-voids were formed due to the entrapment of air towards bottom part and the middle portion of the cab front. Efforts were made using alteration of injection pressure, vent opening operations to reduce the entrapped air. Using injection pressure of 6 atmospheres and all vent pressure 1 atmosphere until 10 minutes had reduced the air entrapment volume from 1% to 0.5% when compared to the volume of the cab front. The curing simulation results showed that the isothermal curing at 390 K decreased the cure time to 55 minutes with the maximum degree of cure.

Figure 6. Effective injection strategy

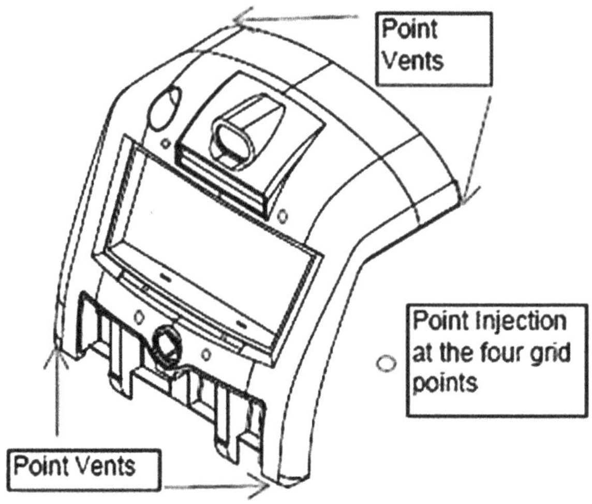

$$t = 162 - \left(39.6 * P_{inj}\right) + \left(3.683 * P_{inj}^{2}\right) - \left(0.1138 P_{inj}^{3}\right) \tag{4}$$

$$t = \left(67.69 * \eta\right) + 0.554 \tag{5}$$

$$t = 3.65 * 10^{-8} * K^{-1} \tag{6}$$

Scaled Down Prototype for Cab Front Using RTM Technology

The scaled down strategy for the radial flow advancement was obtained from Darcy's equation by matching the mould fill time, which is given in Equation 7. If the model is geometrically scaled down by a factor S, then the scaling down strategy is summarized as,

$$P_{inj}(scaled\ down) = \frac{P_{inj}(full\ scale)}{S^{2}} \tag{7}$$

Using this scaled down strategy, it was found that same mould fill time and pressure distribution were obtained for a scaled down model and a full scale model. The details of the full scale and scaled down simulations are tabulated in Table 3.

Experimental Validation of RTM Isothermal Mould Filling Simulations on Scaled Down Prototype

A simplified scaled down cab front mould with a scaling ratio of 1:3 from actual production scale was designed and constructed to validate the isothermal mould filling simulations. A transparent mould was made out of an acrylic material to observe the resin flow in the mould. A digital video recorder was used to record flow advancement. The experiments were performed under constant injection pressure of 2.75 atm at the optimized four injection gates using an unsaturated polyester resin of viscosity 0.25 Pa-sec and orthotropic glass fiber woven roven mat of permeabilities $K_y = 2*10^{-10}$ m^2 and $K_x = 5*10^{-10}$ m^2. An elliptical shape flow fronts were obtained at all the four injection gates for an isotropic preform before the flow fronts merge. The flow fronts advancement obtained from isothermal mould filling simulations and actual experiments on the scaled down cab front model were compared using flow fronts positions at different time interval, which are shown in Figures 7, 8 and 9, and a very close agreement is observed between both the results.

Table 3. Scaled down cab front simulation results

Scaling Factor	P_{inj} (Full Scale), atm	P_{inj} (Scaled Down), atm	Full Scale Model Mould Fill Time, min.	Scaled Down Model Mould Fill Time, min.	Mould Fill Pattern
2	5	1.25	32	32	Same mould fill time distribution
3	5	0.556	32	32	Same mould fill time distribution
4	5	0.3125	32	32	Same mould fill time distribution
3	24.75	2.75	11	11	Same mould fill time distribution

Figure 7. Experimental flow front progression

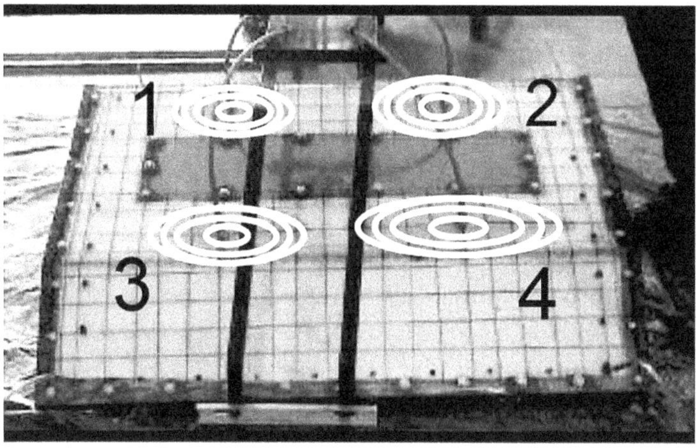

Figure 8. Simulated flow front progression

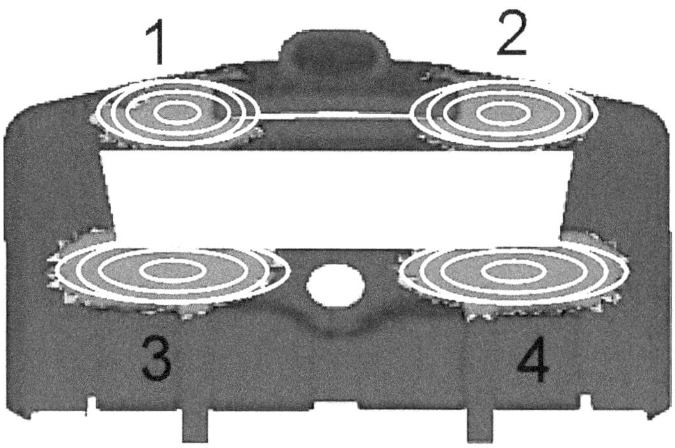

Figure 9. Comparisons of experimental and simulated flow fronts

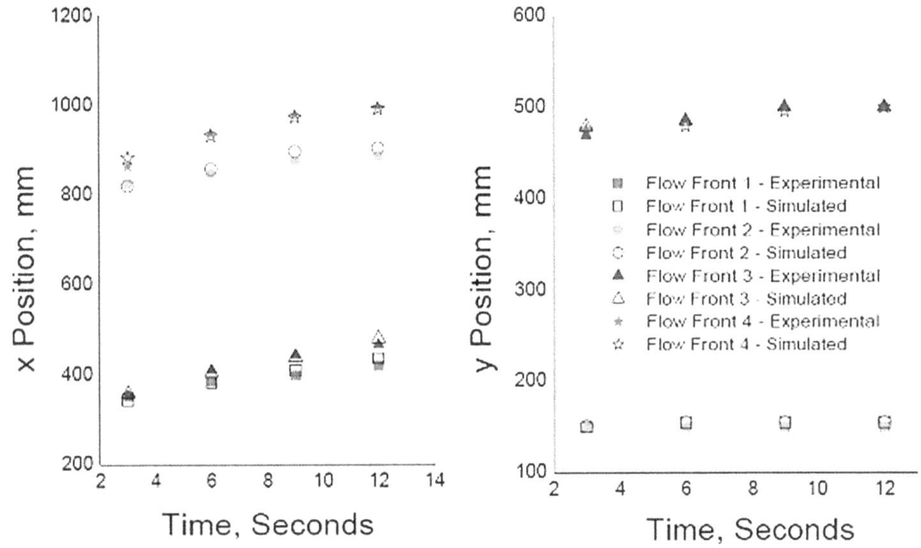

CONCLUSION

In this work, a methodology to develop a RTM technology for manufacturing a complex and large composite structure utilizing the process simulation were demonstrated using high speed cab front as selected component. The methodology will allow the manufacturer to determine the number and position of injection ports and vents, and the pressure distribution throughout the mould. This knowledge can be utilized by the moulders to design the mould avoiding the trial and error method and reduce developmental cost. Also a scaling down methodology based on full scale and scaled down model simulations were also demonstrated that can be used to validate the developed RTM technology as well as to test the product quality by manufacturing the scaled down prototype in the developmental stage.

Contribution of the Book Chapter

The demonstrated methodology for the development of RTM technology and manufacturing a scaled down prototype for process validation will reduce the development cost and will enhance the use of RTM process instead of HLU for manufacturing large and complex composite structures. Use of RTM process instead of HLU will improve the product quality due to higher reinforcement volume fraction, better surface finish and uniformity in the distribution of resin and reinforcement. Being closed mould process, use of RTM process will result in less styrene emission making the process more environment friendly. Moreover, due to lower cycle time product volume will increase. RTM process can also be automated resulting in less waste.

REFERENCES

Advani, S. G. (1994). Resin Transfer Moulding Flow Phenomena in Polymeric Composites. In S. G. Advani (Ed.), *Flow and Rheology in Polymer Composite Manufacturing* (pp. 465–515). Newark, NJ: Elsevier Science.

Bruschke, M. V., & Advani, S. G. (1990). A Finite Element/Control Volume Approach to Mould Filling in Anisotropic Porous Media. *Polymer Composites*, *11*(6), 398–405. doi:10.1002/pc.750110613

Bruschke, M. V., & Advani, S. G. (1994). A Numerical Approach to Model Non-Isothermal Viscous Flow through Fibrous Media with Free Surfaces. *International Journal for Numerical Methods in Fluids*, *19*(7), 575–603. doi:10.1002/fld.1650190704

Calhoun, R. D., Yalvac, S., Wetters, D. G., Wu, C. H., Wang, T. J., Tsai, J. S., & Lee, L. J. (1996). Mould Filling Analysis in Resin Transfer Moulding. *Polymer Composites*, *17*(2), 251–264. doi:10.1002/pc.10610

Demaria, C., Ruiz, E., & Trochu, F. (2007). In-Plane Anistropic Permeability Characterisation of Deformed Woven Fabrics by Unidirectional Injection. Part II: Prediction Model and Numerical Simulations. *Polymer Composites*, *28*(6), 812–827. doi:10.1002/pc.20108

Fracchia, C. A., Castro, J., & Tucker, C. L. (1989). A Finite Element/Control Volume Simulation of Resin Transfer Mould Filling. *Proceedings of the American Society for Composites, Fourth Technical Conference*, 157-166.

Lim, S. T., & Lee, W. I. (2000). An Analysis of the Three-Dimensional Resin Transfer Mould Filling Process. *Composites Science and Technology*, *60*(7), 961–975. doi:10.1016/S0266-3538(99)00160-8

Lin, M., & Hahn, T. (1996). A Robust and Efficient Approach for RTM Simulation. *Advanced Materials*, *74*, 75–76.

Phelan, R. F. (1997). Simulation of the Injection Process in Resin Transfer Moulding. *Polymer Composites*, *18*(4), 460–476. doi:10.1002/pc.10298

Shojaeia, A., & Ghaffariana, S. R. (2003). Modeling and Simulation Approaches in the Resin Transfer Moulding Process: A Review. *Polymer Composites*, *24*(4), 525–544. doi:10.1002/pc.10050

Varma, R., & Advani, S. G. (1994). Three-dimensional Simulations of Filling in Resin Transfer Moulding, Advances in Finite Element Analysis in Fluid Dynamics (ASME). *FED*, *200*, 21–27.

Verrey, J., Wakeman, M. D., Michaud, V., & Manson, A. E. (2006). Manufacturing Cost Comparison of Thermoplastic and Thermoset RTM for an Automotive Floor Pan. *Composites. Part A, Applied Science and Manufacturing*, *37*(1), 9–22. doi:10.1016/j.compositesa.2005.05.048

Wang, H. P., & Lee, H. S. (1989). Fundamentals of Computer Modeling for Polymer Processing. Hanser Publishers.

Chapter 3
Design of Impeller Blade of Mixed Flow Pump:
A Comparative Analysis

Divya Zindani
NIT Silchar, India

Apurba Kumar Roy
Birla Institute of Technology, India

Kaushik Kumar
Birla Institute of Technology, India

ABSTRACT

Mixed flow pumps have got a wide range of application such as that in irrigation, flood control, de-watering and power station cooling systems. A number of design templates have been in use to design the mixed flow pump impeller blades including free vortex theory and mean stream line theory. Here blade of the impeller of the mixed flow pump has been designed using Mean stream line theory and the free vortex theory and stresses arising due to static loading conditions has been obtained for both the cases. Comparison analysis for the equivalent stresses developed due to structural loading using various materials for the blade was undertaken. It was found that the stress for the blade designed using mean stream line theory is always on the higher side in comparison to that designed using free vortex theory. Further, out of the four materials of construction, titanium and copper were found to be suitable for blade designed using free vortex theory and the mean stream line theory respectively.

INTRODUCTION

Pump can be defined as a mechanical system converting mechanical energy to hydraulic energy. Based on the impeller configuration and specific speed the pump is classified traditionally as radial flow, axial flow and mixed flow pump. A mixed flow pump is a centrifugal type of pump which works on the rotary mechanism. In a Mixed flow pump head is developed partially by centrifugal action and partially by lift

DOI: 10.4018/978-1-5225-3401-3.ch003

of the vane on liquid. The liquid enters axially and discharges in radial and axial direction. These pumps have got a wide range of application having been applied to large scale liquid transfer tasks such as irrigation, flood control, dewatering and power station cooling systems. Because mixed flow pumps are inherently versatile and reliable, increased efforts are being made to extend their range. However, their design and performance prediction process is still a cumbersome task, mainly due to the great number of geometric parameters.

Different design templates over the years have been suggested by researchers ranging from simple designs from empirical relations to the relatively complex numerical ones. Few real 3D inverse design methods have been developed (Zangeneh et al.,1996, Xu and Gu, 1992, Borges,1994). These methods are very time-consuming and exhibit some difficulties in correlating the design parameters with the blade geometry (the first two) or convergence problems (the latter).A quasi-3D method had also been proposed (Peng et al., 2002.), which performs a blade-to-blade solution and saves computer time by using only one representative hub-to-shroud surface. Free vortex theory is the widely used design module which stems from the empirical relations and basic Eq.s of the turbomachinery. Another such theory that finds the same root is that of mean stream line theory which however is not so popular. However, no study has been made with regard to comparison of different design modules of the blade. In the present study a mixed flow pump impeller blade has been designed using the free vortex theory and the mean stream line theory. The free vortex theory is used often by the designers for designing of the impeller blade whereas the design of impeller blade using the mean stream line theory is rare. The blade designed using both the theories have been analyzed for von-mises stresses. The blade is under structural loading conditions. Four different materials i.e., Stainless steel, Titanium alloy, Copper alloy, Bronze, which are the most common construction material for the impeller blade and have different advantages under different conditions (stainless steel is best for corrosion etc.) have been used to carry out the comparison analysis for von-mises stress for both the theories. An overall comparison analysis has been made between the two theories of blade design considered.

BACKGROUND

A pump is a device, which converts mechanical energy to hydraulic energy. Based on the impeller configuration and specific speed the pump is classified traditionally as radial flow, axial flow and mixed flow pump. A mixed flow pump is a centrifugal type of pump which works on the rotary mechanism. In a Mixed flow pump head is developed partially by centrifugal action and partially by lift of the vane on liquid. The liquid enters axially and discharges in radial and axial direction. These pumps have got a wide range of application having been applied to large scale liquid transfer tasks. Increased efforts are being made to extend the range of mixed flow pumps. However, their design and performance prediction process is still a cumbersome task, mainly due to the great number of geometric parameters

Over the period of time the "impeller blade design of mixed flow pump" has been carried out according to the various templates. Wislicenus (1947), initiated the design of a mixed flow pump impeller. Myles (1965), Stepanoff (1957), Neumann (1991), Gahlot and Nyiri (1993), have suggested the different design procedures for designing mixed flow pumps. Varchola and Hlbocan (2012), used the experimental results of an internal impeller flow to create a geometrical design of a mixed flow pump.

All the above models are based on the inviscid simplified assumption. John (2009) had recently developed a fast numerical method for flow analysis and blade design of impellers. Li et al. (2012) carried out the optimum design using CFD. They simulated the three-dimensional flow field of the whole flow passage of a mixed-flow pump by using CFD software on the basis of Spalart-Allmaras turbulent model. The reason why the flow rate of this pump cannot reach to the design requirements was found out. The flow simulation of a low-specific-speed high-speed centrifugal pump was carried out by Jafarzadeh et al. (2011). A commercial CFD code was used to solve the governing Eq.s of the flow field. Chaudhari et al. (2013) analyzed the mixed flow impeller with the help of CFD. Desai and Naik (2015), carried out optimum design on mixed flow pump impeller using CFD simulation.

Research had been carried out for studying the effect of various forces on the blade. Mingxiong et al. (2015) presented a numerical investigation of hydrodynamic radial force of mixed-flow pump, through validations and CFD simulations under several flow rates. Şakar and Sabuncu (2008), carried out the dynamic and static stability of an aerofoil cross-section for a rotating blade subjected to axial periodic force. Effect of water drop erosion on a turbine blade was carried out by Li et al. (2008). Kaneko and Hayashi (2014), analyzed the damage in the piping system caused by the pressure pulsations, excited by centrifugal turbo machinery at the blade passing frequency. Experimental analysis of the effect of an inlet swirl on the static pressure acting on the centrifugal pump blades was carried by Kikuyama et al. (1992)

Pei et al. (2013) studied the unsteady flow behaviour to increase the performance of the single flow pump. Pei et al. (2012) investigated numerically and experimentally the unsteady flow-induced impeller oscillations for a single blade pump. The numerical, experimental and theoretical studies for the centrifugal pump as turbine were recently conducted by Wang et al. (2017). Mishra and Mozaffari (2017) used the computational fluid dynamics to analyze the performance of the impeller in solid-liquid mixing.

Another aspect which plays a critical role in the efficient performance of a pump is the analysis of structural and thermal stresses of the system. Ramamurti V. and Balasubramanian P. (1987), Jonker and Essen (1997), Lemeš and Nermina (2002), Bhope and Padole (2004). Arewar and Bhope (2013) contributed much on the stress analysis for highly complex blades of various turbo machines. Methodical design and stress analysis of a mixed flow pump impeller was carried out by Das et al.(2013). Srivastava et al. (2014) also designed the blade of the pump using free vortex theory.

The following salient points summarize the research gap for the research conducted on mixed flow pumps:

- No comparison of the different design templates used in designing the blade.
- Scarce literature on the material selection of the impeller blade material.
- Literature on the FEM analysis for the blade under static loading conditions is scarce.

The proposed work can be divided into four basic sections namely; Design of blade using the free vortex theory and the mean stream line theory, development of three dimensional Model of blade, analysis of von-mises stress using FEM and its validation, comparison of both the theories of blade design modules.

DESIGN METHODOLOGY

Design of Mixed Flow Impeller Blade Using Free Vortex Theory

In this section, a mixed flow impeller blade has been designed using the free vortex theory and the mean stream line theory. The fluid is considered to be incompressible fluid with no change of temperature of the fluid medium. Further, a uniform pressure distribution from hub to tip and inlet to outlet of the blade has been assumed. The blade had been designed for a pump with a non-dimensional specific speed of 1.998 rad/sec.

Assuming,

D_{2i}= 250mm

η= 0.76

P= 15 kW

The specific speed of the pump is a dimensionless quantity which is used to characterize the performance for a pump handling a particular fluid. It is given by the Eq. 1,

$$N_S = \frac{N\sqrt{Q}}{(H)^{3/4}} \tag{1}$$

The values once substituted in Eq. 1, yielded the specific speed of the mixed flow pump to be 105.74 rpm.

Value of e, for specific speed of 105.74 rpm, was found to be 0 .775, (APPENDIX), i.e.

$$e = \frac{D_{2i}}{D_{20}} = \frac{R_{2i}}{R_{20}} = 0.775 \tag{2}$$

Value of D_{2i} once substituted in Eq. 2, yielded D_{20} to be 324 mm.

The mean blade diameter at the outlet (D_{mo}), was then found by the empirical relation given by Stepanhoff, Eq. 3,

$$\frac{D_{2i}}{D_{mo}} = \frac{R_{2i}}{R_{mo}} = 0.9 \tag{3}$$

Therefore, D_{mo}=278 mm.

Stepanhoff gave an empirical relation to find out the outlet blade diameter at the hub (D_{10}). The relation is given by Eq. 4,

$$D_{mo} = \sqrt{\frac{D_{20}^2 + D_{10}^2}{2}} \tag{4}$$

Substituting the values of D_{20} and D_{m2} in Eq. 4, the value of D_{10} was found to be 224 mm.

Figure 1 depicts the different geometrical parameters of the blade.

Table 1. Nomenclature

N_s	Specific speed of the pump	Q	volumetric flow rate, m^3/Kg
H	Head of the pump, m	V_{w0}	Velocity of whirl at oulet, m/s
D_{2i}	Inlet blade diameter at tip	D_{1i}	Inlet blade diameter at hub, mm
D_{10}	Outlet blade diameter at hub	D_{20}	Outlet blade diameter at tip
D_{m0}	Mean blade diameter outlet	D_{mi}	Mean blade diameter at inlet
U_0	Outlet flow velocity, m/s	U_i	Inlet flow velocity, m/s
D_0	Diameter on the outlet side of the blade, mm	D_i	Diameter on the inlet side of the blade, mm
Φ	Outlet fluid angle, degree	θ_i	Inlet fluid angle, degree
V_{f0}	Velocity of flow at outlet, m/s	V_{fi}	Velocity of flow at inlet, m/s
V_{ro}	Relative velocity of flow at outlet, m/s	V_{ri}	Relative velocity of flow at inlet, m/s
V_0	Absolute velocity at outlet, m/s	V_i	Absolute velocity at intlet, m/s
V_s	Velocity of slip, m/s	z	Number of blades
Ψ	Cone angle of the impeller, degree	ϕ'	Modified Outlet fluid angle, degree
α_m	Mean blade angle, degree	D_m	Mean blade diameter, mm
S	Pitch, mm	Θ	Camber angle, degree
H	Manometric efficiency	P	Motor power, kW
C'	Chord length, mm	C_A	Actual Cord Length, mm
L	Blade height from hub to tip, mm	V_m	Mean velocity, m/s
Δ	Difference between old and new stagger angle, degree	Λ	Stagger angle, degree
C_m	Mean stream line chord length, mm	R_m	Mean stream line radius of curvature, mm
T	Torque applied, N-m	R	Radius of curvature of camber line, mm
B	Width of the plate, mm	σ_b	Bending stress, N/mm^2
L	Length of the plate, mm	T	Thickness of the plate, mm
I	Moment of inertia, mm^4	T	Shear stress, N/mm^2
U_d	Distortion energy density	Υ	Poisson's ratio
E	Modulus of elasticity, N/m^2	σ_y	Yield stress, N/m^2
V	Transverse shear force, N	Q	First moment of area
x_m	x coordinate on the mean stream line	y_m	y coordinate on the mean stream line
T	thickness of NACA 10C4 blade profile, mm	x_u	x coordinate on upper surface of the aerofoil
y_u	y coordinate on the upper surface of the aerofoil	x_l	x coordinate on lower surface of the aerofoil
y_l	y coordinate on the lower surface of the aerofoil		

Phenomenon of flow separation should be taken care of during the design of the blade which could cause serious structural failures. Therefore, in order to avoid flow separation in the blade of the pump, Myles suggested that the ratio D_{mo}/D_{mi} should be 1.20. From Figure 2, the ratio was found to be 1.375, and is given by Eq. 5,

Figure 1. Geometrical layout of the mixed flow pump impeller blade

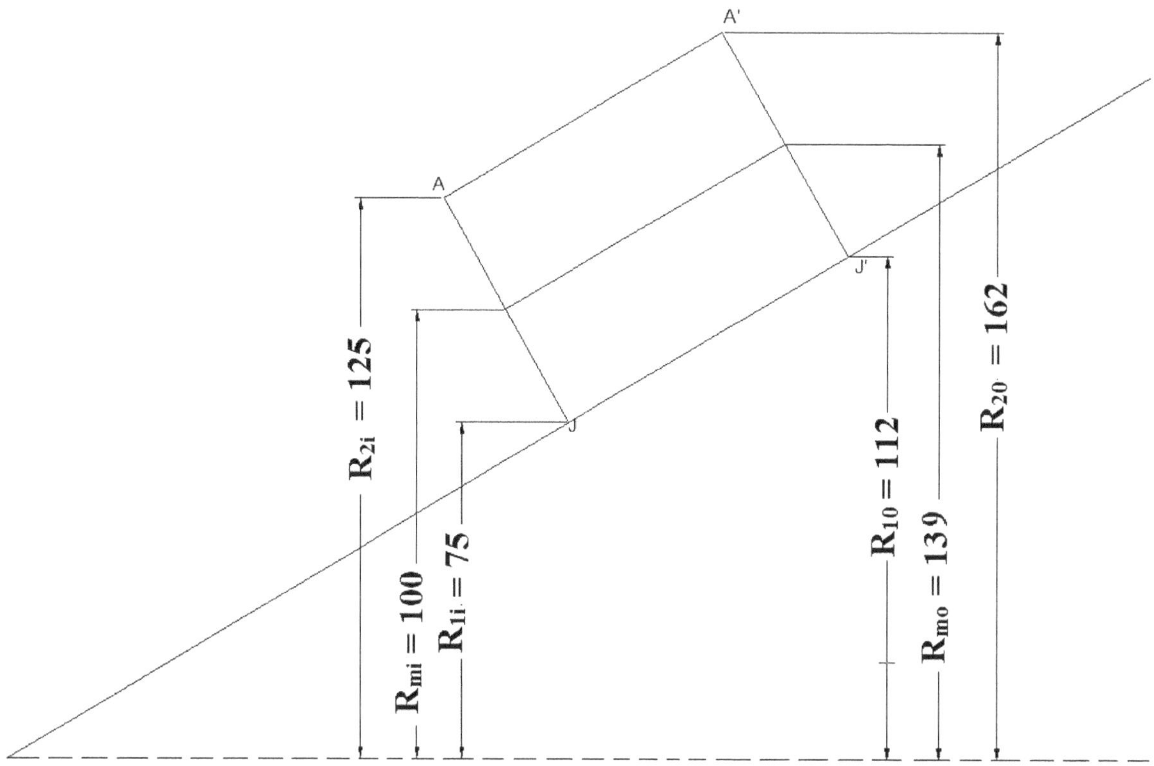

$$\frac{D_{mo}}{D_{mi}} = \frac{R_{mo}}{R_{mi}} = 1.375 \tag{5}$$

Eq. 5 was modified to Eq. 6 to result in ratio value of 1.20.

$$\frac{D_{mo} - X}{D_{mi} + X} = 1.20 \tag{6}$$

X was calculated to be 8 mm, where X is the correction factor taken in account to avoid flow separation.

The final geometrical parameters of the mixed flow pump impeller blade, considering the above discussion was tabulated to result in Table 2.

Table 2 shows the various geometrical parameters of the blade was obtained in AUTOCAD 2013.

The 3D model of the blade was obtained using the following design steps:

The blade layout, Figure 2, was divided into ten different sections as shown in the Figure 3.

Camber curves and camber chords were obtained for different sections of the blade.

Calculations of camber curve's radius of curvature, camber line chord lengths, and the stagger angles for camber lines were carried out.

The inlet and outlet circumferential velocities were obtained using Eqs. (7-8),

Table 2. Final geometrical parameters of the blade

Geometrical Parameter	Dimensions (mm)
Inlet blade diameter at hub (D_{1i})	150 mm
Inlet blade diameter at tip (D_{2i})	274 mm
Mean blade diameter at inlet (D_{mi})	216 mm
Outlet blade diameter at hub (D_{20})	224 mm
Outlet blade diameter at tip (D_{20})	298 mm
Mean blade diameter outlet (D_{mo}) Blade height (L)	260 mm 58 mm

Figure 2. Final geometrical layout of the blade (A-J-J-'A') in AUTOCAD 2013

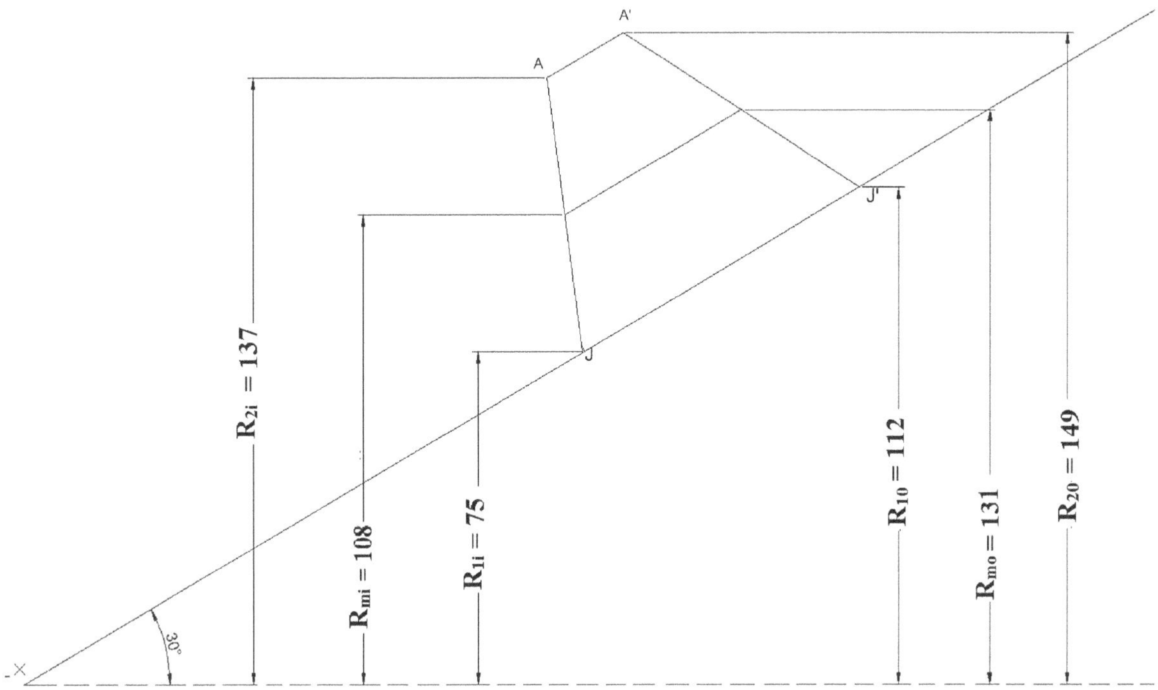

$$U_i = \frac{\Pi D_i N}{60} \tag{7}$$

$$U_0 = \frac{\Pi D_0 N}{60} \tag{8}$$

D_i and D_0 were measured for different sections from Figure 3.
Eq. 9 was derived from Figure 4 to obtain the inlet fluid angle, θ_i

Figure 3. Ten different sections of the blade (A-J-J-'A')

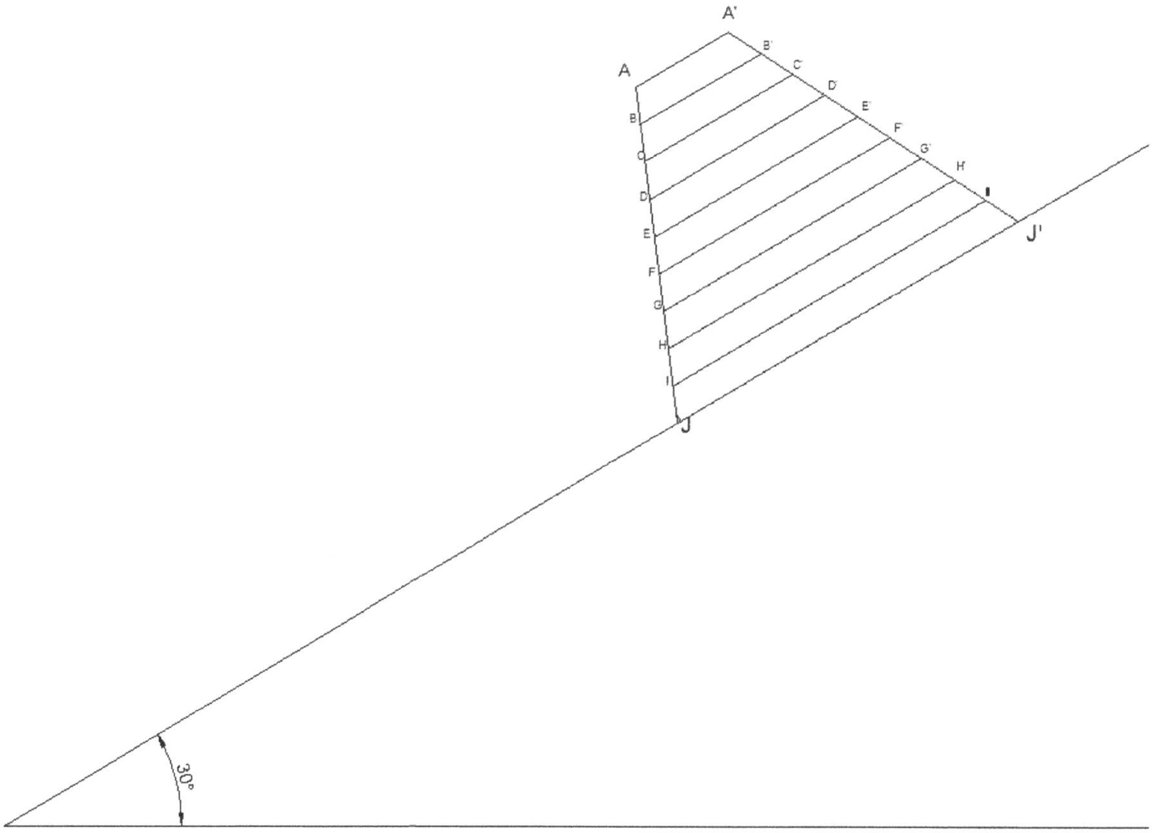

Figure 4. Inlet velocity triangle

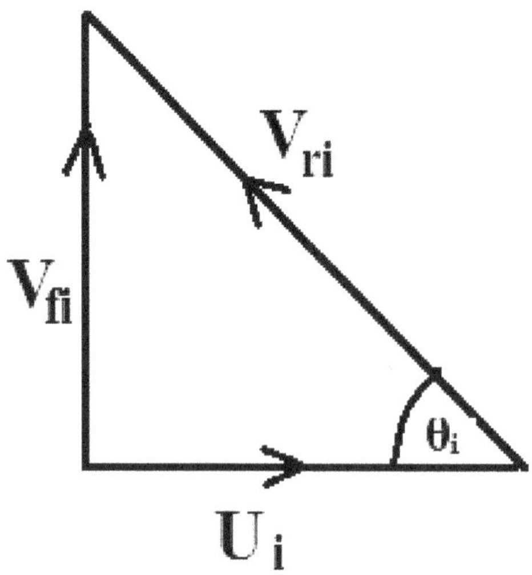

$$\tan\theta_i = \frac{V_{fi}}{U_i} \tag{9}$$

Velocity of flow V_{fi} was obtained using Eq. 10,

$$V_{fi} = \frac{Q}{2\Pi R_{mi} L} \tag{10}$$

The Head imparted to the water by the impeller is given by the Eq. 11,

$$H = \frac{\eta \times V_{wo} \times U_0}{g} \tag{11}$$

Eq. 11 was used to calculate the tangential component of the velocity, V_{wo}.
The outlet fluid angle ϕ was obtained using Figure 5 and the Eq. 12,

$$\tan\varphi = \frac{V_{f0}}{U_0 - V_{wo}} \tag{12}$$

Outlet velocity of flow, V_{f0} was obtained using Eq. 13,

$$V_{f0} = \frac{Q}{2\Pi R_{mo} L} \tag{13}$$

Figure 5. Outlet velocity triangle

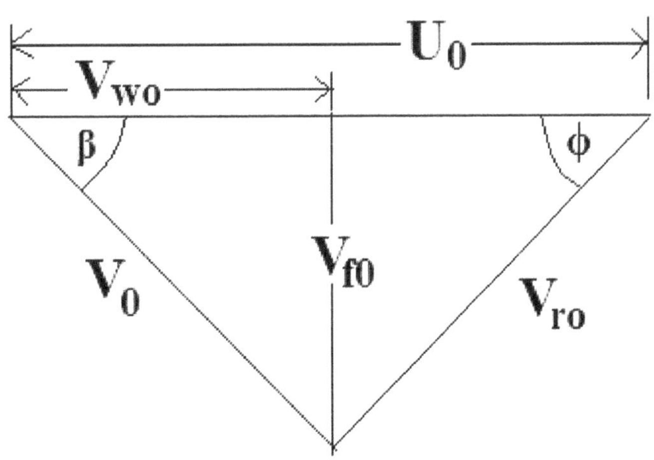

The particles in the impeller channel of the mixed flow pump will have relative angular velocity. This results in relative eddy in each passage, opposite in direction to the rotation of the impeller. The outer edge of the eddy will have a velocity V_S that produces the changed circumferential velocity at the outlet. V_S is known as the slip velocity. According to Balje and Stanitz (2009), for a given impeller tip speed and flow rate, the eddy can be reduced or eliminated by increasing the number of blades. Based on the above recommendations 8 blades were chosen and the slip velocity was calculated based on Stodola (1945) and is given by Eq. 14,

$$V_S = \frac{\Pi \times U_0 \cos\varphi \times \sin\psi}{Z} \tag{14}$$

Thus the outlet fluid angle was finally given by Eq. 15,

$$\tan\varphi' = \tan\varphi - \frac{V_S}{V_{f0}} \tag{15}$$

The mean blade angle, α_m was obtained using Eq. 16,

$$\alpha_m = \tan^{-1}\left[\frac{\tan\theta_i + \tan\varphi'}{2}\right] \tag{16}$$

Pitch S, is given by Eq. 17,

Figure 6. Slip velocity triangle

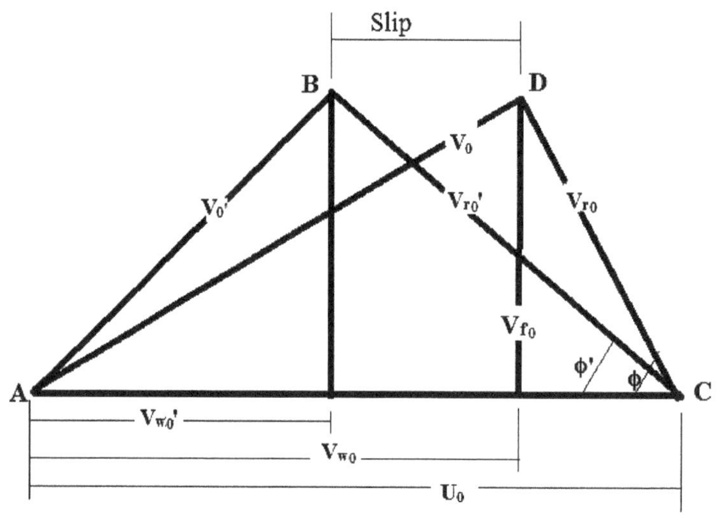

$$S = \frac{\Pi D_m}{Z} \tag{17}$$

where,

$$D_m = \frac{D_i + D_o}{2} \tag{18}$$

Using the results of Table 3 and Eqs. (19-21), an iterative procedure was followed to obtain the camber curve's radius of curvature and camber line chord length for different sections of the blade (A-J-J-'A'). The actual chord length was obtained using Eq. 19,

$$C_A = \frac{C'}{\cos\lambda} \tag{19}$$

where, C' was measured from Figure 3 for different sections of the blade (for instance linear distance J-J' for section J-J') and $\lambda = 90 - \alpha_m$.

Camber angle,θ was calculated using Eq. 20,

$$\theta = \varphi' - \theta_i + \Delta \tag{20}$$

$$\Delta = m \times \theta_i \times \sqrt{\frac{S}{C_A}} \tag{21}$$

'm' is a constant whose value was obtained from the variation with λ, Horlock (1958). (APPENDIX)

Table 3. Diameter blade angles and pitch for different sections of the blade (A-J-J-'A')

Section	Di (mm)	Do (mm)	θi (°)	ϕ' (°)	αm (°)	S (mm)
A-A'	274	298	12.38	13.55	12.94	112.34
B-B'	259	291	13.04	14.25	13.61	108.03
C-C'	243	279	13.87	15.23	14.51	102.88
D-D'	231	273	14.67	16.16	15.38	98.53
E-E'	218	265	15.45	17.2	16.27	94.60
F-F'	203	257	16.45	18.40	17.40	90.28
G-G'	190	247	17.59	19.77	18.61	86.01
H-H'	175	241	18.89	21.42	20.35	81.67
I-I'	161	233	20.39	23.45	21.62	77.35
J-J'	150	224	21.87	26.00	23.77	73.36

The iterative process was carried out for different sections of the blade (A-J-J-'A') till the difference between old stagger angle and the modified stagger angle is 0.01°.

Eq. 22 was used to calculate the camber curve's radius of curvature for different sections of the blade (A-J-J-'A')

$$R = \frac{C_A}{2\sin\frac{\theta}{2}} \quad (22)$$

In order to obtain the profile of a section of the blade (A-J-J-'A'), co-ordinates of the upper and lower surface of the blade section were obtained using Eqs. (23-26). The geometrical interpretation of the same is shown in the Figure 8.

$$x_u = x_m - \frac{t}{2}\sin\gamma \quad (23)$$

$$y_u = y_m + \frac{t}{2}\cos\gamma \quad (24)$$

$$x_l = x_m + \frac{t}{2}\sin\gamma \quad (25)$$

$$y_l = y_m - \frac{t}{2}\cos\gamma \quad (26)$$

Table 4. Actual chord length, stagger angle, camber angle, camber radius for different blade sections

Section	C_A (mm)	Λ	Θ	R (mm)
A-A'	102.60	76.48	1.73	3392
B-B'	119.40	75.90	1.75	3907
C-C'	132.17	74.98	1.91	3963
D-D'	148.18	74.15	2.23	3800
E-E'	159.21	73.19	2.34	3907
F-F'	165.52	72.11	2.79	3402
G-G'	170.27	70.84	3.06	3188
H-H'	176.98	69.34	3.49	2906
I-I'	181.57	67.52	4.12	2525
J-J'	182.23	65.39	5.38	1942

Figure 7. Camber curve and camber chord for section J-J' of the blade (A-J-J-'A')

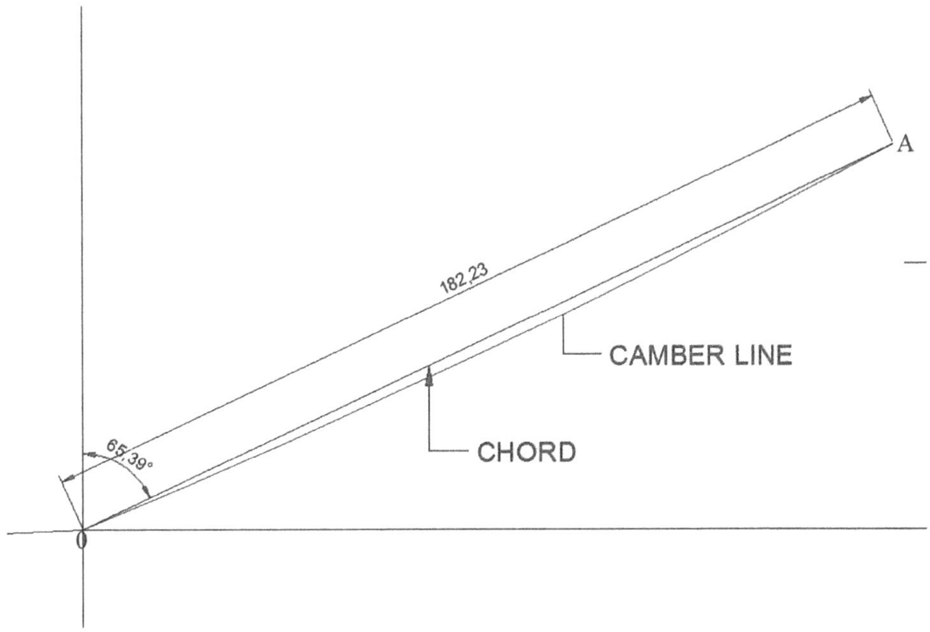

Figure 8. Geometrical description of chord line

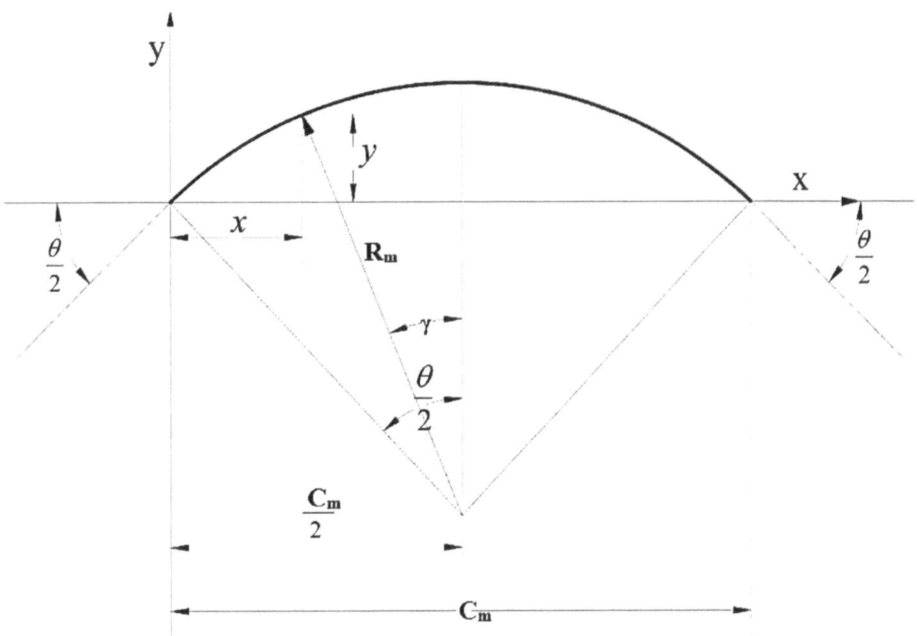

$$\gamma = \sin^{-1}\left[\frac{\frac{C_m}{2} - x_m}{R_m}\right] \tag{27}$$

The thickness, t, for section J-J' of the blade (A-J-J-'A'), has been tabulated in Table 5

3D model of the blade (A-J-J-'A'), was obtained by stacking profiles of different sections of the blade at their respective stagger angles.

Performing the geometrical transformations and connecting the trailing and the leading edges of the blade sections, complete 3D model of the blade was obtained.

Design of Mixed Flow Impeller Blade Using Mean Stream Line Theory

Streamlines are the geometrical representation of the of the flow velocity. The graphical representation is shown in the Figure 10.

Following design procedure was adopted to obtain the streamline.

The chord 'OA' and the camber line obtained in Figure 7 was divided into five equally spaced intervals for section J-J as shown in the Figure 11. The division was based on the pressure distributon across the blade. The entire pressure diagram was divided into equal steps such that $A_1 = A_2 = A_3 = A_4 = A_5$.

Eq. 28 was used to calculate the meridional velocity V_{f0},

$$V_{f0} = \frac{Q}{\pi \times D_{mo} \times L} \tag{28}$$

The relative velocities, V_{ri} and V_{r0}, were then derived using Figure 4 and Figure 5 respectively. Eq.s (29-31) were obtained.

Figure 9. Geometrical description of thickness distribution about chord line

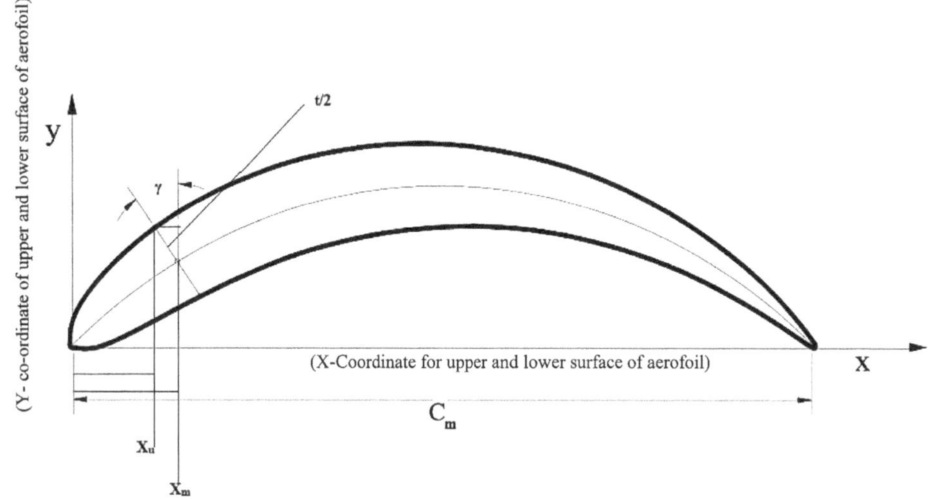

Table 5. Thickness as a percentage of chord length for section J-J' of the blade

10C4 Blade Profile Data		Modelled Blade Profile Data	
Distance From the Leading Edge	**Thickness**	**Distance From the Leading Edge**	**Thickness**
0	0	0	0
1.25	3.3	1.27	0.04
2.5	4.54	2.55	0.11
5	6.16	5.10	0.31
7.5	7.24	7.65	0.55
10	8.04	10.21	0.82
15	9.1	15.31	1.39
20	9.66	20.42	1.97
30	10	30.63	3.06
40	9.78	40.84	3.99
50	9.14	51.06	4.66
60	8.01	61.27	4.90
70	6.74	71.48	4.81
80	5.08	81.69	4.15
90	3.2	91.90	2.94
95	2.12	97.01	2.05
100	0	102.12	0

Figure 10. Graphical representation of streamline

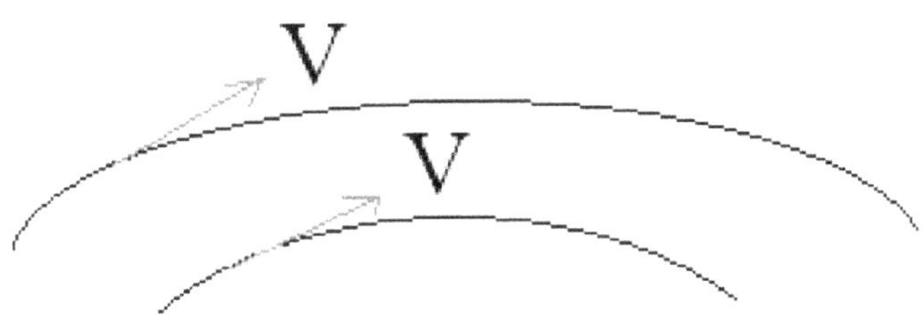

Figure 11. Division of chord line and the camber line for section J-J'

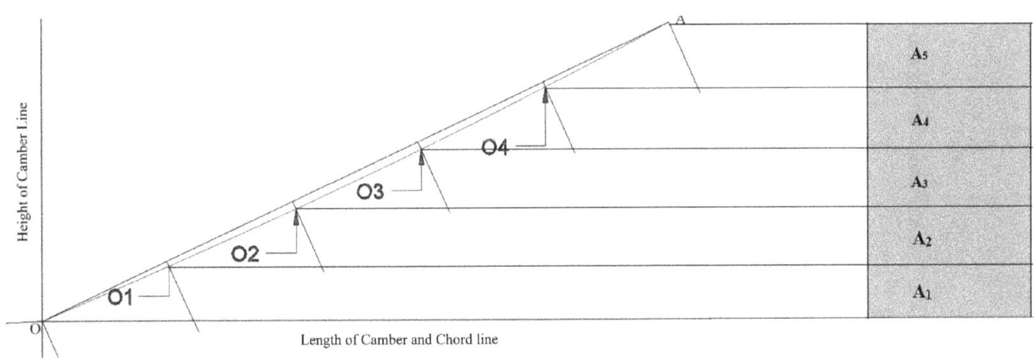

$$V_{ri} = \frac{V_{fi}}{\sin\theta_i} \tag{29}$$

$$V_{r0} = \frac{V_{fo}}{\sin\varphi'} \tag{30}$$

The mean velocity of flow at every section of the blade was obtained using the Eq. 31

$$V_m = \frac{V_{ri} + V_{ro}}{2} \tag{31}$$

The variation of resultant velocity at the inlet (V_{ri}), the relative velocity at outlet (V_{ro}) and the mean velocity of flow (V_m)for different sections of the blade is shown in the Figure 12 below.

The Resultant velocities V_{ri} and V_{r0} were plotted at an angle of 90-θ_i and 90-ϕ' from the vertical, respectively for different sections of the blade. Plot for section J-J' has been shown in the Figure 13.

Mean velocity of flow, V_m, were then plotted at angle of 90-α_m from the vertical, for diffrernt sections of the blade. The end points of the Resultant velocity vectors, V_{ri} and V_{ro}, and the mean velocity of flow vector, V_m were connected by a smooth curve to obtain the mean velocity curve for different sections of the blade. Plot of section J-J' has been shown in the Figure 14.

The change in resultant velocity vector ($\Delta V_r = V_{ro} - V_{ri}$) was determined for different sections of the blade and was then divided into five equally spaced intervals. The divisions intersecting the mean velocity curve gave the stations 0.2, 0.4 etc indicating the end points of various resultant velocity vectors (V_{r3}, Vr4etc…) across the blade section. Sample plot has been shown for the section J-J' of the blade in the Figure 15.

The mean stream line 0, 0.2, 0.4, etc was then obtained graphically by the condition that its direction at the points 0, 0.2, 0.4, etc is equal to the velocity vectors drawn to the corresponding points on the mean velocity curve. Figure 16 shows the mean stream line for section J-J' of the blade.

Figure 12. Relative velocities and mean velocity for different sections of the blade

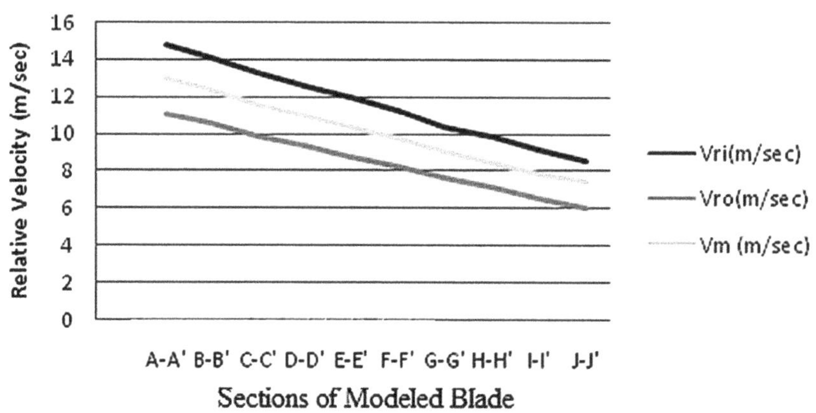

Figure 13. Resultant velocities for section J-J' of the blade (A-A'-J'-J)

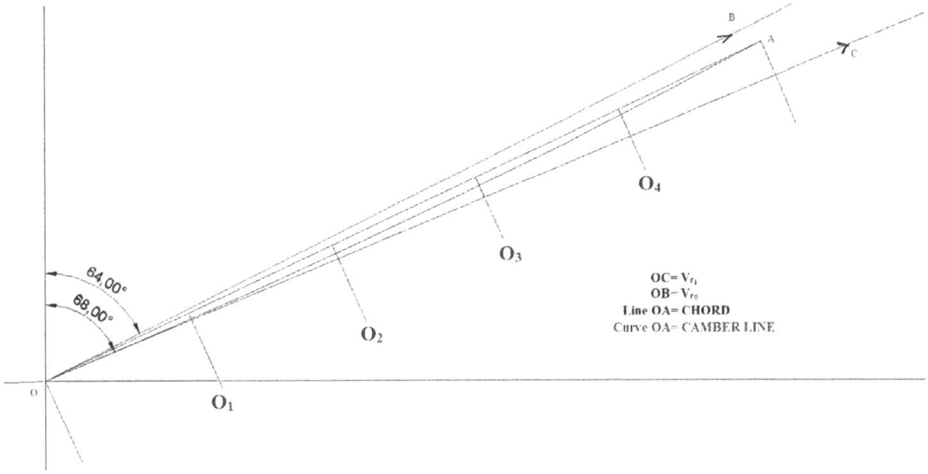

Figure 14. Mean velocity of flow and mean velocity curve for section J-J' of the blade (A-A'-J'-J)

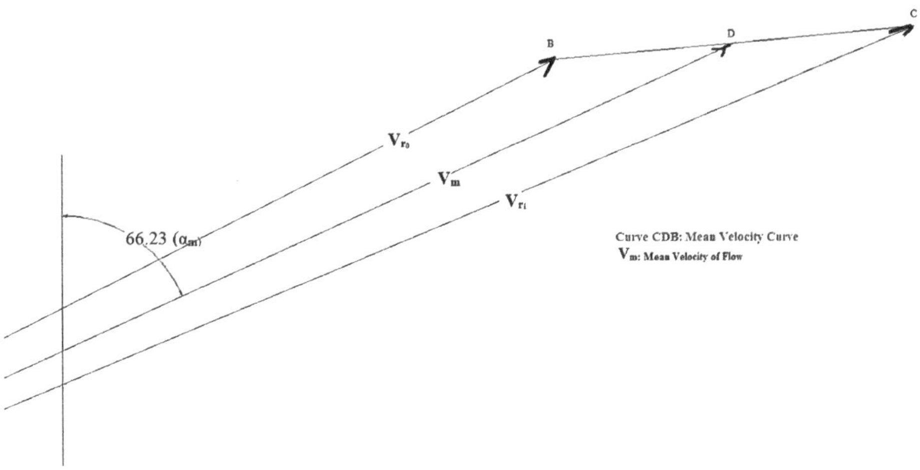

The procedural steps followed, yielded Chord length for mean stream line and its radius of curvature for different sections of the blade. Also the the stagger angle for the mean stream line chord for different sections of the blade were measured from the plot .The variation has been tabulated in Table 7.

The coordinates of the points on the upper and lower surface of the mean stream line were calculated to obtain the complete blade profile of a section of the blade.The co-ordinates on the upper and lower surface of the mean stream were obtained using the following sets of Eq.s (23-27).

Three-Dimensional model of the blade (A-J-J-'A'), for the free vortex theory and the mean stream line theory was obtained by stacking profiles of different sections of the blade at their respective stagger angles. The blade models are shown in the Figure 17 and Figure 18.

Figure 15. Resultant velocity vectors across section J-J' of the blade

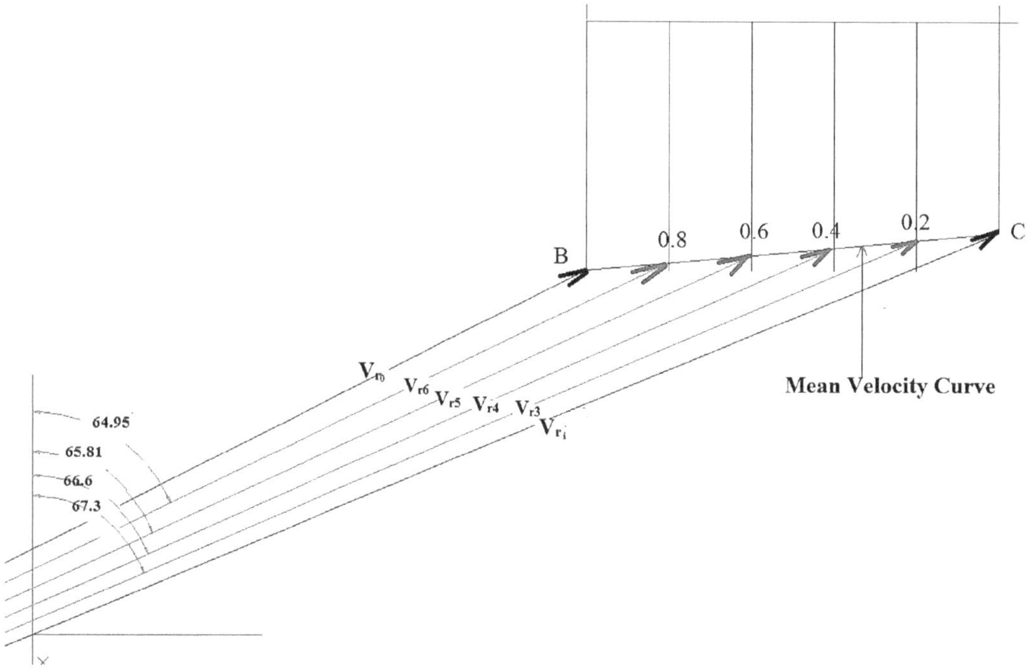

Table 6. Magnitude of resultant velocities and their respective angles

Resultant Velocity (m/s)	α (Degree)
8.51	65.39
8.14	67.33
7.83	66.60
7.62	65.81
7.21	64.95
6.01	64.00

FEM ANALYSIS

Analysis of the Modeled Blade (A-J-J-'A'), Neglecting the Thermal Conditions

Analysis of the modelled blade (A-J-J-'A'), was carried out using different material of constructions.

The analysis was carried out first for an equivalent rectangular plate having the same volume and surface area as the of the complex modeled blade. This was done to obtain the optimum element size for the analysis.

The equivalent plate is of dimension: 58×162.72×10.41 The equivalent model has been shown in the Figure 19.

Torque applied on the plate:

Figure 16. Mean stream line for section J-J'of the blade

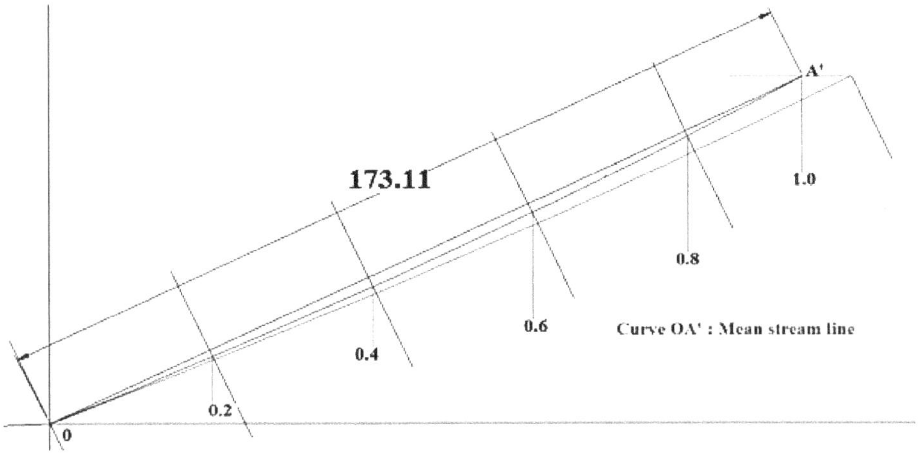

Table 7. Variation of chord length of mean stream line and its radius of curavature after curve fitting

Sections	Mean Stream LIne Chord Length, C_m (mm)	Radius of Curvature, R_m (mm)
J-J'	175.77	1410.50
I-I'	176.33	1803.77
H-H'	174.94	2154.74
G-G'	171.61	2460.10
F-F'	166.33	2716.19
E-E'	159.10	2920.48
D-D'	149.92	3069.23
C-C'	138.78	3159.13
B-B'	125.70	3186.88
A-A'	110.67	3201.15

$$T = \frac{wbl^2}{2} \tag{32}$$

Also,

$$T = \frac{60P}{2\Pi NZ} \tag{33}$$

From Eq.s (32-33), Eq. (34) was obtained,

Figure 17. Impeller blade designed using free vortex theory

Figure 18. Impeller blade designed using mean stream line theory

Figure 19. Equivalent model of the plate

$$w = \frac{60P}{\Pi NZlb^2} \tag{34}$$

where,

P=20kW,
N=1000 rev/min,
Z=8

Therefore,

$w = 87225$ N/m²

For a rectangular beam in pure bending, the shear stress is given by the Eq. (35)

$$\tau = \frac{Vq}{Ib} \tag{35}$$

Simplifying Eq. (35), Eq. (36) was obtained,

$$\tau = \frac{3wl}{t} \tag{36}$$

$\therefore \tau = 1.45 \times 10^6$

The bending stress σ_b was calculated using Eq. (37),

$$\sigma_b = \frac{3wl^2}{t^2} \tag{37}$$

Values once substituted in Eq. (38), resulted in σ_b i.e;

$$\sigma_b = 8.12 \times 10^6 N / m^2$$

The equivalent stress then was obtained using the Distortion energy theory.The distortion energy density associated with yielding is given by the Eq. (38),

$$U_d = \frac{1+\nu}{3E}\sigma_Y^2 \tag{38}$$

According to von Mises's failure criterion, the material under multi-axial loading will yield when the distortional energy is equal to or greater than the critical value for the material:

$$\frac{1+\nu}{3E}\sigma_{VM}^2 = \frac{1+\nu}{3E}\sigma_Y^2 \tag{39}$$

$$\therefore \sigma_{VM} \geq \sigma_Y$$

$$\sigma_{VM} = \sqrt{\sigma_b^2 + 3\tau^2}$$

$$\therefore \sigma_{VM} = 8.5 \times 10^6$$

So, the calculated value of the Von Misses stress at the root of the plate is 8.5x 10^6 N/m^2.

FEM convergence test for the Von Mises stress was carried out to determine the optimum size of the element using ANSYS 14.0 which was found out to be 0.52 mm. Figure 20 shows the convergence test results. Span wise variation of Von Misses stress for the plate for an optimal element size is shown in Figure 21.

A surface force density of 69977 N/m^2 was applied to the impeller blade. The analysis for the span-wise distribution of the equivalent stress obtained for the impeller blade constructed using copper as construction material is shown in the Figure 22.

RESULTS AND DISCUSSION

Analysis was conducted for equivalent stress for the impeller blade modeled using free vortex theory and mean stream line method. Different material of construction was used for the analysis. Loading conditions and boundary conditions for both the design templates were kept similar i.e. the blade was considered to be cantilever beam bearing a surface force density of 69977 N/m ^{2} on its surfaces. The analysis was then conducted using the ANSYS 14.0 software for the equivalent stresses generated in the modeled

Figure 20. Convergence history for copper alloy plate

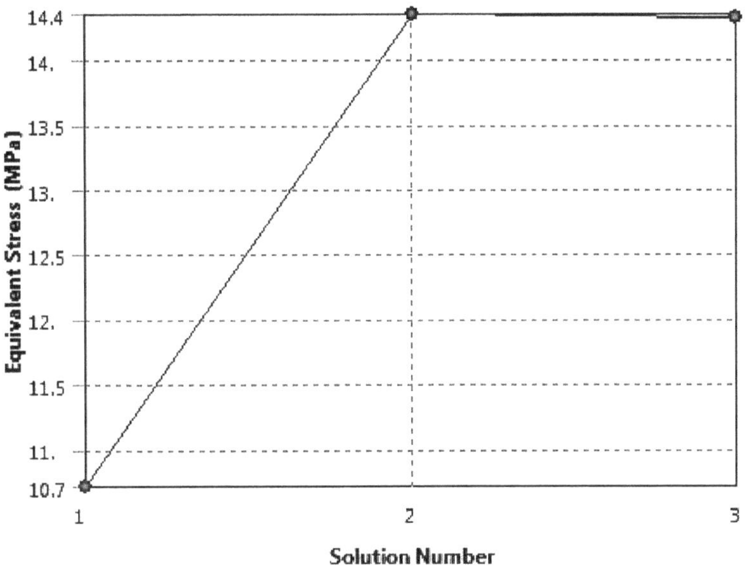

	Equivalent Stress (MPa)	Change (%)	Nodes	Elements
1	10.717		609388	141180
2	14.351	28.994	238304	122655
3	14.321	-0.21142	541636	326291

Figure 21. Span-wise variation of Von Misses stress for copper alloy plate

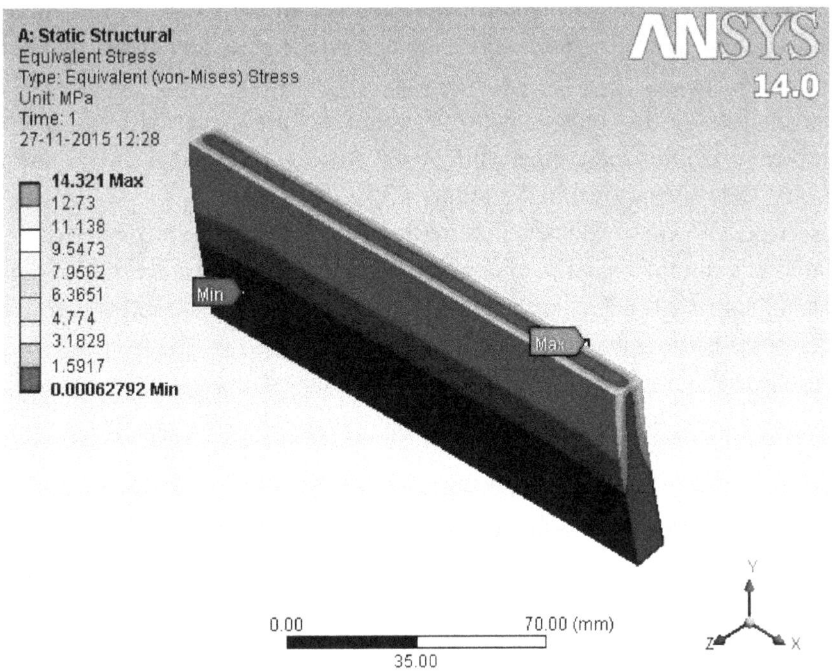

Figure 22. Span-wise variation of Von Misses stress for copper alloy modelled impeller blade

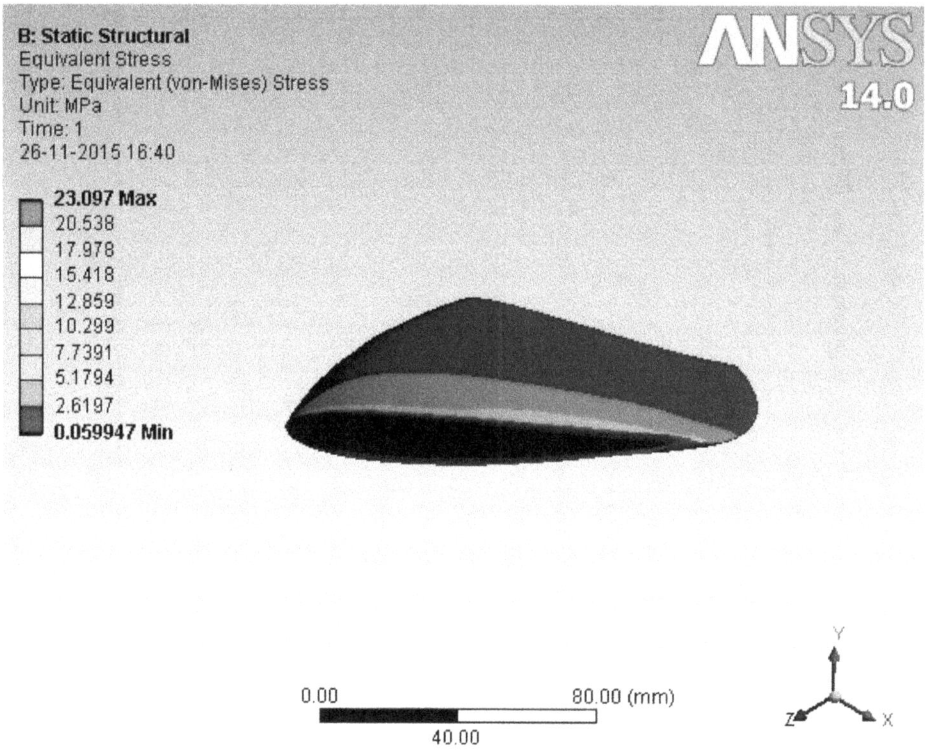

blade for both the design templates i.e., mean stream line theory and the free vortex theory. The mesh size used for the complex modeled blade was determined from the FEM convergence test conducted for the equivalent modeled plate. The test was conducted on the equivalent plate owing to the complex geometry of the impeller blade.

A comparison analysis for the equivalent stress generated for different material of construction i.e., stainless steel, bronze, titanium and copper alloy, for the mean stream line theory and the free vortex theory modeled blade has been shown in the Figure 23.

The comparison analysis clearly depicts that for the mean stream line theory modelled blade bronze could be considered as a suitable material of construction out of the four materials of construction used. The same material of construction i.e., bronze turns out to be the most suitable material of construction when the impeller blade is designed using the free vortex theory.

Further, the comparison analysis shows that the equivalent stress for the mean stream line theory is always on the higher side than that for the free vortex method. Although mean stream line method yields results at par with the free vortex method, still free vortex method is considered for designing of the impeller blade for mixed flow pump. This may be due to free vortex theory being relatively simpler in approach to designing of the impeller blade than the mean stream line theory.

Figure 23. Comparison of equivalent stress for both the design templates

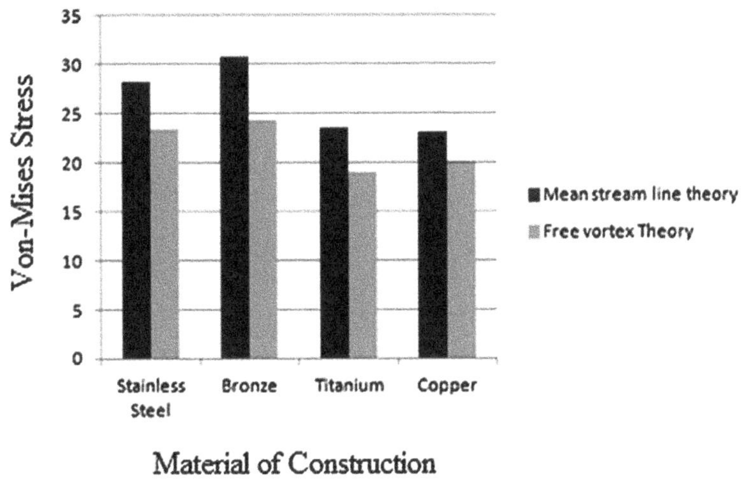

CONCLUSION

The comparison analysis between the two design templates of design concludes the following:

1. The Equivalent stress value for mean stream line method is always on the higher side than that for the free vortex method. This may be attributed to the difference in geometric profiles of the modeled impeller blade.
2. Figure 24 shows the comparison on the basis of chord length for the mean stream line method and the free vortex method. The comparison concludes that the variation of camber line chord length for different sections of the blade for both the design templates is almost similar.

Figure 24. Variation of camber line chord length for both the design templates

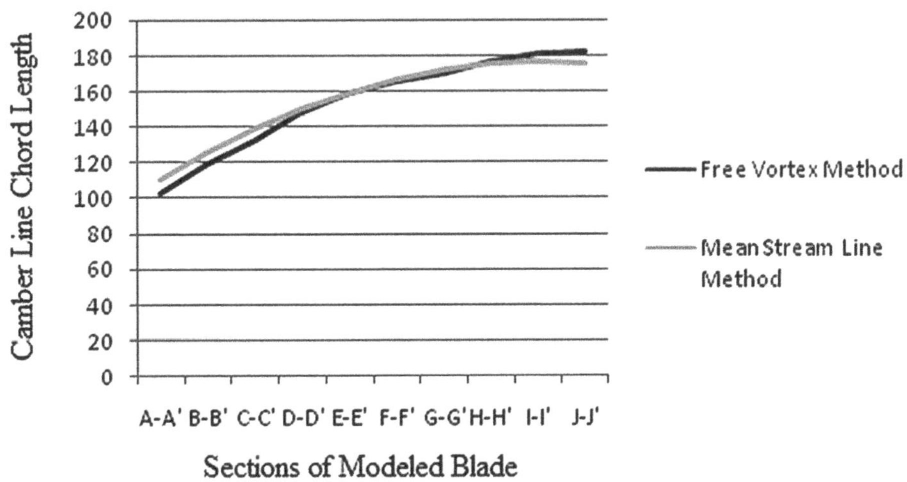

3. The mean stream line method although gives results at par with the free vortex method, but being relatively cumbersome is less preferred over the free vortex method.
4. Out of the considered four different material of construction for the impeller blade, titanium could be selected as the construction material if free vortex method is chosen as the design template; whereas copper may be preferred over the others if design template is mean stream line method.

FUTURE SCOPE

The present work has considered the comparison of two design templates of blades. Since there are other design templates of design, a future research direction could be to consider the other design templates and then to make an overall comparison to come out with a suitable design template for the mixed flow pump impeller blade. Experimental analysis is another area where the research could be made.

ACKNOWLEDGMENT

The authors sincerely acknowledge the comments and suggestions of the reviewers that have been instrumental for improving and upgrading the paper in its final form.

REFERENCES

Arewar, A. P., & Bhope, D. V. (2013). Stress analysis of axial flow fan impeller. *International Journal of Engineering Research and Applications*, *3*(1), 2086–2090.

Bhope, D. V., & Padole, P. M. (2004). Experimental and theoretical analysis of stresses, noise and flow in centrifugal fan impeller. *Mechanism and Machine Theory*, *39*(12), 1257–1271. doi:10.1016/j.mechmachtheory.2004.05.015

Borges, J. E. (1994). A three-dimensional inverse method for turbo machinery: Part-1Theory. *ASME Trans J Turbo*, *112*(3), 346–354. doi:10.1115/1.2927666

Chaudhari, S. C., Yadav, C. O., & Damor, A. B. (2013). A comparative study of mix flow pump impeller cfd analysis and experimental data of submersible pump. *International Journal of Research in Engineering & Technology*, *1*(3), 57–64.

Das, A., Roy, A. K., & Kumar, K. (2013). Design and Stress Analysis of a Mixed Flow Pump Impeller. *International Journal of Mechanical Engineering and Computer Applications*, *1*(5), 1–7.

Desai, S. M., & Naik, B. R. (2015). Optimum Design on Impeller of Mixed Flow Pump using CFD Simulation. *International Journal of Research in Engineering & Technology*, *4*(8), 535–537.

Gahlot, V. K., & Nyiri, A. (1993). Impeller Pumps. Theory and design. M.A.C.T.

Horlock, J. H. (1958). *Axial Flow Compressor –Fluid mechanics and thermodynamics*. London: Butterworth Scientific Publications.

Jafarzadeh, B., Hajari, A., Alishahi, M. M., & Akbari, M. H. (2011). The flow simulation of a low-specific-speed high-speed centrifugal pump. *Applied Mathematical Modelling*, *35*(1), 242–249. doi:10.1016/j.apm.2010.05.021

John, A. (2009). Fast numerical method for flow analysis and blade design in centrifugal pump impellers. *Computers & Fluids*, *32*, 284–289.

Jonker, J. B., & Essen, V. G. T. (1997). A finite element perturbation method for computing fluid induced forces on a centrifugal impeller, rotating and whirling in a volute casing. *International Journal for Numerical Methods in Engineering*, *40*(2), 269–294. doi:10.1002/(SICI)1097-0207(19970130)40:2<269::AID-NME63>3.0.CO;2-G

Kaneko, S., & Hayashi, I. (2014). Pressure pulsations in piping system excited by a centrifugal turbo machinery taking the damping characteristics into consideration. *Journal of Fluids and Structures*, *45*, 216–234. doi:10.1016/j.jfluidstructs.2013.11.012

Kikuyama, K., Hasegawa, H., & Maeda, T. (1992). Unsteady pressure change in centrifugal pump impeller passages due to inlet swirl. *Journal of Fluids and Structures*, *6*(3), 337–351. doi:10.1016/0889-9746(92)90013-S

Lemeš, S., & Zaimović-Uzunović, N. (2002). Mode shapes of centrifugal pump impeller. *Trends in the Development of Machinery and Associated Technology B&H*, 18-22.

Li, N., Zhou, Q., Chen, X., Xu, T., Hui, S., & Zhang, D. (2008). Liquid drop impact on solid surface with application to water drop erosion on turbine blades, Part I: Nonlinear wave model and solution of one-dimensional impact. *International Journal of Mechanical Sciences*, *50*(10-11), 1526–1542. doi:10.1016/j.ijmecsci.2008.08.001

Li, Z., Zeng, Y., Liu, X., & Wang, H. (2012). Optimum design on impeller blade of mixed-flow pump based on CFD. *Procedia Engineering*, *31*, 187–195. doi:10.1016/j.proeng.2012.01.1011

Mingxiong, O., Weidong, S., Weidong, J., & Qiang, F. (2015). Numerical simulation and experimental validation on hydrodynamic radial force of mixed-flow pump impeller. *Nongye Gongcheng Xuebao (Beijing)*, *9*, 71–76.

Mishra, P., & Mozaffari, F. E. (2017). Using computational fluid dynamics to analyze the performance of the Maxblend impeller in solid-liquid mixing operations. *International Journal of Multiphase Flow*, *91*, 194–207. doi:10.1016/j.ijmultiphaseflow.2017.01.009

Myles, D. J. (1965). *A design method for mixed flow fans and pumps*. Report No. 117 National Engineering Laboratory.

Neumann, B. (1991). *The interaction Between Geometry and Performance of a centrifugal Pump*. London: Mechanical Engineering Publications.

Pei, H., Dohmen, H. J., Yuan, S. Q., & Benra, F. K. (2012). Investigation of unsteady flow-induced impeller oscillations of a single-blade pump under off-design conditions. *Journal of Fluids and Structures*, *35*, 89–104. doi:10.1016/j.jfluidstructs.2012.08.005

Pei, J., Yuan, S., Yuan, J., & Wang, W. (2013). The influence of the flow rate on periodic flow unsteadiness behaviors in asewage centrifugal pump. *Journal of Hydrodynamics*, *25*(5), 702–709. doi:10.1016/S1001-6058(13)60415-1

Peng, G., Cao, S., Ishizuka, M., & Hayama, S. (2002). Design optimization of axial flow hydraulic turbine runner: Part I – an improved Q3D inverse method. *International Journal for Numerical Methods in Fluids*, *39*(6), 517–531. doi:10.1002/fld.342

Ramamurti, V., & Balasubramanian, P. (1987). Steady state stress analysis of centrifugal fan impellers. *Computers & Structures*, *25*(1), 129–135. doi:10.1016/0045-7949(87)90224-0

Şakar, G., & Sabuncu, M. (2008). Dynamic stability analysis of pretwisted aerofoil cross-section blade packets under rotating conditions. *International Journal of Mechanical Sciences*, *50*(1), 1–13. doi:10.1016/j.ijmecsci.2007.06.001

Srivastava, S., Roy, A. K., & Kumar, K. (2014). Design of a mixed flow pump impeller blade and its validation using stress analysis. *Procedia Materials Science*, *6*, 417–424. doi:10.1016/j.mspro.2014.07.053

Stepanoff, A. J. (1957). *Centrifugal and Axial flow pumps*. New York: John Wiley and sons.

Stodola, A. (1945). *Steam and gas turbine*. New York: Peter Smith.

Varchola, M., & Hlbocan, P. (2012). Geometory design of a mixed flow pump using experimental results of an internal impeller flow. *Procedia Engineering*, *39*, 168–174. doi:10.1016/j.proeng.2012.07.021

Wang, T., Wang, C., Kong, F., Gou, Q., & Yang, S. (2017). Theoretical, experimental, and numerical study of special impeller used in turbine mode of centrifugal pump as turbine. *Energy*.

Wislicenus, G. F. (1947). *Fluid mechanics of turbomachinery*. New York: McGraw-Hill.

Wright, T., & Gerhart, P. (2009). *Fluid Machinery application, selection and design*. Boca Raton: CRC press. doi:10.1201/b15836

Xu, J. Z., & Gu, C. W. (1992). A numerical procedure of three-dimensional design problem in turbo machinery. *ASME Trans J Turbomach*, *114*(3), 548–582. doi:10.1115/1.2929178

Zangeneh, M., Goto, A., & Takemura, T. (1996). Suppression of secondary flows in a mixed flow pump impeller by application of three-dimensional inverse design method: Part 1 – design and numerical validation. *ASME Trans J Turbomach*, *118*(3), 536–543. doi:10.1115/1.2836700

KEY TERMS AND DEFINITIONS

ANSYS: Analysis software.

Free Vortex Theory: A design template for design of mixed flow pump impeller blade.

Impeller Blade: The rotating part which increases the flow and pressure of fluid.

Mean Stream Line: A curve on which the resultant velocity vectors of the fluid are tangential.

Mean Stream Line Theory: A design template for design of impeller blades on the basis of mean stream lines.

Mixed Flow Pump: Pump in which fluid experiences both radial acceleration and lift.

Von-Mises Stress: The equivalent stress undergone by a structure under complex state of loading.

APPENDIX

Figure 25. Impeller Constants as a function of Sp. speed
Stepanoff, 1957.

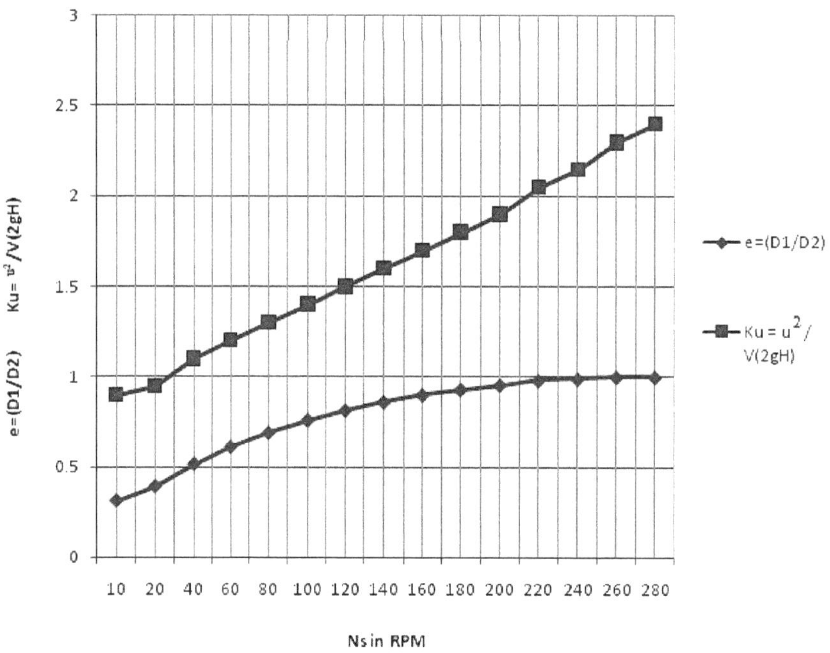

Figure 26. Deviation rules
A.D.S. Carter, Courtesy: George Newnes.

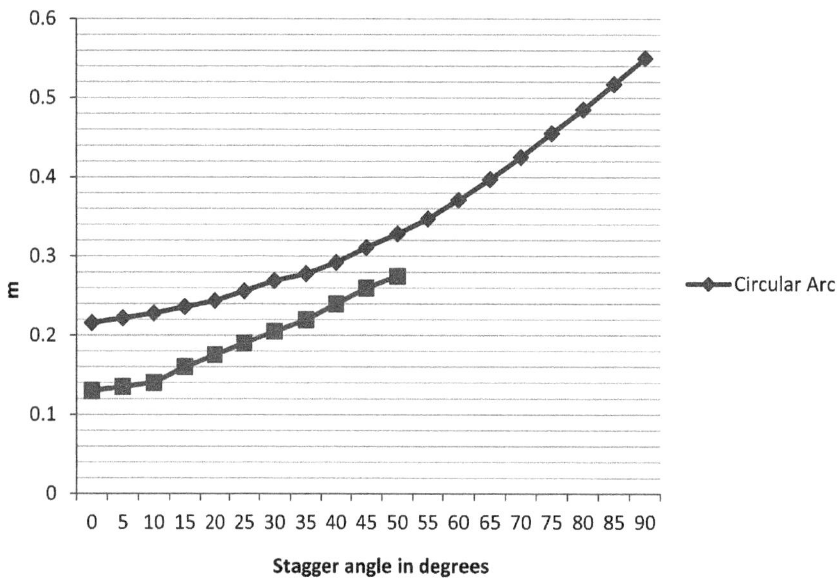

Chapter 4
Design of Base Isolation System for Buildings

Govardhan Bhatt
National Institute of Technology Raipur, India

D. K. Paul
IIT Roorkee, India

Shubhankar Bhowmick
National Institute of Technology Raipur, India

ABSTRACT

Seismic base isolation is one of the most widely implemented and accepted seismic protection systems and is a relatively recent and evolving technology. The most common isolation system used is Laminated Lead Rubber Bearings (LLRB). They combine the function of isolation and energy dissipation in a single compact unit, giving structural support, horizontal flexibility, damping, and a re-centering force in a single unit. The force deformation behavior of LLRB is modeled as bilinear system with viscous damping. In this paper, a comprehensive design of LLRB is presented. Accurate evaluation of the structural properties and precise modeling of isolation devices are of utmost importance in predicting the response of the structure during the earthquakes. Base isolation has now been used in numerous buildings in countries like Italy, Japan, New Zealand, and USA. Base isolation is also useful for retrofitting of important buildings (like hospitals and historic buildings).

INTRODUCTION AND BACKGROUND

Past earthquakes have indicated, especially in developing countries, that major loss of life often occurs due to the collapse of poorly constructed buildings. If the level of seismic demand on these buildings is reduced through a simple but reliable engineering solution, this would result in much safer design. Seismic base isolation, which is aimed at reducing the seismic demand instead of increasing the earthquake-resistant capacity of the structure, is an attractive alternative to conventional earthquake-resistant design methods. Seismic base isolation is one of the most widely implemented and accepted seismic protection

DOI: 10.4018/978-1-5225-3401-3.ch004

systems. Seismic isolation is a relatively recent and evolving technology. It has seen an up-rise in use since 1980s, and has been well evaluated and reviewed internationally. The two most common types of base isolation systems adopted in practice utilize either rubber bearings or sliding systems between the foundation and superstructure for the purpose of isolation from ground motions in the buildings as well as bridges. Base isolation has now been used in numerous buildings in countries like Italy, Japan, New Zealand, and USA. Base isolation is also useful for retrofitting of important buildings (like hospitals and historic buildings). By now, over 1000 buildings across the world have been equipped with seismic base isolation. A four storey hospital building at Bhuj, India has been constructed using base isolation technique where the isolators were brought from New Zealand (Govardhan and Paul, 2016).

The isolation system reduces the effects of an earthquake by partially isolating the superstructure and its contents from potentially damaging ground motion. It is very essential to understand the different parameters affecting the response of base-isolated structure when used for seismic protection of the structures. Especially in case of the base-isolated structures, that houses sensitive equipment, determination of acceleration imparted and associated peak displacement are the key issues for the design engineer (Stanton and Roeder, 1991). Accurate evaluation of the structural properties and precise modeling of isolation devices are of utmost importance in predicting the response of the structure during the earthquakes. The most common isolation system used is Laminated Lead Rubber Bearings (LLRB). They combine the function of isolation and energy dissipation in a single compact unit, giving structural support, horizontal flexibility, damping, and a re-centering force in a single unit. The force deformation behavior of LLRB is modeled as bilinear system with viscous damping. Deb (2004) has given an overview of seismic isolation and also a three dimensional nonlinear analysis procedure. Effects of soft soil and near faults on performance of base isolated buildings are also reported. Skinner et al., (1999) proposed the design procedures for seismically isolated structures. Using a logical progression, seismic isolation along with the concepts of earthquake structural dynamics underlying the isolation theory is reported. Methods discussed provide the basis for continuing development and refinement of seismic isolation design. Mkrtychev et.al. (2014) reported the usefulness of non-linear LRB isolation system in buildings having different heights using an explicit scheme in LS-DYNA. The effect of mechanical property variation of LRBs on the seismic response of base isolations under input ground motions is reported under different acceleration to velocity ratio (A/V ratio) by Choun et.al. (2014). In a recent work (Islam et.al, 2013), the insertion of Lead rubber bearing (LRB) with bilinear hysteretic behavior and high damping rubber bearing (HDRB) on corresponding structural bases is reported along with the check of structural feasibility of the insertion. For a bidirectional site-specific earthquake, the linear static, free vibration and dynamic frequency-domain analyses for both isolated and non-isolated buildings is reported and effectiveness of isolation for multistoried buildings under seismic responses is revealed. As approximate methods are popular due to reduced computational time, equivalent linear (EL) method has evolved to be one of the best alternative to NLTH analysis wherein the non-linear response of the isolator is modeled by a fictitious viscously damped elastic structure. Zordan et.al. (2014) proposed an improved expression for equivalent linearization of LRB supported structures wherein based on secant stiffness, the optimal damping ratios are calculated and these lead to the estimation of an equivalent damping ratio. The proposed model is reported to have better agreement of displacement with that obtained using NLTH analysis. Ras and Boumechra (2017) investigated the seismic response of LRB fitted multi-storey steel building subject to bidirectional seismic ground motions using a 3D numerical investigation based upon Bouc–Wen nonlinear model.

The seismic isolation systems are designed based on the recommendations of the FEMA, ASCE/SEI, British Standard, AAHSTO and NHERP guidelines wherein the mechanical properties of the LRB isolation system comply with recommendations of the standard guidelines. However, none of the standard provides the complete procedure for design of isolation bearings. Each standard code covers some certain aspect of the design but doesn't focus on the design and selection of the bearing as a whole. The present chapter proposes a unified procedure complying all the standard guidelines and includes the safety checks of the complete system including material as well as the validation of the design calculations. Literature review indicates towards the non-availability of unified code and hence has prompted towards the preparation of one as presented in this chapter.

MODELING OF LLRB ISOLATORS

The rubber-based bearings isolation system consists of layers of rubber and steel with the rubber being vulcanized to the steel plates which provide horizontal flexibility and high vertical stiffness. This isolator consists of a lead-plug insert (Figure 1) which provides its hysteretic energy-dissipation characteristic as shown in Figure 2. Therefore, LLRB system is able to support the structure vertically, to provide necessary horizontal flexibility together with the restoring force, and provide the required hysteretic damping. These specific isolators provide hysteretic damping through the yielding of the lead core. The isolation system can be modeled as a linear or nonlinear element. The linear model of an isolation system is modeled by effective stiffness and effective damping at the design displacement. The behavior of LLRB isolators is simulated by the use of the bilinear hysteretic model of Park et al., (1986) as proposed in Nagarajaiah et al., (1991) and implemented in finite element program taking into account the following mechanical properties.

Properties of Isolation System

1. For nonlinear analysis of a base isolated structure requires the properties of the seismic isolation system to be specified. So, a trial design of the seismic isolation system is performed.

Figure 1. Details of LLRB

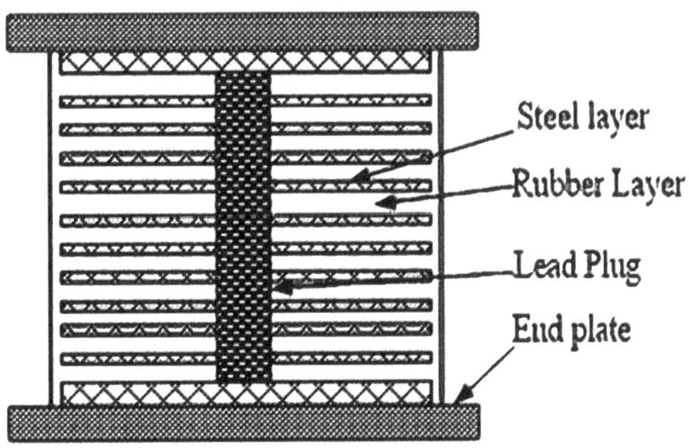

2. Obtain the initial design properties of the isolator based on the target displacement, effective stiffness and assumed effective damping of the system a non-iterative procedure has been developed by Govardhan *et al.*, (2014).

The effective stiffness, design displacement and global energy dissipation per cycle are calculated from the equations (1), (2) and (3) respectively. The bilinear properties of the hysteretic curve are obtained from the non-iterative products such as K_{eff}, Δ_D and W_D.

The procedure to derive the relationship for the non-linear parameters of the bilinear hysteretic curve is described as follows (Govardhan *et al.*, 2014):

Step 1: The effective horizontal stiffness is given by

$$K_{eff} = \frac{W}{g}\left(\frac{2\pi}{T_b}\right)^2$$

where W is the Axial load from the column in kN, g is the acceleration due to gravity in m/sec^2.

T_b is the natural time period of the base isolated structure.

Step 2: The Design Displacement is given by

$$\Delta_D = \frac{C_V}{B}\left(\frac{g}{4\pi^2}\right)T$$

where $C_V = \frac{S_a}{g}$ from the demand spectra for Time period =1sec,

B = Damping coefficient and is given by

Figure 2. Bilinear idealization of the hysteretic behaviour of an isolator

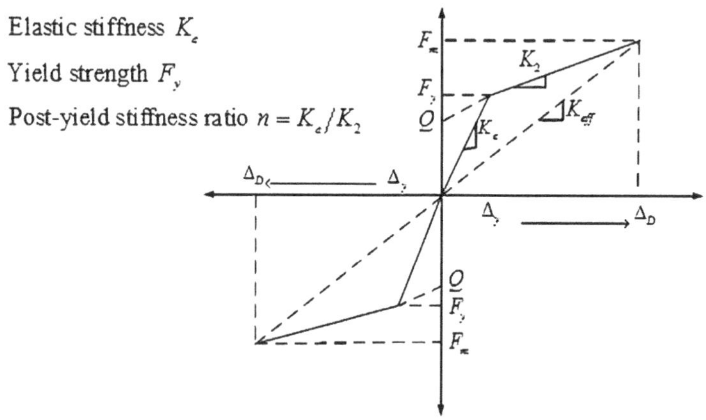

$$\frac{1}{B} \approx 0.25\left(1 - \ln \xi_{eff}\right)$$

Step 3: The Global Energy dissipation per cycle is given by

$$W_D = 2\pi K_{eff} \Delta_D^{\;2} \xi_{eff}$$

where ξ_{eff} =Damping ratio required (It is around 15% for LLRB) for $\Delta_D \gg \Delta_y$, we have,

$$K_{eff} = K_2 + \frac{Q}{\Delta_D}$$

and also,

$$W_D = 4Q\left(\Delta_D - \Delta_y\right)$$

$$\Delta_y = \frac{Q}{\left(K_e - K_2\right)}$$

For laminated lead rubber bearings as an approximation, we take $K_e \approx 10K_2$. From equations (1) to (6), it is clear that the K_2, Q and Δ_y are interdependent on each other. The values of the K_2, Q and Δ_y can be accurately calculated only by iteration. Therefore equations are proposed to evaluate their values directly. P_G is a constant and is given by

$$P_G = \left(4\Delta_D^2 K_{eff} - W_D\right)\left(n-1\right)$$

The yield displacement is given by,

$$\Delta_y = \frac{P_G - \sqrt{P_G^{\;2} - 16\Delta_D^2 K_{eff}\left(n-1\right) W_D}}{8\Delta_D K_{eff}\left(n-1\right)}$$

For LLRB, n=10,

$$\Delta_y = \frac{12\Delta_D^2 K_{eff} - 3W_D - \sqrt{144\Delta_D^4 K_{eff}^2 - 88\Delta_D^2 K_{eff} W_D + 9W_D^2}}{24\Delta_D K_{eff}}$$

The post yield stiffness K_2 is given by,

$$K_2 = \frac{1}{4}\left(4K_{eff} - \frac{W_D}{\Delta_D\left(\Delta_D - \frac{P_G - \sqrt{P_G^{\,2} - 16\Delta_D^2 K_{eff}\left(n-1\right)W_D}}{8\Delta_D K_{eff}\left(n-1\right)}\right)}\right)$$

The characteristic strength Q, is given by

$$Q = \frac{W_D}{4\left\{\Delta_D - \frac{P_G - \sqrt{P_G^{\,2} - 16\Delta_D^2 K_{eff}\left(n-1\right)W_D}}{8\Delta_D K_{eff}\left(n-1\right)}\right\}}$$

DESIGN OF THE LAMINATED LEAD RUBBER BEARINGS

After finalizing the specification and number of bearings and the target displacement of base isolated building, individual isolator characteristics are determined. The important points to be considered for the design of a base-isolation system are (i) the vertical loads carrying capacity, (ii) flexible fundamental time period, (iii) High vertical stiffness to reduce the amplification in vertical direction and difficulties arising due to rocking, (iv) energy dissipation capacity to maintain displacements at the isolation level within acceptable limits, and (v) minimum initial stiffness to maintain serviceability criteria.

The isolators are designed based on the recommendations of the FEMA, ASCE/SEI, British Standard, AAHSTO and NHERP guidelines. The mechanical properties of the LLRB isolation system are set to comply with recommendations of the above mentioned guidelines. However, none of the standard provides the complete procedure for design of isolation bearings. Each standard code covers some certain aspect of the design but doesn't focus on the design and selection of the bearing as a whole.

Design Procedure

Design displacement and bilinear properties of base isolation system for both Maximum Credible Earthquake and Design Basis Earthquake cases are obtained as discussed above. Using these values, design of individual isolator characteristics are determined and checked for design limits for all load cases. The stages involved in the design of LLRB are as follows:

1. Specify the design vertical load (W) on the individual isolator, the design period for the isolated structure T_b and the site conditions. The load, W on individual isolator, is computed for the dead load and live load combination.
2. Find corresponding horizontal stiffness of the bearing, K_{eff} using

$$K_{eff} = \frac{W}{g}\left(\frac{2\pi}{T_b}\right)^2$$

3. Assign Young's modulus, E. In general, shear strain "γ" is considered to be in the range of 100 to 150%. Shear Modulus G for the rubber can be taken from design specifications provided by the manufacturer. The G value for above mentioned range of shear strains varies from 0.69 to 0.86 MPa, depending on the rubber composition (Kelly, 1997).
4. Consider the effective equivalent damping in the isolation system and allowable pressure on the isolator "p" for vertical loads. Usually "p" for LLRB can be taken as about 6.9 N/mm^2 (Kelly, 1997). Damping can be taken as 15 to 20%, subject to verification.
5. The plan area of the rubber A_r in isolator is mainly controlled by the design vertical load W and allowable pressure p; and can be expressed using;

$$A_r = W/p$$

6. The total thickness of rubber layers, T_r; in the isolator is evaluated based on bilinear properties of the hysteretic curve using design displacement, effective stiffness and effective damping from the following equation: (Govardhan *et al.*, 2014)

$$T_r = \frac{GA_r}{K_2}$$

Total thickness T_r should not be less than Δ_D/γ:

7. Evaluate the shape factor S for the desired value of ratio of vertical stiffness K_v and horizontal stiffness K_{eff} using following expressions:

For circular isolators,

$$6S^2 \approx f_v/f_h \approx K_v/K_h$$

For Square isolators,

$$6.73S^2 \approx f_v/f_h \approx K_v/K_h$$

where, f_v is vertical frequency and is assumed to be 10Hz and $f_h = 1/T_b$. When the shape factor of a single layer of elastomer becomes large, the rubber starts to become more compressible. Thus, in design it is common to set the maximum value of the shape factor around 10.

8. The ratio of the cross-sectional area of the isolator to force-free area of a single layer of rubber is called as shape factor, the thickness of each rubber layer can now be determined using shape factor by

For circular isolators,

$$t_r = \frac{\phi}{4S}$$

For square isolators,

$$t_r = \frac{L}{4S}$$

where, ϕ is the diameter of isolator and L is the side for isolator. Knowing the values of T_r and t_r; rubber layers n_r can be computed.

9. Adjust the values of A_r, t_r and T_r such that the value of K_2 is adjusted.
10. The area of lead plug, A_{pb} can be determined using;

$$A_{pb} = Q/\sigma_{pb}$$

in which yield strength of the lead σ_{pb} in shear ranges between 8 to 10 MPa (Skinner et al., 1993). Q is characteristic strength of the isolator. Lead plug should be checked also for height:

$$1.25 \leq \frac{h_{pb}}{d} \leq 5.0$$

where, h_{pb} is the lead plug effective height and d is lead plug diameter (Yang *et al.*, 2003).

Design Checks

The isolator which is designed must be checked to satisfy the following conditions also.

The Vertical Stiffness

The isolator vertical stiffness is,

$$K_v = \frac{E_c A_r}{T_r}$$

where, E_c is the laminated rubber-steel composite compression modulus. It is evaluated as $E_c = 6GS^2$ for circular isolator and $E_c = 6.73GS^2$ for square isolator.

Safety Against Buckling

When low-shear stiffness of the rubber is used in a multi-layered elastomeric isolator, buckling type of instability will occur in isolator. The usual conditions for isolator in an isolation system are constrained against rotation at both ends and can move freely at the top in lateral direction. The result for the critical buckling load P_{crit}, with the above boundary conditions, is the solution of the equation:

$$P^2 + PP_S - P_S P_E = 0$$

P_E is the Euler load for the standard column and P_S is the load due to combined bending and shear stiffness of the isolator composite system

$$P_E = \frac{\pi^2 E_c I h}{3h^2 T_r}$$

$$P_S = \frac{G A_r h}{T_r}$$

where, h is the height of the isolator which includes the steel shims, I is the moment of inertia of a steel plate and T_r is total thickness of rubber layers. From the above equations, we have

$$P_{crit} = \sqrt{P_S P_E}$$

$$P_{crit} = \frac{\sqrt{2}\pi GAS}{T_r} r$$

where, r is radius of gyration. For a square bearing $r = L/2\sqrt{3}$ and for a circular bearing $r = \phi/4$ In the design of all of the isolators it is ascertained that the load carried by the bearing is less than its critical load, P_{crit}.

Under gravity loading only, as per ASCE, AASHTO and NEHRP recommendations for service loads (W), a factor of safety of three should be used for maximum shear strain, γ_S. The γ_S must satisfy the condition as given below:

$$\gamma_S = 6S\frac{W}{E_C A} \leq \frac{\varepsilon_b}{3}$$

where, ε_b is the minimum tensile strain at which rubber breaks and E_C is compression modulus for the isolator and is given by $E_C = E\left(1 + k'S^2\right)$ where E and k', for a given value of rubber hardness.

In case of ultimate conditions that include the effect of seismic loads, AASHTO recommends a factor of safety of 1.33. Hence, maximum shear strain γ_u should not exceed $\varepsilon_b/1.33$ due to combined effect of seismic load γ_{eq}, compression γ_c and torsion γ_t; and γ_t is given by

$$\gamma_t = \frac{L^2}{2t_r T_r}\frac{12\Delta e}{\left(B_b^2 + L_b^2\right)}$$

where, B_b and L_b are plan dimensions of building and 'e' is the eccentricity.

γ_c ($= 6SW_E/E_C A_r'$) can be computed for W_E (vertical load and effect of earthquake load). In this A_r' is the reduced area due to applied lateral displacement Δ. A_r' is determined using the following equation:

For circular isolator,

$$A_r' = \frac{1}{2}\left[\phi^2 \sin^{-1}\left(\frac{\sqrt{\phi^2 - \Delta^2}}{\phi} - \Delta\sqrt{\phi^2 - \Delta^2}\right)\right]$$

For square isolator,

$$A_r' = A_r\left(1 - \frac{\Delta}{b}\right)$$

Roll Out Stability

Maximum lateral displacement should not exceed the roll-out displacement $\delta_{roll-out}$ given by

$$\delta_{roll-out} = \frac{W_E L}{W_E + K_{eff} h}$$

where, L is size or the diameter of isolator, and h is the height of the isolator as shown in the Figure 3.

The check should be made for isolator with bolted connections also, so as to avoid the development of tensile stresses. The rubber shear modulus is considered based on the required hardness.

The flow chart in Figure 4 gives the design of an isolation system for a building.

CASE STUDY OF A BUILDING

Details of the Superstructure

A (Ground plus Four) Reinforced Concrete moment-resisting frame building, including continuous columns and their adjacent girders, is shown in Figure 5 and Figure 6 respectively. The column locations are grouped into three categories based of the vertical load, and a lead laminated rubber bearing is designed for each of the category. The two types of isolator units, their design vertical loads and the corresponding maximum and minimum vertical loads as obtained for the isolator design are tabulated below. Total number of columns is 21. Center to center height of the column is 3m. For both fixed base and base isolated buildings, the column sections are 300x450 mm; and the beam sections are 300x450 mm in the first floor and foundation, 300x300 mm for top floor designed as per IS 456:2000. The characteristic strength of concrete is taken as 25N/mm^2. URM infill is considered and is modeled as a diagonal strut element. The structure is assumed to be located in seismic Zone IV according to IS: 1893-Part-I: 2002 with zone factor of 0.24. The buildings are assumed to be resting on the medium soil. The buildings are

Figure 3. Stability against roll out of an isolator

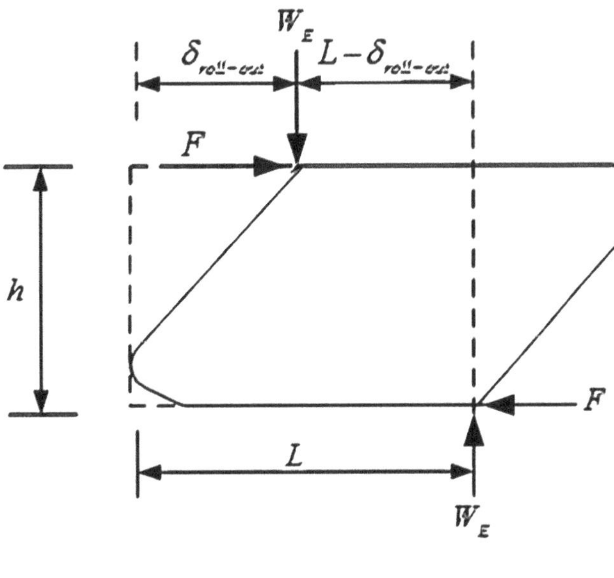

Figure 4. Design of an isolation system for a building

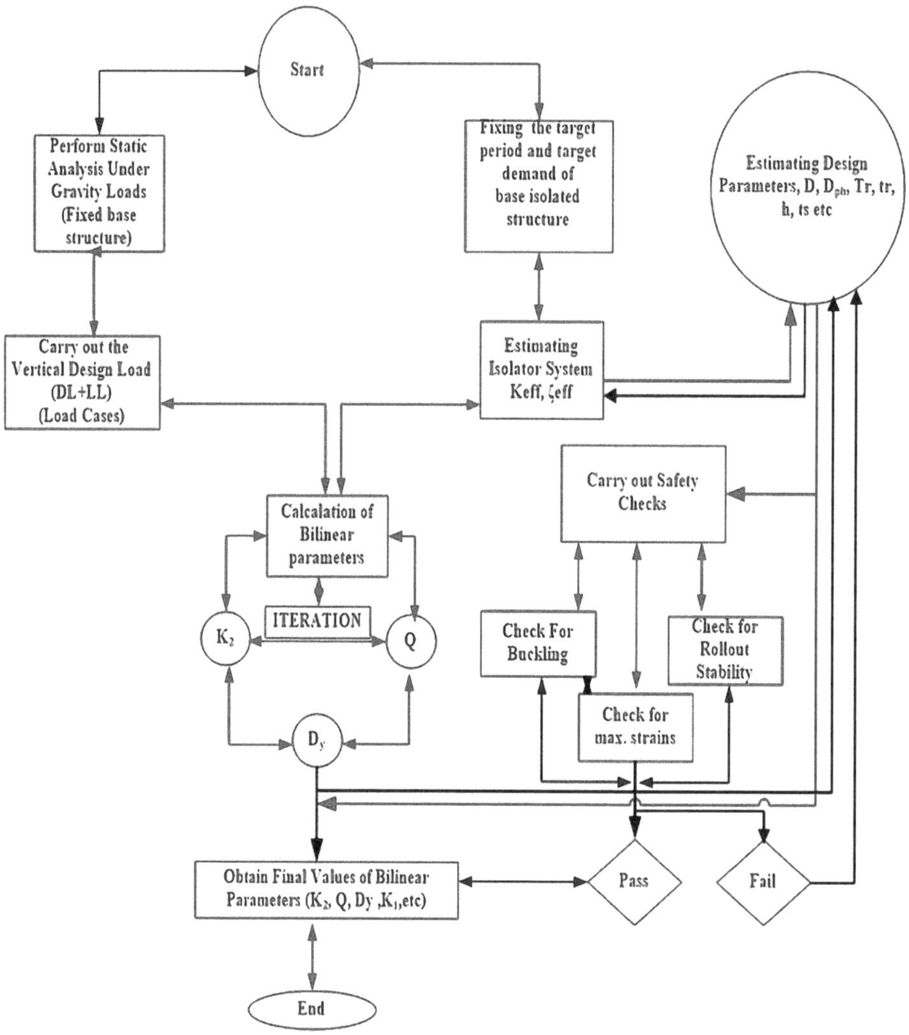

designed for 1% drift. Lateral forces are applied in proportion to the storey masses and the square height of the floor for the pushover analysis (Nonlinear Static Analysis).

Design of LLRB Isolator System

The design vertical loads as obtained for the isolator design are given in Table 2 (a). The columns are grouped into two categories based on the vertical load, and lead laminated rubber bearing are designed for each of the load category. The two capacities of isolator units, their design vertical loads and the corresponding maximum and minimum vertical loads as obtained for the building are tabulated in Table 2 (a). Initially the base isolated time period is considered as 2.0 sec and effective damping assumed as 15%.

Figure 5. Plan of the building

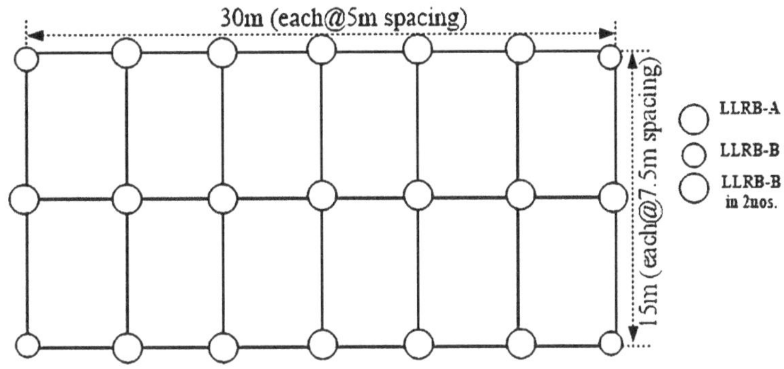

Figure 6. Elevation of the building

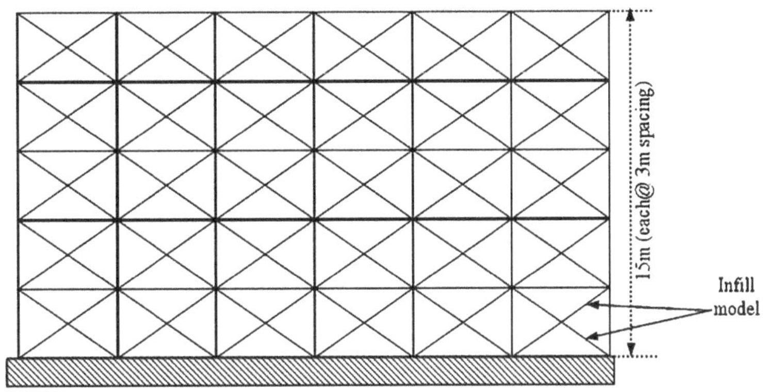

Table 1. Design parameters

Design Parameters	DBE	MCE
Time period of the Base isolated structure (s)	1.87	2.26
Design Displacement (mm)	87.5	206.2

CONCLUDING REMARKS

A thorough study has been made on design of elastomeric bearings based on different available codes such as FEMA, NEHRP, AASTHO, ASCE and British Standards. The mechanical properties of the LLRB isolation system are set to comply with recommendations of the above mentioned guidelines. However, none of the standard provides the complete procedure for design of isolation bearings. Each standard code covers some certain aspect of the design but doesn't focus on the design and selection of the bearing as a whole. From the study, a complete design procedure for a LLRB isolation system including

Table 2. Properties, design parameters and design checks for LLRB 1 and LLRB 2

Load Case	LLRB-A	LLRB-B
Bilinear Input Properties of the Isolator		
Min Vertical Load	800	550
Design Vertical Load	1000	750
Maximum (DL + LL + EQ)	1200	950
Individual Isolator Characteristics		
Post yield stiffness (K_2) in *kN/m*	769.70	527.60
Initial stiffness (K_e) in *kN/m*	7697.00	5276.00
Characteristic strength (Q)in *kN*	50.26	35.30
Effective stiffness (K_{eff}) in *kN/m*	1012.54	704.43
Design Parameters		
Bearing diameter (*mm*)	550	415
Lead plug diameter (*mm*)	80	67
Rubber layer thickness (*mm*)	10	10
Number of rubber layers (*mm*)	13	11
Shim thickness (*mm*)	3	3
End plate thickness (*mm*)	25	25
Bearing height (*mm*)	216	190
Design Checks for the LLRB		
Factor of safety under compression (>1)	2.16	1.48
Roll out displacement (*mm*) (> Design Displacement)	410	338
Maximum design strain of Isolator (<7)	1.91	3.01
Design shear strain due maximum horizontal displacement (<2.5)	1.66	1.66

important steps in all the codes has been presented and explained with the help of a flow chart. A case study has been presented to illustrate the design procedure wherein the columns of a G + 4 reinforced concrete moment-resisting frame building are grouped into two categories based on the vertical load, and lead laminated rubber bearing are designed for each of the load category. The design parameters of the LLRB are tabulated. The proposed design methodology could serve as a complete system design code for structural bearings for buildings throughout the world.

REFERENCES

Choun, Y. S., Park, J., & Choi, I. K. (2014, October). Effects of mechanical property variability in lead rubber bearings on the response of seismic isolation system for different ground motions. *Nuclear Engineering and Technology*, *46*(5), 605–618. doi:10.5516/NET.09.2014.718

Deb, S. K. (2004). Seismic base isolation – an overview. *Current Science*, *87*(10), 1426–1430.

Govardhan & Paul, D. K. (2016). Effect of Lead in Elastomeric Bearings for Structures Located in Seismic Region. *Procedia Technology, 25*, 146 – 153.

Govardhan, Paul, D.K., & Singh, Y. (2014). Non-iterative bilinear properties of structural bearings, IALCCE 2014 symposium. Tokyo, Japan: Waseda University.

Islam, A. B. M. S., Jumaat, M. Z., Hussain, R., & Alam, M. A. (2013). Incorporation of rubber-steel bearing isolation in multi-storey building. *Journal of Civil Engineering and Management*, *19*(sup1), 33–49. doi:10.3846/13923730.2013.801904

Kelly, J. M. (1997). *Earthquake-Resistant Design with Rubber* (2nd ed.). London: Springer-Verlag. doi:10.1007/978-1-4471-0971-6

Mkrtychev, O. V., Dzhinchvelashvili, G. A., & Bunov, A. A. (2014). Study of Lead Rubber Bearings Operation with Varying Height Buildings at Earthquake. *Procedia Engineering*, *91*, 48–53. doi:10.1016/j.proeng.2014.12.010

Naeim, F., & Kelly, J. M. (1999). *Design of seismic isolated structures: From theory to practice*. New York: John Willey and Sons Inc. doi:10.1002/9780470172742

Nagarajaiah, S., Reinhorn, A. M., & Constantinou, M.C. (1991). *3D-BASIS: Nonlinear dynamic analysis of three dimensional base isolated structures: Part II*. Technical Report NCEER-91-0005. National Center for Earthquake Engineering Research, State University of New York at Buffalo.

Park, Y. J., Went, Y. K., & Ang, A. H. S. (1986). Random vibration of hysteretic systems under bi directional ground motions. *Earthquake Engineering & Structural Dynamics*, *14*(4), 543–557. doi:10.1002/eqe.4290140405

Ras, A., & Boumechra, N. (2017Dissipation's Capacity Study of Lead–Rubber Bearing System in Seismic Steel Structures Design. *Arabian Journal for Science and Engineering*. doi:10.1007/s13369-017-2489-z

Skinner, R. I., Robinson, W. H., & McVerry, G. H. (1999). *An Introduction to Seismic Isolation*. New York: John Wiley & Sons.

Stanton, J., & Roeder, C. (1991). Advantages and limitations of seismic isolation. *Earthquake Spectra*, *2*(2), 301–322. doi:10.1193/1.1585630

Yang, Y. B., Chang, K. C., & Yau, J. D. (2003). *Base Isolation Earthquake Engineering Handbook*. Washington, DC: CRC Press.

Zordan, T., Liu, T., Briseghella, B., & Zhang, Q. (2014). Improved equivalent viscous damping model for base-isolated structures with lead rubber bearings. *Engineering Structures*, *75*, 340–352. doi:10.1016/j.engstruct.2014.05.044

APPENDIX: NATIONAL AND INTERNATIONAL STANDARDS USED

American Association of State Highway and Transportation Officials (AASHTO), "AASHTO M 251 Standard Specification for Plain and Laminated Elastomeric Bridge Bearings," 2006.

American Society of Civil Engineers. (2005). Minimum Design Loads for Buildings and Other Structures. ASCE/SEI 7-05. Reston, Virginia: ASCE.

American Society of Civil Engineers. (2010). Minimum Design Loads for Buildings and Other Structures. ASCE/SEI 7-10. Reston, Virginia: ASCE.

American Society of Civil Engineers. (2013). Seismic Evaluation and Retrofit of Existing Buildings. ASCE/SEI 41-13. Reston, Virginia: ASCE.

Applied Technology Council (ATC). Proceedings of ATC-17-1 Seminar on Seismic Isolation: Passive Energy Dissipation and Active Control, San Francisco, California, 1993.

BS EN 15129 (2009), "Anti-seismic devices", British Standards Institution London, UK.

BS ISO 22762-3 (2005), "Elastomeric seismic-protection isolators. Applications for buildings. Specifications", British Standards Institution London, UK.

Federal Emergency Management Agency FEMA-450. (2003). NEHRP Recommended Provisions for Seismic Regulations for New Buildings and Other Structures; Design Examples. Report. Washington, DC: Building Seismic Safety Council.

Federal Emergency Management Agency FEMA-451. (2006). NEHRP Recommended Provisions for Seismic Regulations for New Buildings and Other Structures; Design Examples. Report. Washington, DC: Building Seismic Safety Council.

IS – 1893 (Part 1). Criteria for earthquake resistant design of structures – Part 1: General provision and buildings, Bureau of Indian Standards, New Delhi, INDIA; 2002.

Chapter 5
Designing of a Twin Tube Shock Absorber:
A Study in Reverse Engineering

Dheeman Bhuyan
Girijananda Chowdhury Institute of Management and Technology, India

ABSTRACT

Reverse engineering has become the one of the most relevant concepts in modern design doctrines. Advances in technology demand shorter lead time in the overall product development stage, especially in the automobile sector. Hence as a study in reverse engineering, the author has reverse engineered a twin tube shock absorber. The process involved the obtaining of subassemblies of the damper mechanism to generate a 3D CAD model of the damper in PTC CREO 2.0. The model was used to conduct static structural and CFD analysis of the same using ANSYS 15.0 Workbench. The data obtained was used as the datum for the design modifications and performance enhancement of the part. It was seen that the design of the piston valve was optimum hence modifications to the base valve were done. Following the generation of the datum, similar analyses were conducted on the modified assemblies. The results were compared to the datum for the selection of the most appropriate design. Four designs are analyzed and compared with the datum and the set with four orifices in the valve disk was found to be optimum.

INTRODUCTION

With ever shortening lead times and the demand for accuracy of designs and manufacture, the path of least resistance for the redesign of a product is that of reverse engineering. As innocuous as it may seem, the term reverse engineering in itself encompasses multiple fields and specializations. It involves the reproduction of anything based on extracted information (Chikofsky & Cross, 1990; Eilam & Eldad, 2005). It has proved to be one of the best methods for data acquisition for thc purpose of redesigning any product. It has come to be considered one of the technologies that have been able to provide business benefits by enabling a shorter product development cycle (Raja, 2008).

DOI: 10.4018/978-1-5225-3401-3.ch005

The demand for higher quality standards in the automotive industry has created the need for higher performance of shock absorbers. Noise reduction and vibration damping have become the thrust areas for designs.

Figures 1(a) and 1(b) show the designs of modern shock absorbers.

The function of the shock absorber is often misconstrued to be the absorption of vibrations in the system. However that job is done by the spring. The shock absorber provides the damping force necessary to dissipate the energy absorbed by the spring while neutralizing the original shock.

BRIEF BACKGROUND

The design, manufacture, assembly and maintenance of products, systems or both, is the essence of the engineering process. In the pursuit of these goals, who paths are generally available to engineers: forward engineering and reverse engineering. The progression from high level abstractions and logical designs to the physical implementation of the system is termed as forward engineering. However, in certain scenarios, the physical product/system may be available readily, albeit without the accompanying technical details such as design specifications, drawings, etc. The reproduction of the part or system in such cases, in the absence of technical specifications is known as reverse engineering. The process of obtaining a geometric CAD model from 3-D points acquired by scanning or digitizing the extant system can also be called Reverse Engineering (RE).

Figure 1. (a) Mono tube shock absorber; (b) twin tube shock absorber
PartInfo.co.uk, n.d.

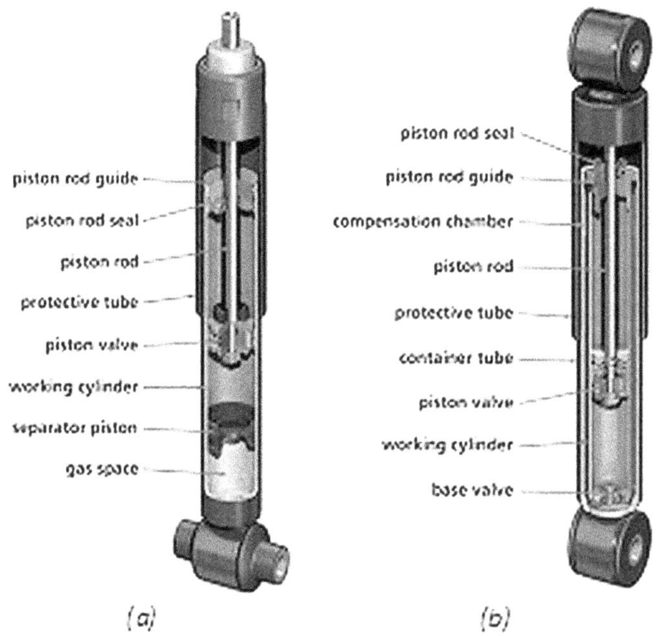

Often defined by researchers according to their specific task, the process of digitally capturing the physical entities of a component is most commonly referred to as reverse engineering (Motavalli & Shamsaasef, 1996). RE has also been described as, "the basic concept of producing a part based on an original or physical model without the use of an engineering drawing" (Abella et al., 1994).

The mono-tube shock absorber presents a shorter stroke length due to the gas charge at the end. Hence it is most appropriate for surfaces which are relatively smooth. Motor sports such as Formula racing present such surfaces. Hence the mono-tube design finds widespread application in racing and associated motorsports. The twin tube design on the other hand provides greater stroke length. This enables the vehicle to accommodate larger bumps and allows the vehicle to navigate rougher surfaces. Therefore production cars and off-road vehicles use the twin-tube design.

Performance optimization of production cars has become a major thrust area for research and development. Despite the advances made in the field, the surface has barely been scratched. The potential for development in the field is still mostly untapped. With diminishing fossil fuel reserves and increasing global carbon footprint coupled with ever increasing quality of life and economic development, the need to improve the performance characteristics such as fuel efficiency is a persistent and major challenge, especially in the Indian Subcontinent given that the suspension system of a vehicle affects its fuel economy (United States Environmental Protection Agency, 1980; Gyenes & Mitchell, 1994). Hence optimization of performance characteristics of suspension elements is required.

Shock absorbers are a critical part of a suspension system, connecting the vehicle to its wheels. The need for dampers arises because of the roll and pitches associated with vehicle manoeuvring, and from the roughness of roads. Technical advancements in the 19th century led to the need for vibration control and damping in automobiles.

The role of shock absorbers in vehicle dynamics is a critical one. In addition to reducing the noise in the passenger compartment of the vehicle and increasing the comfort of the ride, the shock absorbers are involved in reducing the relative motion of the wheels and the body of the car and providing better handling and safety.

The need for damping of vibrations produced by the movement of the vehicle over a road surface has been felt for as long as the automobile industry has existed. The first production shock absorber was the Telesco Shock Absorber exhibited in the 1912 Olympia Motor Show. Technology has progressed by leaps and bounds over the following years to get to where we stand today.

Initial designs of dampers utilized friction as the damping force. Hydraulic damping was introduced early in the 20th century but did not catch on. Most modern shock absorbers are based on viscous damping. Fluid is forced through valves and orifices and the viscous reaction of the fluid dissipates the energy in the form of heat.

Houdaille (Sept. 9 1908) was granted a patent for improvements to hydraulic shock absorbing apparatus for automobiles.

Houdaille (Sept. 7 1909) was granted a patent for a shock absorbing apparatus.

Chai and Mason (September 1996) studied landing gear integration in aircraft conceptual design. They examined documented design procedures and analyses to determine their applicability, and to ensure compliance with current practices and regulations. Using information obtained during industry survey, the analyses were modified and expanded to accommodate the design criteria associated with the advanced large subsonic transports.

Lang (1997) studied the dynamic behaviour of the automotive shock absorber with particular attention given to its frequency dependant behaviour. This study identified each of the elements that comprise of the typical shock absorber and describes the manner in which these elements function to produce the desired force-velocity relationship.

Talbott (2002) presented a mathematical model of a gas-charged mono-tube racing damper. The model was exercised to show the effects of tuning on damper performance. Results obtained from this exercise were that the pressure variation on the compression side of the piston is insignificant relative to that on the rebound side because of the gas charge, valve shim stiffness can be successfully modeled using stacked thin circular plates, bleed orifice settings dominate the low speed regime, and shim stack stiffness dominates the high speed regime.

Atherden (2004) has considered the design and manufacture feasibility of a damper customised to a Formula SAE style vehicle. Research was done into the history of dampers, racing dampers and the concept of the damper in Formula SAE style vehicles, governing the direction of the new design. Vehicle dynamics theory and incompressible fluid flow theory were applied through Simulink and Matlab models for specific component design and 3D CAD of the proposed design was developed.

Dixon (2007) provided a comprehensive coverage of design, testing, installation and use of the damper. The work provides a thorough treatment of the issues surrounding the design and selection of shock absorbers.

Czop et al. (2011) proposed a finite element fluid flow model, which can be used in order to reduce the velocity of fluid flow through a cavity of a shock absorber valve. High flow velocity can cause high-content frequency vibrations and, in turn, audible noise. The model was used for initial screening of new valve concepts and on the other hand to improve the currently used ones.

Martande, Jangale, and Motgi (2013) studied and developed new correlated methodologies that will allow engineers to design components of Shock Absorbers by using FEM based tools. The structural design for Hero Motorcorp shock absorbers was validated and commercially available FEA engine ANSYS was successfully used in the design validation.

Patel, Gehlot and Gupta (2014) reviewed the implementation of a parametric model of an automotive damper. The goal of this research was to create a damper model to predict accurately damping forces to be used as a design tool for damper design.

Suh (2014) carried out studies with the aim of redesigning the tricycle main landing gear based on the reference data of a Boeing 777 aircraft under the landing conditions. The oil-pneumatic type shock absorber of the shock strut was replaced by a spring shock absorber to verify its feasibility as an alternative solution. Comparisons between the two types of shock absorbers were made by using vibration equation and durability of the landing gear was verified.

A review of the published literature on the design and performance of shock absorbers lends an insight into the importance of the shock absorbers or dampers and their design implementation. Dampers have been found to be an important part of vehicle assemblies, providing control, stability and better comfort for vehicle and its occupants. The mono-tube design is seen to be more effective on racing vehicles while twin tube shock absorbers are more suited to production cars meant for the consumer market. Also, it has been seen that hydraulic damping is an effective tool for vibration damping in automotive applications, the characteristics of which have been extensively studied using numerical models.

The literature has also shown that 3D CAD modeling has been done successfully using Pro/ENGINEER or its successor, CREO, CATIA and other modeling suites. Also commercially available FEA engines like ANSYS and Abaqus FEA and CFD engines like FLUENT and OpenFOAM have been shown to have provided accurate and reliable results.

Based on thorough literature review it is seen that most of the studies concentrate on mono-tube designs of shock absorbers and that optimization and performance enhancement studies on twin-tube designs are rare and not easily available. Also, reverse engineering of dampers is not a well documented field of study.

REVERSE ENGINEERING OF VALVE COMPONENTS

In this section, the reverse engineering of the valve components of a commercially available shock absorber is described. The reverse engineering process involves product teardown followed by generation of 3D CAD models of the various components. Parametric modelling suite PTC CREO 2.0 was used for the generation of the models for the various parts and assembly of the same.

Product Teardown and Modelling of Base Valve

Product teardown is a bench marking process, which adds value to the design and generates new concepts. It helps to understand the current technology and functional behaviour of parts, and identifies current product strengths, weaknesses, and opportunities for new product design. Product teardown also helps in finalizing the specifications for a new product, based on the analysis done in the competitor product.

A commercially available shock absorber was acquired and dismantled to obtain the base valve. Figure 2 shows the dismantled shock absorber and the exposed base valve.

Figure 3 shows the inlet and outlet of the base valve respectively.

The base valve was disassembled and the dimensions of each component were recorded and the data generated was used to generate 3D models of the same in PTC Creo 2.0.

A parametric 3D modelling software suite, PTC CREO was used to generate the 3D model of the valve assembly. CREO is the successor to the ProEngineer modeling software as used by Pavani et al (2014, pp 988 – 995).

The base valve consists of seven parts. The assembled valve is shown by Figure 4(a) and the exploded view of the same is shown by Figure 4(b).

Figure 2. Dismantled shock absorber and exposed base valve

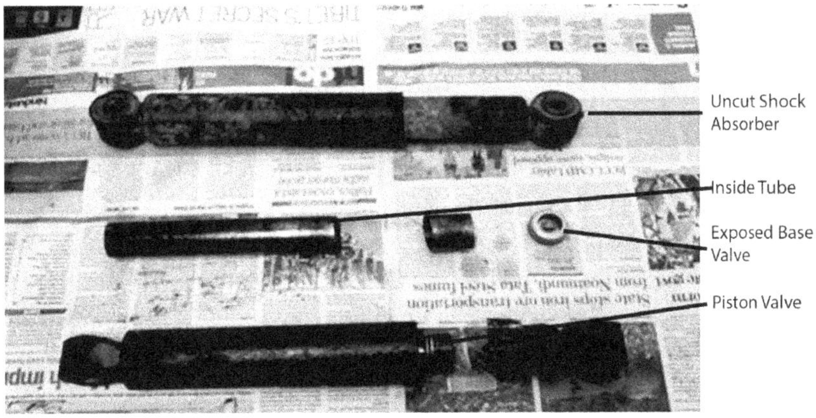

Figure 3. (a) Base valve inlet; (b) base valve outlet; (c) base valve to scale

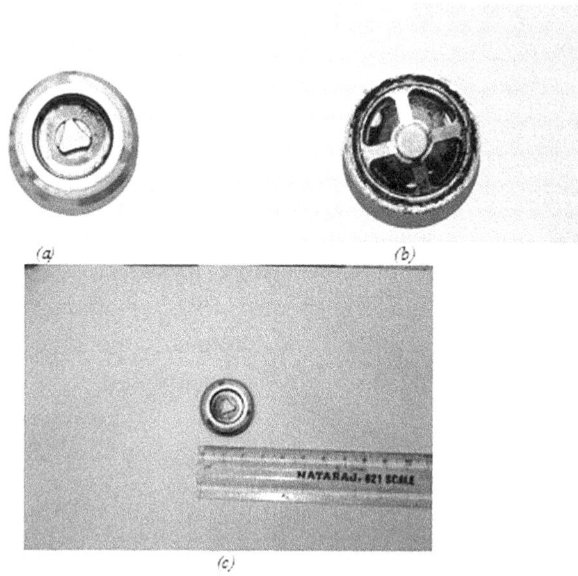

Figure 4. (a) Base valve modelled in Creo; (b) exploded view of the assembly

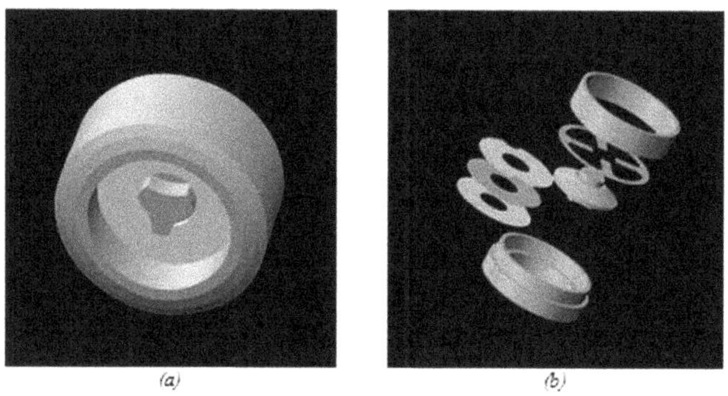

The valve is pressure actuated. There are three orifices for the hydraulic fluid to flow through in open condition. The various parts are shown by Figure 5.

Product Teardown and Modelling of Piston Valve

The piston valve assembly comprises of the piston which has orifices and valve discs. The kinematically forced movement of the piston valve causes the fluid to move through the orifices. This gives rise to viscous damping force which dissipates the energy of the shock.

A parametric 3D modelling software suite, PTC CREO® 2.0 was used to generate the 3D model of the valve assembly. CREO is the successor to the ProEngineer modeling software as used by Pavani et al (2014, pp 988 – 995).

Figure 6 shows the various parts of the piston valve assembly.

Figure 5. (a)(i) Body of the valve, (ii) cross section of body; (b) washer; (c) slotted washer; (d) valve disc; (e) diaphragm spring

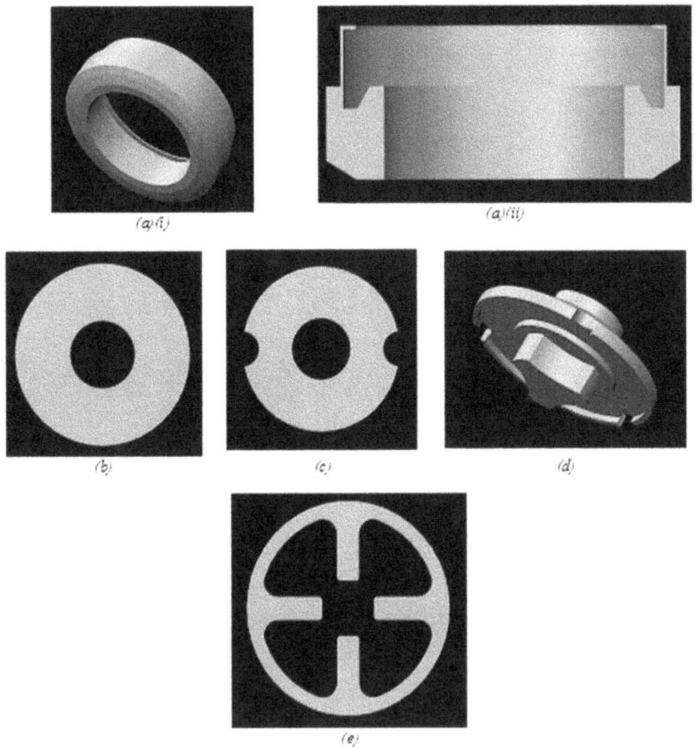

STRUCTURAL ANALYSES

Structural analysis of the components is important to determine the performance of the mechanical components. The general-purpose finite element analysis (FEA) software package ANSYS® Workbench was used for the design validation through FEA analysis of the spring. Pavani, et al (2014, pp 988 – 995.) have done a similar analysis of wave springs. Banerjee, & Pohit (2014, pp 188 – 195) have used the same for vibration analysis. Kumbhalkar et al. (2015, pp 331 – 343), have done static structural analysis on helical springs using ANSYS 12.0.

Static Structural Analysis of Base Valve

The stress analysis of the valve was done in ANSYS 15.0 Workbench. The only moving part of the assembly is the diaphragm spring and it is observed that the failure of the valve is caused mainly by the failure of the failure of the spring. Hence the stress analysis has been done on the spring.

Table 1 shows the geometry details of the spring while Table 2 shows details of the meshing. The dimensioning has been kept true to the original.

The results of the analysis are shown by Figure 7, Figure 8 and Figure 9.

Figure 6. (a)(i) Top view, (ii) front view, (iii) cross-sectional front view of piston; (b)(i) perspective view, (ii) front view of diaphragm spring; (c) piston ring; (d) slotted washer; (e) washer; (f) bleeder; (g) washer used to mount assembly on piston rod; (h) spacer for spring; (j) fully assembled piston valve

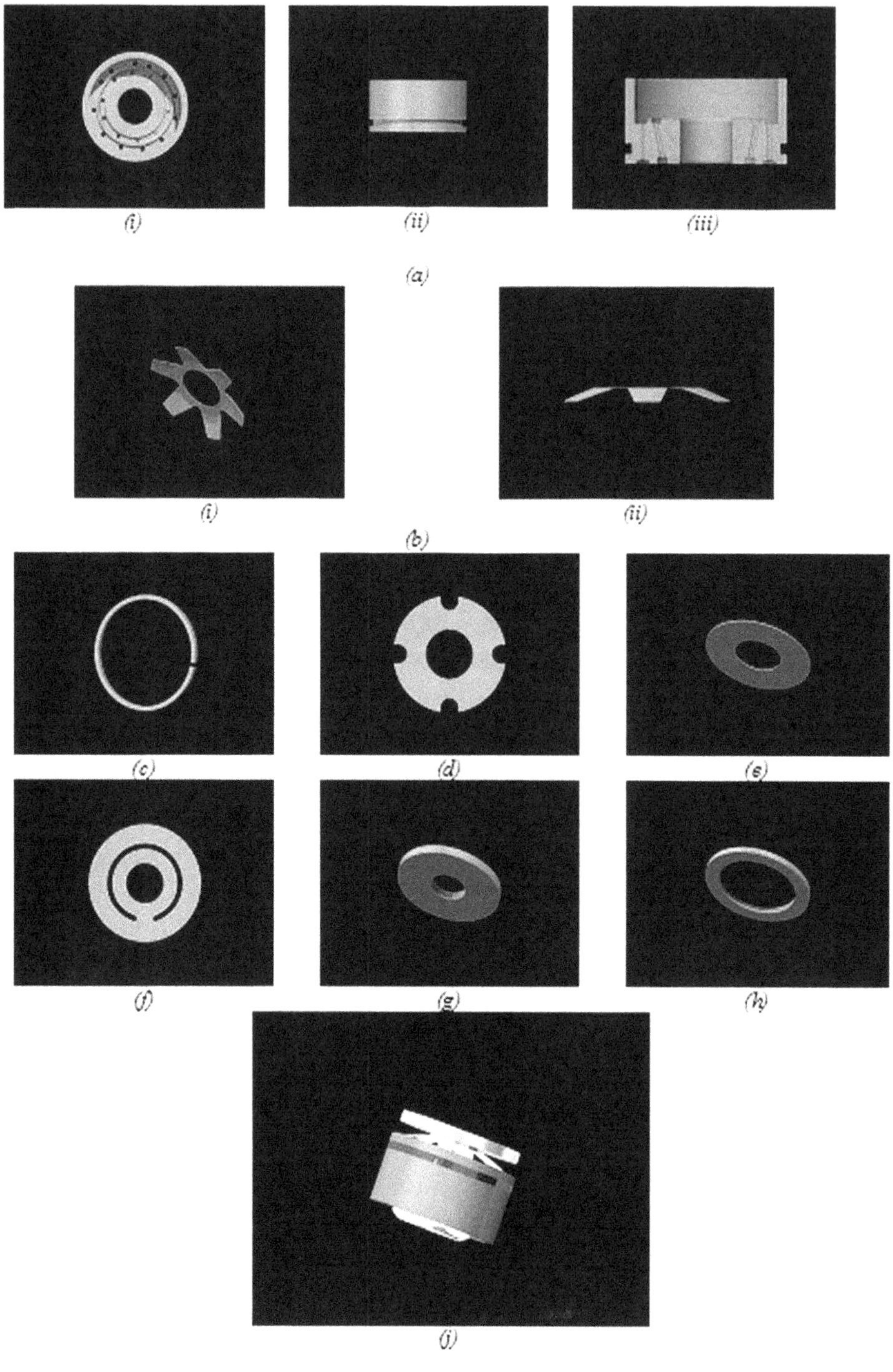

Table 1. Geometry details

Bounding Box	
Length X	23. mm
Length Y	23. mm
Length Z	0.25 mm
Properties	
Volume	44.862 mm^3
Mass	3.521e-004 kg
Scale Factor Value	1.
Statistics	
Bodies	1
Active Bodies	1
Nodes	292185

Table 2. Mesh details

Defaults	
Physics Preference	Mechanical
Sizing	
Use Advanced Size Function	Off
Relevance Center	Coarse
Element Size	0.01 mm
Initial Size Seed	Active Assembly
Smoothing	Medium
Transition	Fast
Span Angle Center	Coarse
Minimum Edge Length	0.250 mm
Inflation	
Use Automatic Inflation	None
Inflation Option	Smooth Transition
Transition Ratio	0.272
Maximum Layers	5
Growth Rate	1.2
Inflation Algorithm	Pre
Patch Conforming Options	
Triangle Surface Mesher	Program Controlled
Patch Independent Options	
Topology Checking	Yes

Figure 7. (a) Equivalent stress; (b) equivalent strain

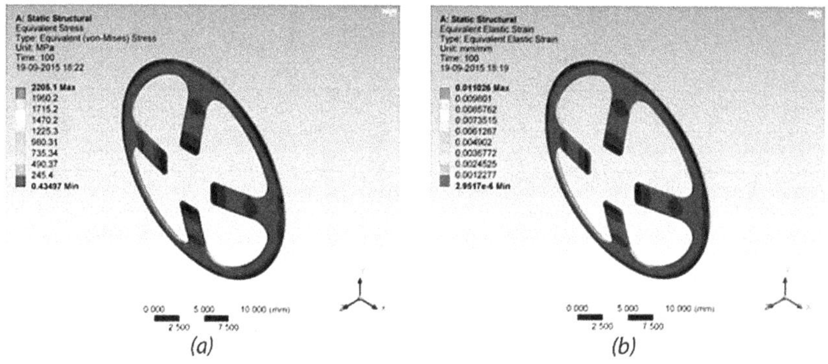

Figure 8. (a) Maximum shear stress; (b) maximum shear elastic strain

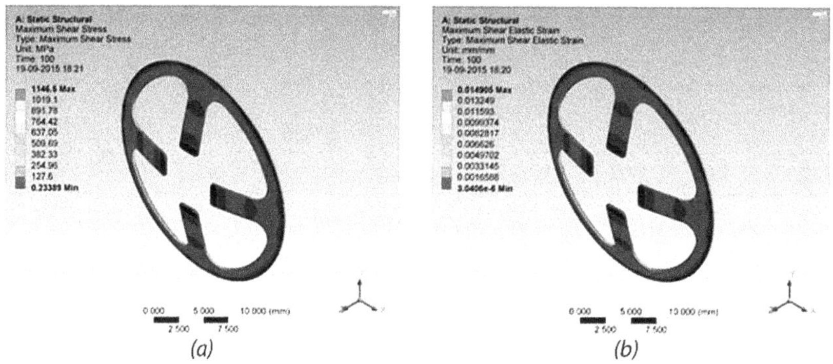

Figure 9. (a) Total deformation; (b) strain energy

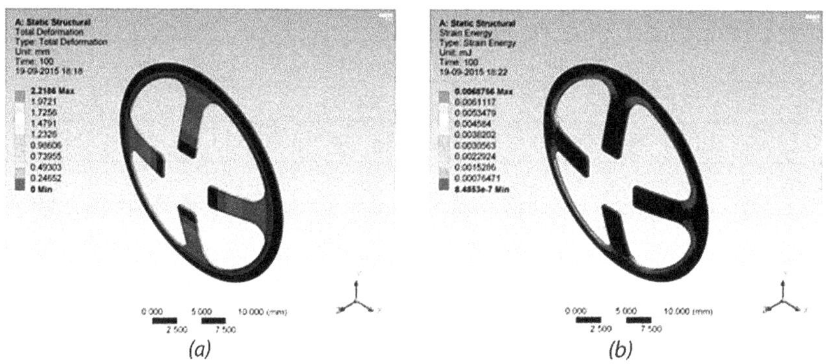

Static Structural Analysis of Piston Valve

ANSYS Workbench was used to carry out the simulation. Material selection was set to structural steel. The forces acting on the system were obtained from CFD analysis of the valve (Bhuyan &. Kumar, In Press).

Table 3 and Figure 10 show the results of the analysis.

Table 3. Results of simulation

Object Name	Total Deformation	Equivalent Elastic Strain	Equivalent Stress
Results			
Minimum	0. mm	3.5844e-017 mm/mm	2.1161e-012 MPa
Maximum	2.7319e-008 mm	3.4693e-010 mm/mm	6.9251e-005 MPa
Minimum Value Over Time			
Minimum	0. mm	3.5844e-017 mm/mm	2.1161e-012 MPa
Maximum	0. mm	3.5844e-017 mm/mm	2.1161e-012 MPa
Maximum Value Over Time			
Minimum	2.7319e-008 mm	3.4693e-010 mm/mm	6.9251e-005 MPa
Maximum	2.7319e-008 mm	3.4693e-010 mm/mm	6.9251e-005 MPa

Figure 10. (a) Undeformed spring; (b) total deformation; (c) equivalent elastic strain; (d) equivalent (von-Mises) stress

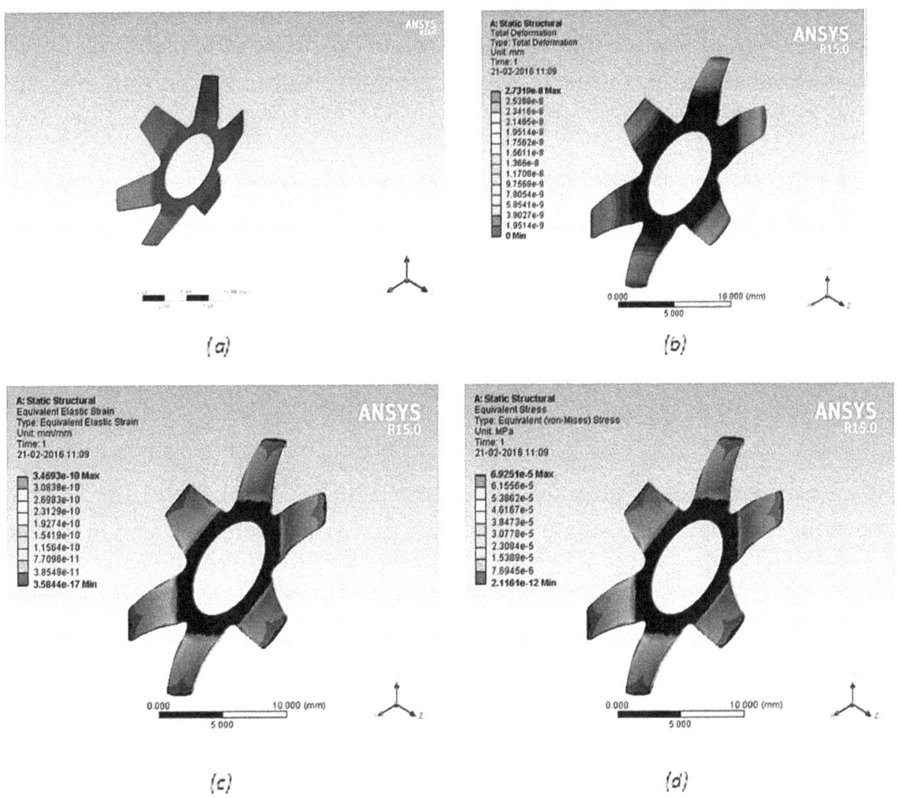

FLOW ANALYSES

Commercially available CFD code FLUENT, integrated into ANSYS Workbench was used to conduct the flow analysis. Meshing of the assembly was done using ANSYS Workbench 15.0. Shukla and Harsha (2015, pp.2056 – 2063) have used a similar modelling and analysis regimen where they did the solid modelling in ProE and analysis was carried out using ANSYS. Kumar, Balakrishna and Chandra (2015, pp.1427 – 1434) have used ANSYS 12 for meshing of non linear bodies. Kumar and Bhingole (2015, pp.2314 – 2322) did CFD analysis of turbine blades using ANSYS Fluent.

SAE10W to SAE 30W oils are prescribed as general purpose hydraulic fluid for shock absorbers, depending on temperature conditions. The fluid considered was SAE20W mineral oil. The properties of ISO grade oils are shown by Table 4.

Flow Analysis of Base Valve

The base valve separates the cylinder space from the reserve chamber. The transfer of fluid from the reserve chamber to the working cylinder happens due to the movement of the piston rod. A shock absorber utilizes two types of valves: intake valves and control valves. Intake valves are check valves which provide low resistance to flow in one direction and, upon reversal of the differential pressure, prevent flow in the opposite direction. A valve spring is used to preload control valves and to prevent their opening until a specified pressure differential has built up across the valve. The height of the opening is controlled by stiffness of the valve spring.

Figure 11 shows the meshing of the valve assembly.

The section in blue is the inlet and the section in red is the outlet defined by named selections.

Figure 12 shows the results of the simulation.

Table 5 shows the total forces in the direction of flow. Table 6 shows the mass flow rate through the valve.

Flow Analysis of Piston Valve

The CFD model was created using Fluent integrated into ANSYS Workbench 15.0. The model consists of 1351129 nodes for minimal disc opening. The number of nodes increases with increased opening of the discs due to enlarged volumes. Figure 13 shows the mesh.

Table 4. Hydraulic fluid properties at 40°C

ISO Grade	Equivalent SAE Grade	Density (kg/m3)	Dynamic Viscosity (Pa-s)
32	10W	857	0.02765
46	20	861	0.03927
68	20W	865	0.05856
100	30	869	0.08681

EngineeringToolbox.com, 200?.

Figure 11. Meshing of assembly for CFD analysis

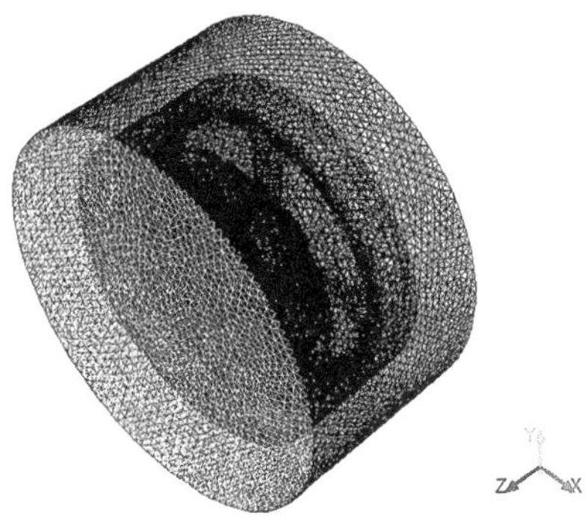

Figure 12. (a) Histogram of static pressure (inlet vs outlet); (b)velocity vectors coloured by velocity magnitude (m/s); (c) pathlines coloured by particle ID; (d) static pressure; (e) turbulent kinetic energy

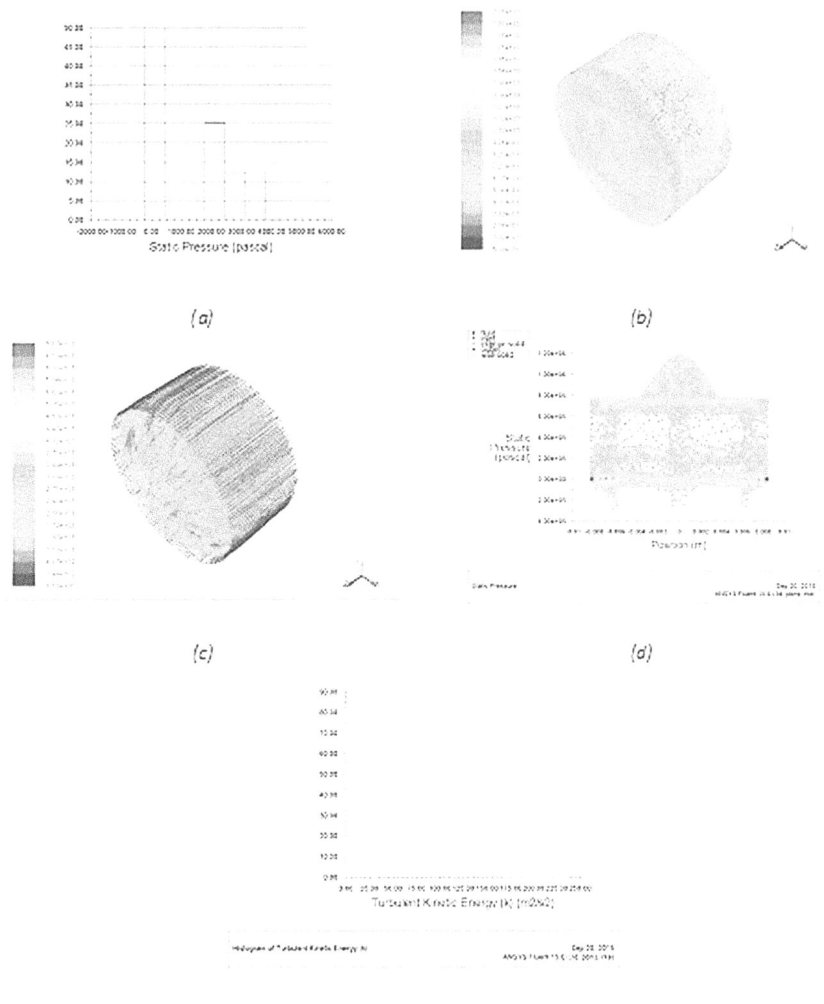

Table 5. Forces along z-direction

Zone	Forces (N)			Coefficients		
	Pressure	Viscous	Total	Pressure	Viscous	Total
Wall-solid	-1.340	-0.389	-1.729	-2.188	-0.635	-2.823
Net	-1.340	-0.389	-1.729	-2.188	-0.635	-2.823

Table 6. Mass flow rate (kg/s)

Inlet	0.30225933
Outlet	-0.30225906
Net	2.682209e-07

Figure 13. Grid of finite elements

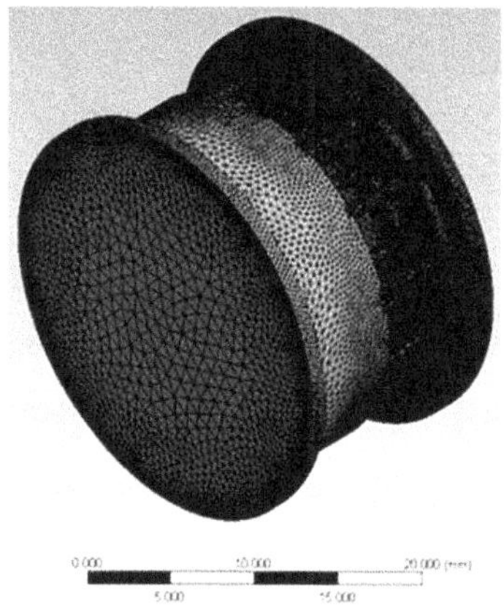

Figure 14 shows the flow pathlines of the fluid in the case of minimal disc opening. The flow pattern through the valve shows the dissipation of energy absorbed by the system.

DESIGN MODIFICATION

The design modification was carried out on the base valve assembly. This was done after consideration of results generated. The piston valve assembly in a twin tube damper is similar to that of a mono tube damper. The design of the piston valve is already highly optimized. There was however, scope for improvement in the performance characteristics of the base valve assembly. Stress concentration in the diaphragm spring of the valve and relatively low flow rate through the valve present scope for improvement.

The changes to the present design were made in the form of changes to the valve disk and the bleeders. The number of orifices was changed and 3D CAD models of the changed design were generated. Stress analysis and flow analysis was conducted on the changed designs.

Figure 14. (a) Flow pathlines through valve; (b) close up of flow pathlines for orifice outlet

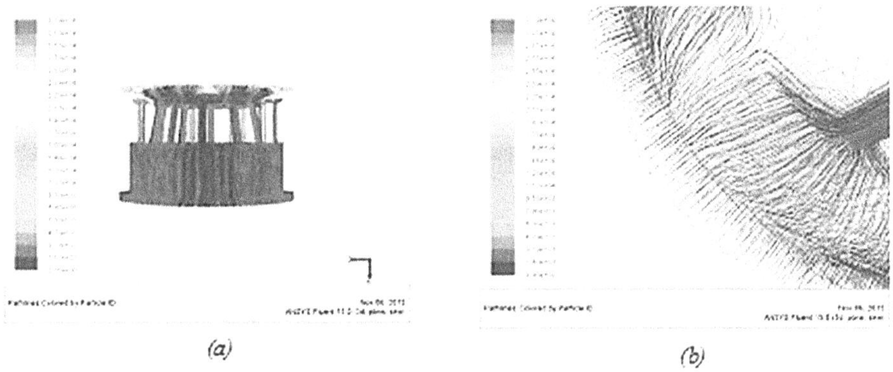

Figure 15 shows the 3D CAD models generated of the modified components.

As can be seen in the above figures, the shape of the bleeder has been kept constant. The only variables are the number and dimensions of the orifices in the valve disk.

RESULTS

The results of the analyses carried out on the modified designs are presented in this section. The modified designs were prepared in sets numbered from 1 to 4. Set 1 was such that the valve disk had two orifices. With every proceeding set, the number of orifices was increased. However, the solution failed to converge when the number of orifices was seven. It was also seen that greater opening of the discs causes greater flow velocity in the piston valve as observed by Czop et al. (2011). Hence there are four sets with two, four, five and six orifices each. The three orifice design being the design obtained by reverse engineering, serves as the datum.

Set 1: Two Orifices

Figure 16 shows the results for the simulations.

Table 7 and Table 8 give the details of the forces in the system and the mass flow rate.

Set 2: Four Orifices

Figure 17 shows the results for the simulations. The system did not converge for this setup.

Set 3: Five Orifices

Figure 18 shows the results for the simulations.

Table 9 and Table 10 give the details of the forces in the system and the mass flow rate.

Figure 15. (a) Set 1 for flow analysis: (i) valve disk (ii) bleeder; (b) set 2 for flow analysis: (i) valve disk (ii) bleeder; (c) set 3 for flow analysis: (i) valve disk (ii) bleeder; (d) set 4 for flow analysis: (i) valve disk (ii) bleeder

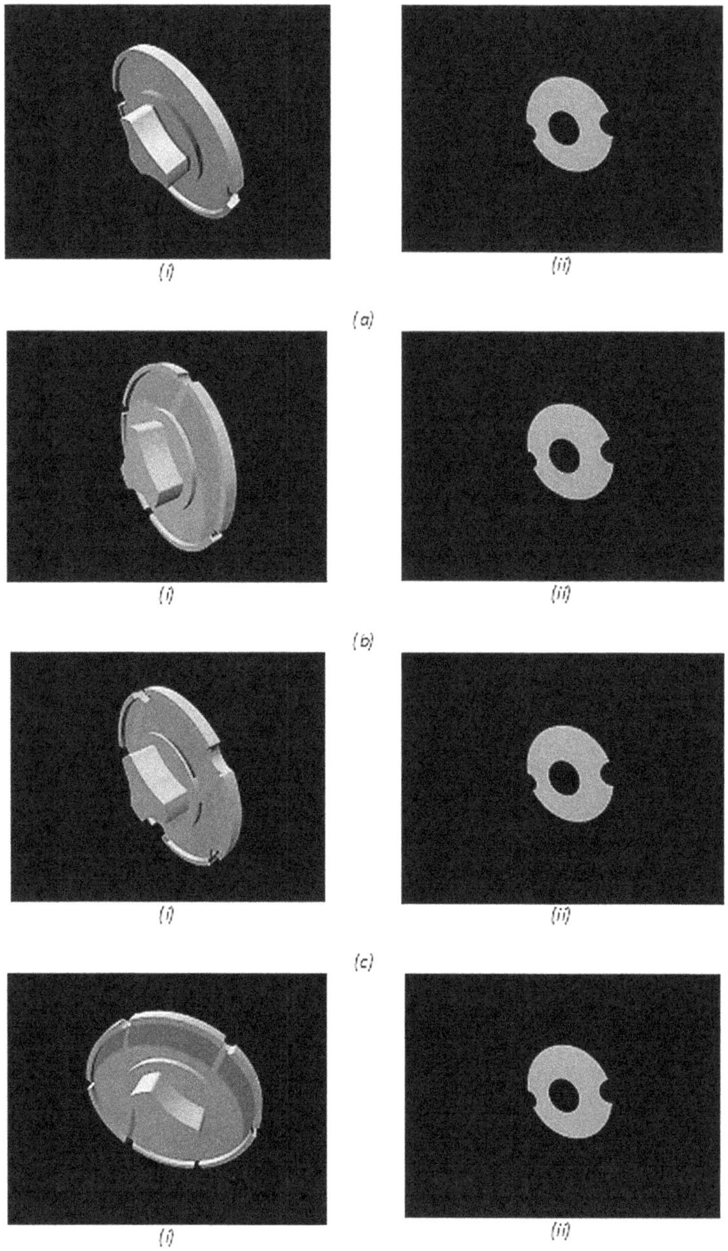

Figure 16. (a) Mesh; (b) velocity vectors (m/s); (c) pathlines coloured by particle ID(d) scaled residuals

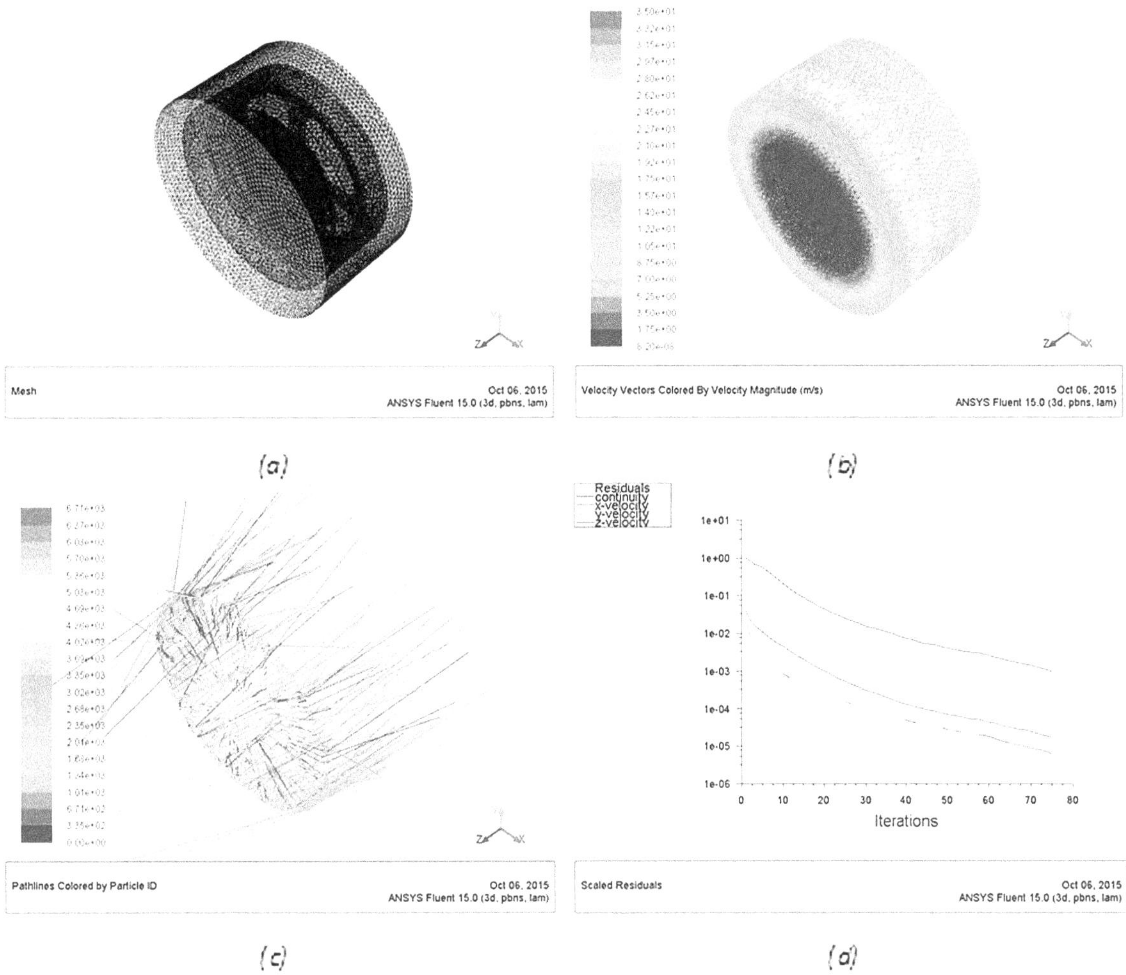

Table 7. Forces in (0,0,1) direction

Zone	Forces (N)			Coefficients		
	Pressure	**Viscous**	**Total**	**Pressure**	**Viscous**	**Total**
Wall-solid	-0.0324	-0.0031	-0.0355	-0.0529	-0.0051	-0.0580
Fluid	0.0882	0.0029	0.0911	0.1440	0.0047	0.1488
Net	0.0558	-0.0002	0.0555	0.0911	-0.0003	0.0907

Table 8. Mass flow rate (kg/s)

Inlet	3.3878248
Outlet	-3.3878236
Net	1.1004557e-06

Figure 17. (a) Scaled residuals; (b) pathlines coloured by particle ID

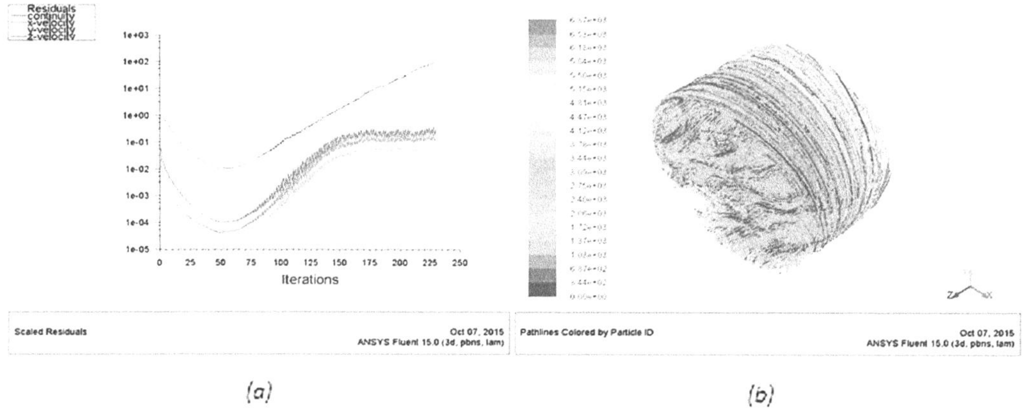

Figure 18. (a) Mesh; (b) velocity vectors (m/s); (c) pathlines coloured by particle ID; (d) scaled residuals

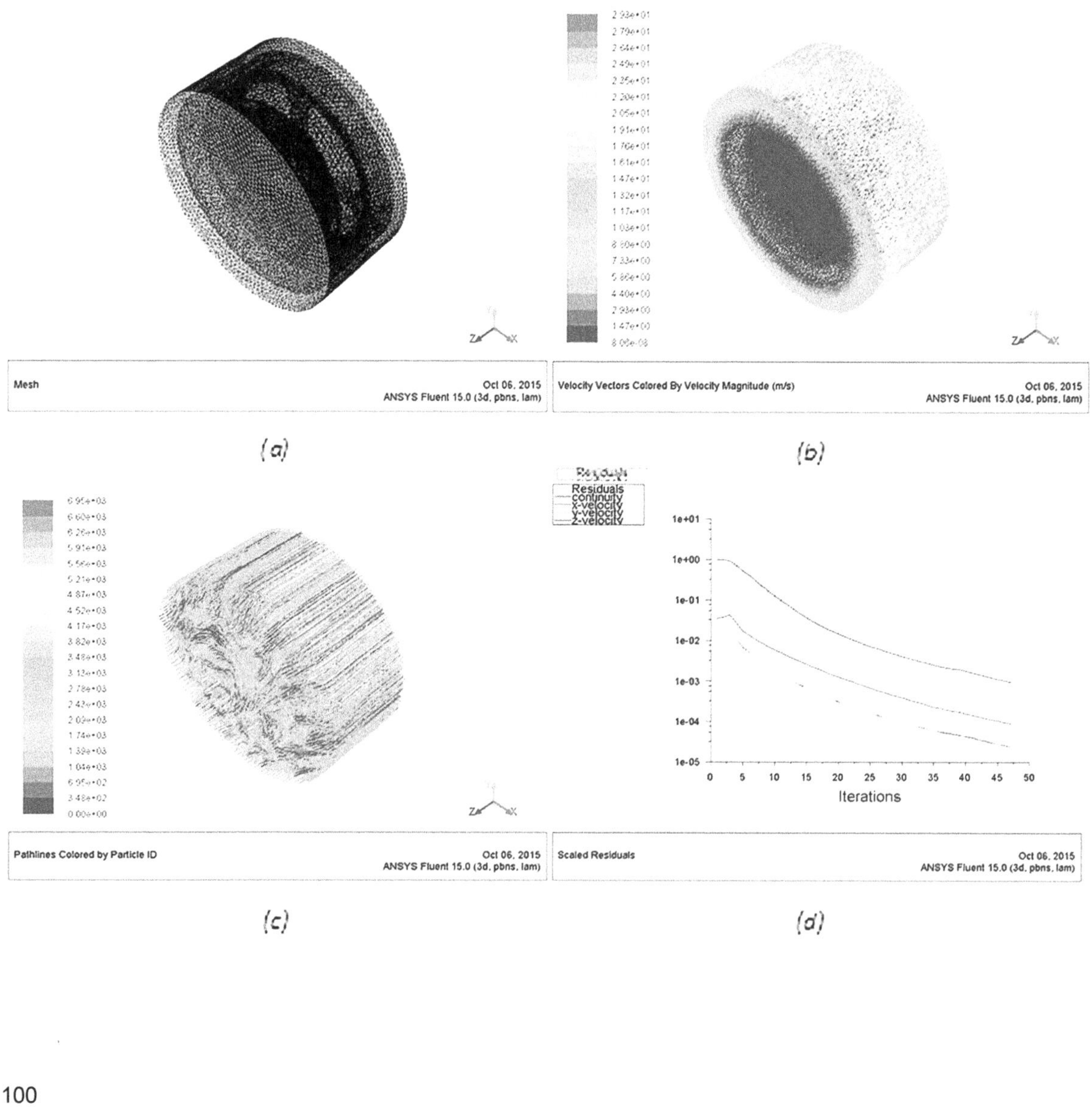

Table 9. Forces in (0,0,1) direction

Zone	Forces (N)			Coefficients		
	Pressure	Viscous	Total	Pressure	Viscous	Total
Wall-solid	-166.1339	-4.9124	-171.0463	-271.2390	-8.0203	-279.2593
Fluid	0.0094	-8.2101	-8.2007	0.0154	-13.4043	-13.3889
Net	-166.1244	-13.1225	-179.2470	-271.2236	-21.4246	-292.6482

Table 10. Mass flow rate (kg/s)

Inlet	1.4472516
Outlet	-1.4472504
Net	1.1484768e-06

Set 4: Six Orifices

The results are shown by Figure 19.

Table 11 and Table 12 give the details of the forces in the system and the mass flow rate.

CONCLUSION AND FURTHER SCOPE OF WORK

From the results of the initial analysis, it can be seen that the maximum stresses generated in the spring in the base valve assembly range from 0.233 MPa to 127 MPa. These stresses are well within the allowable limits for Structural steel, which has yield strength of 250 MPa. It can hence be safely stated that structural steel can be used to manufacture the diaphragm spring without excessive risk.

It is also seen that the three orifice design of the disc exerts considerable forces on the mechanism. Under the operating conditions wherein the maximum operating pressure inside the tube is 0.3 MPa, the maximum deflection produced is 2.21mm which is enough to allow the hydraulic fluid to flow from the compensation tube to the working cylinder. There is however, cause for concern about stress concentration in the spring. There is need for design modifications to reduce this effect. However, taking into account factors of safety as well as the conditions presented in the Indian subcontinent, the mechanism can function safely.

It has been seen that the flow velocity at the outlet is much greater than the inlet. This causes a greater dissipation of energy due to the large pressure gradient created. The results also indicate the possibility of noise generation in the system. The yield strength of structural steel being 250MPa, the spring operates under very safe conditions and does not run the risk of failure. Also, the maximum stress is localized due to stress concentration in the body. However, the magnitude of the stress is not large enough to warrant any design modifications.

Figure 19. (a) Mesh; (b) velocity vectors (m/s); (c) pathlines coloured by particle ID

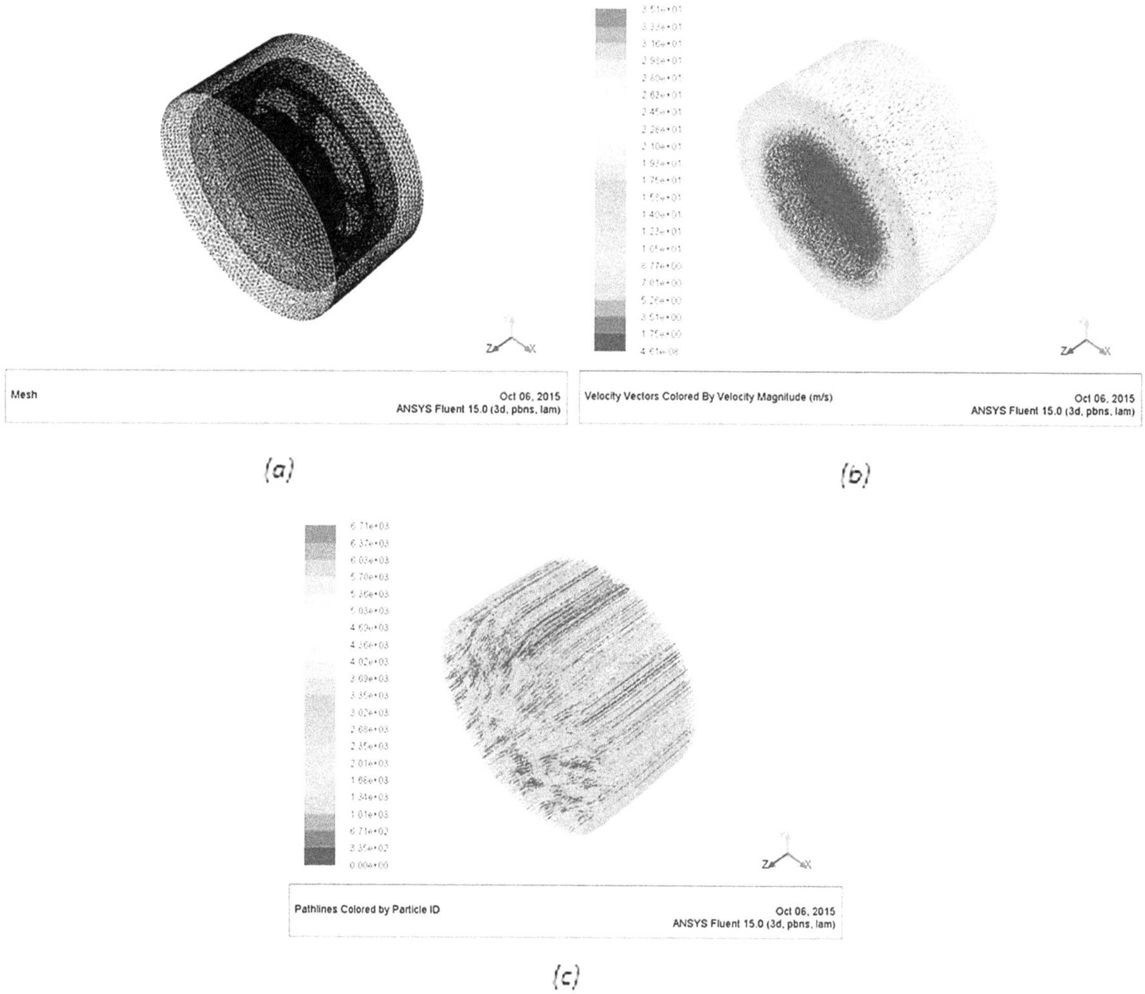

Table 11. Forces in (0,0,1) direction

Zone	Forces (N)			Coefficients		
	Pressure	**Viscous**	**Total**	**Pressure**	**Viscous**	**Total**
Wall-solid	-158.6951	-3.2876	-161.9827	-259.0940	-5.3675	-264.4616
Fluid	0.0036	-9.6902	-9.6866	0.0059	-15.8208	-15.8148
Net	-158.6914	-12.9778	-171.6693	-259.0881	-21.1883	-280.2765

Table 12. Mass flow rate (kg/s)

Inlet	3.3857985
Outlet	-3.3857996
Net	-1.2826779e-06

It has also been seen that the same models can be used to design and analyse further improvement on this system and the performance for various combinations of input parameters.

Following design modification, the forces exerted in the five orifice design are well within safety limits. Also, the net flow rate of hydraulic fluid is high enough for proper functioning of the assembly.

As observed above, the models can be used for further study of the systems. The designs may be validated using rapid prototyping techniques.

REFERENCES

Abella, R. J., Daschbach, J. M., & McNichols, R. J. (1994, April). Reverse Engineering Industrial Applications. *Computers & Industrial Engineering*, *26*(2), 381–385. doi:10.1016/0360-8352(94)90071-X

Atherden, M. J. (2004). *Formula SAE Shock Absorber Design. Bachelor of Engineering*. University of Queensland.

Banerjee, A., & Pohit, G. (2014). Crack Investigation of Rotating Cantilever Beam by Fractal Dimension Analysis. *Procedia Technology*, *14*, 188–195. doi:10.1016/j.protcy.2014.08.025

Bhuyan, D., & Kumar, K. (in press). 3D Cad Modelling and Computational Fluid Analysis of Piston Valve of Twin Tube Shock Absorbers. *Materials Today: Proceedings*.

Chai, S. T., & Mason, W. H. (1996). Landing Gear Integration in Aircraft Conceptual Design. Virginia Polytechnic Institute and State University.

Chikofsky, E. J., & Cross, J. H. (1990). Reverse Engineering and Design Recovery: A Taxonomy. *IEEE Software*, *7*(1), 13–17. doi:10.1109/52.43044

Czop, P. et al.. (2011). A Computational Fluid Flow Analysis of a Disc Valve System. *Journal of KONES Powertrain and Transport*, *18*(1).

Dixon, J. C. (2007). *The Shock Absorber Handbook* (2nd ed.). West Sussex, UK: Professional Engineering Publishing Ltd. and John Wiley and Sons Ltd. doi:10.1002/9780470516430

Eilam, E. (2005). *Reversing:Secrets of Reverse Engineering*. Indianapolis, IN: John Wiley & Sons.

EngineeringToolbox.com. (n.d.). *ISO Grade Oils - Viscosities and Densities* [Table]. Retrieved August 2015, from http://www.engineeringtoolbox.com/iso-grade-oil-d_1207.html

Gyenes, L., & Mitchell, C. G. B. (1994). *The effect of vehicle-road interaction on fuel consumption. Vehicle-Road Interaction*. ASTM International.

Houdaille, M. (1908). *Improvements in or Relating To Shock-Absorbing or Dash-Pot Apparatus for Motor Vehicles and the Like*. France Patent number FR394081-A.

Houdaille, M. (1909). *Shock-absorbing apparatus*. United States patent number US933076 A.

Kumar, D., & Bhingole, P. P. (2015). CFD Based Analysis of Combined Effect of Cavitation and Silt Erosion on Kaplan Turbine. *Materials Today: Proceedings*, *2*(4-5), 2314–2322. doi:10.1016/j.matpr.2015.07.276

Kumar, E. A., Balakrishna, M. V., & Chandra, M. R. (2015). Effect of Shim Percentage on Non-Linear Static Behavior of Flex Seal of Rocket Nozzle. *Materials Today: Proceedings*, *2*(4-5), 1427–1434. doi:10.1016/j.matpr.2015.07.063

Kumbhalkar Manoj, A., Bhope Dr, D. V., & Vanalkar Dr, A. V. (2015). Material and Stress Analysis of Railroad Vehicle Suspension: A Failure Investigation. *Procedia Materials Science*, *10*, 331–343. doi:10.1016/j.mspro.2015.06.064

Lang, H. H. (1997). *A Study of the Characteristics of Automotive Hydraulic Dampers at High Stroking Frequencies. Ph.D.* Ann Arbor, MI: University of Michigan.

Martande, S., Jangale, Y. N., & Motgi, N. S. (2013). Design and Analysis of Shock Absorber. *International Journal of Application or Innovation in Engineering & Management*, *2*(3), 195–199.

Motavalli, S., & Shamsaasef, R. (1996). Object-Oriented Modeling of a Feature-Based Reverse Engineering System. *International Journal of Computer Integrated Manufacturing*, *9*(5), 354–368. doi:10.1080/095119296131463

Patel, K. N., Gelot, B. N., & Gupta, A. H. (2014). A Review Study – Design and Analysis of Suspension System. *International Journal for Scientific Research & Development*, *2*(3).

Suh, J. (2014). *Redesign and Stress Analysis of Simplified Landing Gear* (BEng thesis). University of Exeter.

Talbott, M. S. (2002). *An Experimentally Validated Physical Model of a High Performance Automotive Damper. M.S.* Purdue University.

United States Environmental Protection Agency. (1980). *Passenger Car Fuel Economy: EPA and Road*: a *Report to the Congress in Response to the National Energy Conservation Policy Act of 1978, Public Law 95-619, Title IV, Part 1, Section 404.* Author

PartInfo. (n.d.). *Boge Monotube Shock* [Electronic Image]. Retrieved 2 August 2015 from http://www.partinfo.co.uk/articles/74

Pavani, P. N. L., Prafulla, B. K., Rao, R. P., & Srikiran, S. (2014). Design, Modeling and Structural Analysis of Wave Springs. *Procedia Materials Science*, *6*, 988–995. doi:10.1016/j.mspro.2014.07.169

Raja, V. (2008). Introduction to Reverse Engineering. In V. Raja & K. J. Fernandes (Eds.), *Reverse Engineering: An Industrial Perspective* (pp. 1–9). London: Springer Verlag. doi:10.1007/978-1-84628-856-2_1

Shukla, A., & Harsha, S. P. (2015). An Experimental and FEM Modal Analysis of Cracked and Normal Steam Turbine Blade. *Materials Today: Proceedings*, *2*(4-5), 2056–2063. doi:10.1016/j.matpr.2015.07.191

Chapter 6
CFD-Supported Designing of a Louvered Cylindrical Mixing Pipe for Maximum Air Entrainment

Sameer Ranjan Sahu
Birla Institute of Technology, India

Dipti Prasad Mishra
Birla Institute of Technology, India

ABSTRACT

Conservation equations of mass, momentum and energy have been solved numerically along with a two-equation-based k-ε turbulent model to determine the air entrainment into a mixing pipe. Two different louvered pipe configurations (one with entrance face closed to atmosphere while the other was opened) were used to entrain air into the mixing pipe and by changing various design parameters, the resulting effects were studied. For both the configurations, the opening area and the number of nozzles (assumed exhaust pipe) had significant effect in bringing down the exhaust temperature especially the configuration with its entrance face opened to atmosphere. Temperature as low as 309 K (approx.) was successfully achieved in the analysis. Nozzle protrusion length also proved to be an important parameter among other parameters like nozzle pitch circle diameter, inclination angle and pipe diameter.

1. INTRODUCTION

Transportation is an important part of the daily life. During ancient times, horses were the mode of transportation. From the invention of wheel, the mode of transportation took a whole new turn and got inclined towards carts. Since then, many developments have taken place to make this process more comfortable. All those modes of transportation used human or animal power. Moreover, the travelling also took time. With the passage of time, considerable research works were conducted by many investigators so that the travelling time can be reduced making the journey more comfortable. Thus vehicles came into action.

DOI: 10.4018/978-1-5225-3401-3.ch006

Brilliant minds like Nikolaus A. Otto, Sir Dougald Clerk, George Brayton, Gottlieb Daimler, Rudolf Diesel and Daimler made various engine designs so that they could be employed on these vehicles.

Technology, nowadays have advanced to an even higher level. Compare to earlier times, the engines are now more fuel efficient and reliable. In the current times, vehicles have been the easiest mode of transportation for every individual and thus have become an important part of today's lifestyle. They have made the world easier in eve ry aspect. But every coin has two sides. On one hand, the life style has become easier, while on the other hand a huge price is being paid for it. The heavy and continuous usage of vehicles has led to deterioration of not only the environment but also one's health. Every year the temperature of the planet is increasing causing many undesirable environmental changes. Global warming is being one of the most dangerous one. All these just because of the hot exhaust gases from the vehicles running all over the world every day and night. Jie (2011) studied the effects of exhaust gases on local as well as global scale showing their effect on the ecosystem. Locally viewing, these gases enter the respiratory systems which may become fatal to asthma patients and in the long run will affect the rest too in one way or another. These gases have a tendency to rise up, thereby creating adverse effects on the environment like ozone layer depletion and smog formation. The climatic conditions are also changing creating an inhospitable environment for all the life forms in the planet. So, a major challenge is faced by all of us. Therefore, some major steps have to be taken so as to control this degradation.

A simple step of cooling the exhaust gases prior to its release into the atmosphere can be applied to take care of such environmental issues. If fresh atmospheric air, which is at a lower temperature than the exhaust gases, is sucked in without using any external power or source, then it can be used for cooling of these exhaust gases. A louvered pipe surrounding the exhaust pipe will do the purpose of entraining the atmospheric air into the pipe. This works on Bernoulli's principle creating low pressure inside the louvered pipe by high speed jet action of the exhaust gas. Since, the exhaust gases travel at a greater speed while coming out of the exhaust pipe, they create a low pressure inside the pipe which thereby sucks in atmospheric air into the ventilated pipe as shown in the Figure 1. As the exhaust gases are hot, the exhaust pipe material should have high thermal resistance or should be stable at high temperatures. Now if theses exhaust gases are cooled down to a lower temperature, relatively lower price material can be employed for making the exhaust pipe. This in turn will cut the fabrication cost of the vehicles for large scale production. If sufficient cooling of these gases is achieved, then there is a chance of reducing the rate of increment in the local environmental temperature. The present work was numerically done so that an optimum design can be made that would suck maximum air from the atmosphere into the louvered pipe. Practically, it can be applied to any vehicle of any size.

The numerical analysis was aimed to obtain an optimum design which would effectively help in bringing down the temperature of the exhaust gas. For this purpose, two different pipe configurations were used. The first design was named as Configuration A. In this design, the air was admitted to enter the pipe only through the louvers on the pipe. The second was Configuration B, where the air entered the pipe via entrance face of the pipe as well as the louvers.

For designing the mixing pipe, various parameters have to considered that would affect its performance under different conditions. These parameters are the opening area (louver as well as total), protrusion length of nozzle, number of nozzles, nozzle inclination angle, pipe diameter and pitch circle diameter of nozzles on the air suction rate. Some assumptions were made to carry out this numerical analysis. They are as follows:

Figure 1. Entrainment of atmospheric air into the louvered pipe for configuration A and configuration B

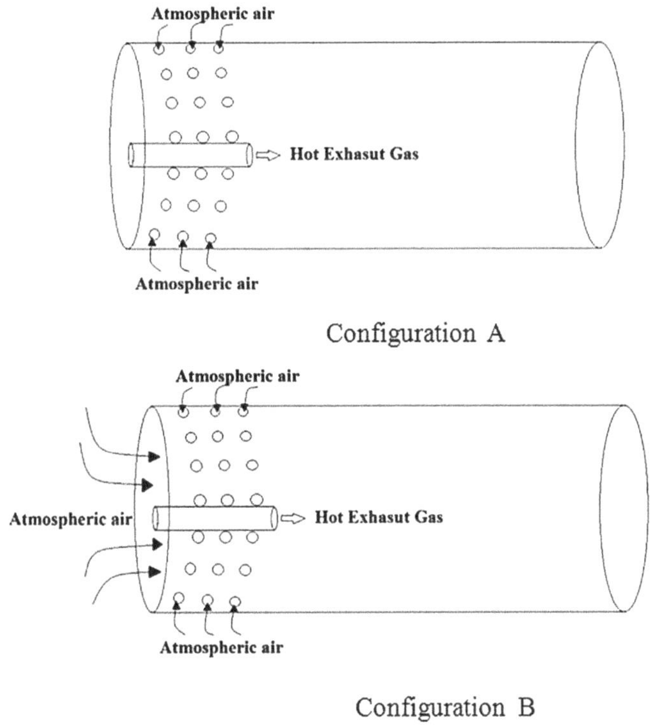

1. Flow from the exhaust pipe (assumed as nozzle) is taken as continuous and at a constant velocity.
2. Hot air is taken as nozzle fluid and treated to be incompressible.

The flow characteristics due to air entrainment were studied by many researchers both numerically and experimentally. The flow inside the pipe becomes turbulent because of the air suction created by the high velocity jet.

2. BACKGROUND

Considerable works based on experiment and numerical have been carried out to understand the fluid flow and heat transfer characteristics in cylindrical pipes. The mass flow measurement inside a sucking pipe is really difficult and becomes more cumbersome if the fluid density varies with temperature at different locations inside the pipes. A novel technique was described by Ricou and Spalding (1961) through which it might be comfortable to measure the axial mass flow rate formed inside a source of non-circulating/non-flowing air kept at unvarying pressure by injecting a turbulent jet. Their novel technique for measuring the entrainment rate was additionally applicable to jets of non-varying density. Through their experiments, the entrainment rate was raised by 30% when the gas got burnt out within the jet.

As mentioned earlier, turbulent flow characteristics are somewhat difficult to visualize. Hill's (1972) experiments came as an aid in this complicated field. The main focus was on the region occurring at the beginning of axi-symmetric turbulent air jets and thus, experiments were conducted for measuring the local entrainment rate. During the experiments, he found that the local entrainment rate was dominated by the axial distance from the injection site and the nozzle Reynolds number lost its dependency when it crossed the value of 6×10^4. In the initial part of the region, the entrainment coefficient was found to increase with axial length. But later, in the fully developed region, it almost became unvarying in nature.

Further adding to the entrainment study, Han and Mungal (2001) used Particle Image Velocimetry (PIV) turbulent jet co-flow, with and without reaction. As per their observation, the rise in co-flow speed for both reacting and non-reacting jets led to reduction in the entrainment. Heat transfer effects were also found to play some role in this entrainment area. The result of this heat transfer effect near the field of reacting jets was to reduce the entrainment by a factor of 2.5. The authors also found that the effect of buoyancy enhanced the entrainment which in turn compensated for the heat release effect in the reacting jet.

It is known that turbulence creates better mixing and an enhanced effect of heat transfer can be achieved. So, if some arrangements are made which can enhance mixing effect, then the required heat transfer or flow effects can be achieved. But conducting experiments is not always possible for every variation in design. In such cases, numerical analysis helps us significantly. Mishra and Dash (2010a, 2010b, 2011) conducted such numerical investigations. Their model had louvers cut on a funnel to suck in air by using a high velocity jet placed at centre. Their experiments were aimed to reduce the exhaust temperature of the naval ships. In order to do that, the hot gases released from the funnel were to be cooled by sucking the surrounding air. They increased the louver opening area and found that it enhanced the air suction rate. Although, the number of louvers per row had no significant effect. They also varied the protrusion of the funnel in the louvered pipe. The maximum suction effect from the bottom wall was found at around 0.5 metres. They also changed the design of louvered pipe and made it into a frustum keeping the volume same as that of the cylindrical funnel previously used. With this design, the maximum suction was found by keeping the diameter ratio at 0.8 and it was comparatively more than that sucked in by the cylindrical funnel. They realized that there may be some other factors that could affect this suction effect. The requirement was to entrain air at maximum rate. So there must be some other parameters that would affect the suction rate. By varying parameters, they found that when the normalised ratio of funnel diameter to nozzle diameter had the most prominent effect on entrainment rate. When this ratio was increased from 2.2 to 4.55, the entrainment rate was found to increase by 130% at an operating temperature of 573 K.

Extending their research for finding out more parameters that would affect the air suction, Mishra et al. (2010c, 2014) analysed different protrusion length and funnel diameter. Assuming isothermal conditions, the funnel inclination did not play any significant effect on air mass suction. Its effect was visualised when the nozzle fluid velocity was low and the temperature of the fluid was high. The entrainment rate was found to be mostly affected by the ratio of the mixing pipe length and the nozzle diameter. This ratio when changed from 5 to 20 altered the entrainment rate by 180%.

Barik et al. (2014a, 2014b, 2015) conducted somewhat similar experimental and numerical study on air entrainment. Firstly, its effect for infra-red suppression was studied. Here, the movement of nozzle away from the funnel bottom decreased the mass entrainment. Moreover, by changing the ratio of funnel overlap height to nozzle diameter from 0 to 1.6, the entrainment got reduced by 19.2%. They also studied flow recirculation in the suction zone. At some particular Reynolds number, the entrainment due to hot jet was comparatively greater than the isothermal jet because of buoyancy. Irrespective of the nozzle jet temperature, the mass ingress of air was found to be increasing with the nozzle protrusion. When the louvers opening area is increased from 116. 7 to 584, there was an enhancement of 70% in air entrainment. New correlations were found for air entrainment into infra-red suppression device for the prediction of the funnel exit temperature and mass ingress of air into IRS device.

The before mentioned researchers used finite volume method to analyse entrainment characteristics on various geometries and their applications. Singh et al. (1999), by the use of finite element model, predicted the mixing and entrainment of a confined turbulent jet. By modifying Prandtl mixing length model, they achieved the turbulence closure. They concluded that the main factors influencing the entrainment and mixing were the tube diameter and the air density inside the mixing pipe and the outside free stream. The positioning of the jet from the inlet also had an effect on entrainment. For short distances, the moving of the jet aside from the inlet raised the entrainment but concentrated when going for longer distances. Experiments were also conducted by using different jet shapes (Singh et al., 2003). Here they used confined and semi confined circular and non-circular jets. Non-circular jets proved to be more efficient in providing higher suction and mixing with surrounding fluid as compared to the circular jets. Even the entrainment ratio for all non-circular jets was more when compared to circular jet. Again the moving of the jet location helped in enhancing the entrainment by 30%.

Markatos (1986) diversified the study of the flow phenomena which involved turbulence in various practical applications. Computational methods were used to furnish details of the turbulent motion for various process including engineering, environmental, bio medical, etc. For the k-ω model, the varying of the distance from the wall was required to fit the logarithmic wall law. Another turbulent model used was k-ε model which proved to be more suitable to use as ε equation can be derived from N-S equation.

3. MAIN FOCUS OF THE CHAPTER

The previous works have thrown a light on the turbulent entrainment behaviour. Both numerical and experimental studies have been done on various shapes and geometries so that they can be applied for real world applications. The works cited above have not covered applications where air suction is required in horizontal louvered pipe using of CFD methodology. Thus, the present numerical study was based on methods of designing the mixing pipe for highest mass suction for different operating parameters. Since, this work has been done numerically; the results were validated against the experimental analysis conducted by Singh et al. (2003).

4. METHODOLGY

In the present work, D_p was taken as the diameter of the cylindrical pipe having a length L_p (Figure 2). As mentioned in Section 1, the Configuration A of the pipe will be closed at the entrance end and will be open to the surrounding atmosphere at the exit end. In case of Configuration B, the pipe was kept opened to the atmosphere at both the ends. The nozzle diameter was taken as D_n, which was having an extrusion length of L_{Pt} within the pipe. The nozzle was directed at the centre of the pipe's entrance face. Hot air is supplied by using this nozzle. The diameter (D_{CD}) and the length (L_{CD}) of the computational domain selected was about 6 times of D_p and about 3 times of L_p respectively and was placed around the pipe for considering the surrounding atmosphere. Accordingly, the boundary conditions were applied on this computational domain for the air suction to take place by the louvers cut around the pipe. The circular shaped louvers, each of diameter D_h = 0.03 m, are placed in a stack one over the other by maintaining a distance of 0.5 m between each stack. Each stack is having a total of 12 louvers. More louvers could have been made but keeping in mind the material strength, only 12 louvers per stack was made. The initial stack of louvers was kept 0.3 m apart from the bottom of the pipe. With this setting, the number of rows of louvers was varied from 8 to 9 with 12 columns of louvers used in the pipe. The flow field in the domain being three-dimensional and the fluid assumed as incompressible, accordingly Navier-Stokes equations were solved along with energy equation. As the flow is turbulent, a two equation based k-ε turbulence model was selected to account for the turbulent flow. The Reynolds number (Re_n) based on nozzle diameter was varied from 2.4×10^4 to 2.6×10^4. The purpose of selecting the two equation k-ε turbulence model is that it provides better results for such Re flows and also matched well with the existing experimental results. The nozzle fluid is taken as air with a flow rate of 0.02 kg/s (assumed as incompressible) with a temperature (temperature of the exhaust gas) of 343 K. For all the numerical investigations, these values were kept constant. The exhaust gas temperature and its mass flow rate were calculated from a twin cylinder vertical 4 stroke Petter Kirloskar Diesel CI Engine. The engine was having the following specifications; 1500 rpm, 10 hp, compression ratio 17:1, cylinder diameter 80mm, stroke 110mm, exhaust pipe diameter 5 cm and for the numerical simulation purpose, for all the wall defined determined, Steel was chosen as the solid material.

Figure 2. Schematic view of computational domain and the boundary conditions applied to it

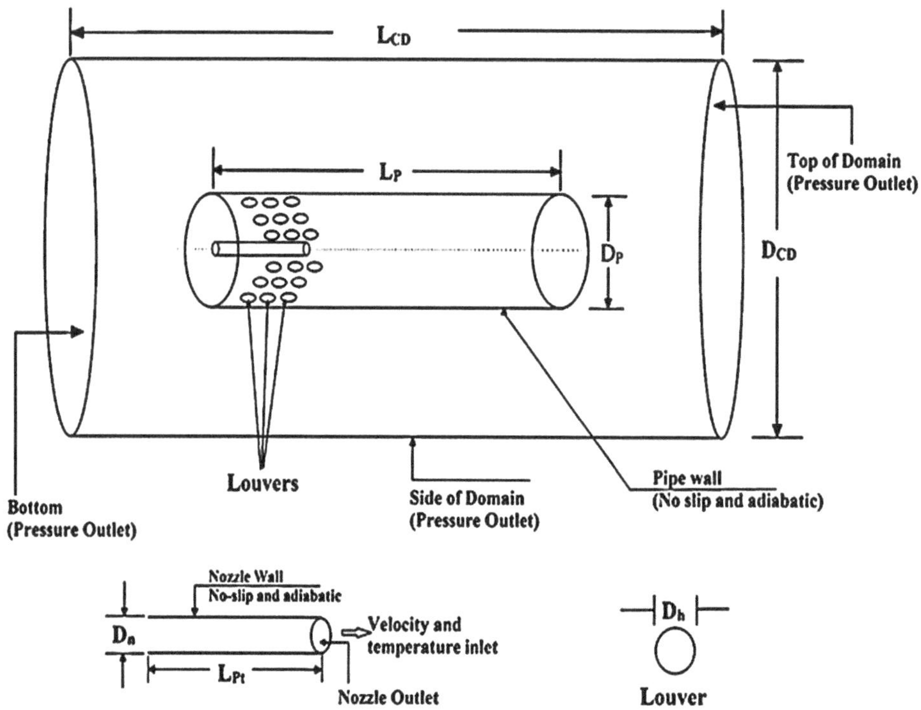

4.1 Governing Equations

The following governing equations were used or rather solved using finite volume method (ANSYS FLUENT 14.0) for the present analysis.

Continuity equation:

$$\frac{\partial}{\partial x_i}(\rho U_i) = 0 \tag{1}$$

Momentum equation:

$$\frac{D(\rho U_i)}{Dt} = -\frac{\partial p}{\partial x_i} + \frac{\partial}{\partial x_j}\left[\mu\left(\frac{\partial U_i}{\partial x_j} + \frac{\partial U_j}{\partial x_i}\right) - \rho\overline{u_i u_j}\right] \tag{2}$$

$$\frac{p}{\rho} = RT \tag{3}$$

As per the ideal gas law equation (3), the density ρ of the fluid was taken as a function of temperature, while keepingits viscosity and thermal conductivity constant.

Energy equation:

$$\frac{D(\rho T)}{Dt} = \frac{\partial}{\partial x_i}\left[\left(\frac{\mu}{\Pr} + \frac{\mu_t}{\Pr_t}\right)\frac{\partial T}{\partial x_i}\right] \tag{4}$$

Turbulence kinetic energy, k:

$$\frac{D}{Dt}(\rho k) = D_k + \rho P - \rho\varepsilon \tag{5}$$

Dissipation rate of k:

$$\frac{D}{Dt}(\rho\varepsilon) = D_\varepsilon + C_{1\varepsilon}\rho P\frac{\varepsilon}{k} - C_{2\varepsilon}\rho\frac{\varepsilon^2}{k} \tag{6}$$

$$\overline{u_i u_j} = \frac{2}{3}k\delta_{ij} - v_t\left(\frac{\partial U_i}{\partial x_j} + \frac{\partial U_j}{\partial x_i}\right) : v_t = 0.09\frac{k^2}{\varepsilon} \tag{7}$$

$$D_{\phi} = \frac{\partial}{\partial x_j}\left[\left(\mu + \frac{\mu_t}{\sigma_{\phi}}\right)\frac{\partial \phi}{\partial x_j}\right] \quad (8)$$

$$P = -\overline{u_i u_j}\frac{\partial U_i}{\partial x_j} \quad (9)$$

σ_k and σ_ε are the Prandtl numbers for k and ε.
R is characteristic gas constant whose value is 0.287 kJ/kg-K.
These k-ε equations have certain constants mentioned below

$C_{1\varepsilon} = 1.44$, $C_{2\varepsilon} = 1.92$, $C_\mu = 0.09$, $\sigma_k = 1.0$, $\sigma_\varepsilon = 1.3$, $Pr_t = 1$.

Generally, the above mentioned constants are found to be used for internal flows. But they can also be applied for flows involving mixing i.e., occurrence of both internal and external flows, similar to the present case under study.

4.2 Boundary Conditions

Figure 2 shows the boundary conditions applied on the louvered pipe. For the solid walled pipe and nozzles, no-slip boundary condition was applied. For considering the surrounding atmosphere, a computational domain was taken whose top, side and bottom surface were given a pressure outlet boundary condition. The nozzle from which the hot air is released into the louvered pipe was given a velocity inlet condition along with the required temperature. To compute the velocity, the local pressure field was used, which in turn satisfied the continuity equation at the pressure outlet boundary. The rest of the variables like T, k and ε were calculated from zero gradient condition. At the nozzle inlet, the turbulent intensity was set at 2% and for the back flow; it was set as 5% for all pressure outlet boundaries.

4.3 Numerical Solution Procedure

There was a variety of choices for discretisation of the domain. The information available from previous numerical work conducted by different researchers show us that the use of tetrahedral cells for discretizing the computation domain gave much better results. A finite volume method was used for discretization of the three dimensional mass, momentum, energy and turbulent equations over a control volume for obtaining a set of algebraic equations. To solve these algebraic equations under the imposed boundary conditions, an iterative method using multi grid solver of ANSYS FLUENT 14.0 was used. Two different types of solvers are available in ANSYS FLUENT. First one was the pressure based and the second one was the density based. Density based solvers are useful when working with compressible flows. But the flow condition here is an incompressible one. So, the Pressure based solver was found to be more appropriate to be used in such type of conditions. Momentum and turbulent equations were discretised using First Order Upwind scheme. The choice of domain size and number of grids was an important factor because changing them might change the output. So, its effect was studied in detail which have been separately discussed in the later sections.

As the momentum equation suggests, the pressure and velocity are coupled in the equation. To solve this, the SIMPLE algorithm with Standard scheme was used. For satisfying the convergence of all the variables, under relaxation factors for pressure, momentum, energy, k and ε were taken as 0.3, 0.7, 1, 0.8 and 0.8 respectively. For getting more accurate results, finer grids were applied at the nozzle entrance and louver openings. For the cylindrical computational domain, tetrahedral cells were used for meshing as mentioned earlier. To satisfy the convergence criteria, the field residual for all the variables appearing in the discretization equation was taken as 10^{-3} except for energy equation, for which the criteria was taken as $10^{-6.}$

5. RESULTS AND DISCUSSIONS

5.1 Validating with Other Experiments

As the present work was a numerical investigation of the mass ingress into a sucking pipe, therefore it was important to validate the results with some previously available experimental results. It has been shown in the Figure 3, the matching of experimental results with present CFD results. When compared with Pritchard's (1977) analytic solution and experimental results from Singh et al. (2003), the present numerical k-ε model was found to match well and within the allowable percentage error range. It can be seen from the plot that the results obtained from the present CFD with matched well with the analytic solution developed by Pritchard (1977) and existing experiment results from Singh et al. (2003).

Figure 3. Comparing suction ratio for different D_p/D_n for present CFD model with Pritchard's relation (1977) and available experiment

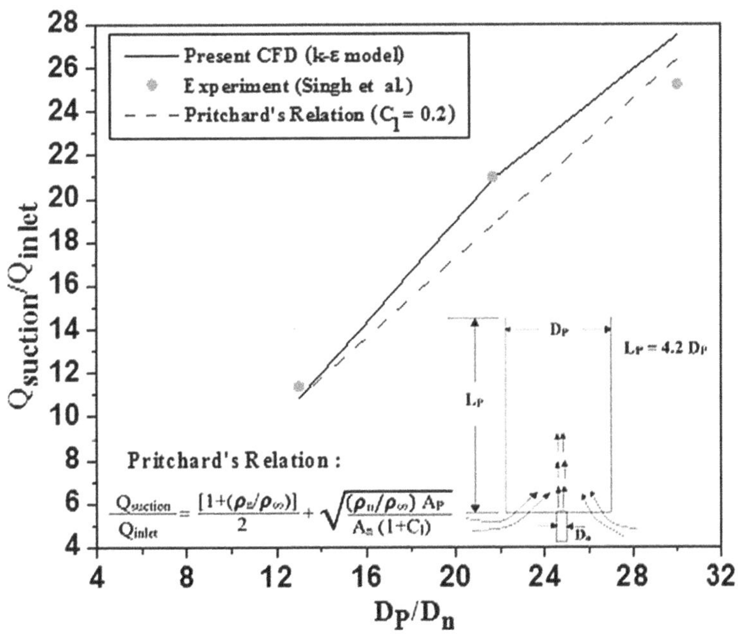

Figure 4 shows the radial profile of mean axial velocity of a wall-constraint. The ratio of centreline velocity u(x, 0) and the radial distance scaled to the jet half width ($y_{0.5}$) was used for normalising the mean axial velocityu(x, y).The transverse distance between the jet axis and the position where mean axial velocity u(x, y) was found to be half of the centreline velocity u(x, 0) was the required jet half width. As proposed by Singh et al. (2003) and Becker et al. (1963), the axial velocity profile can be reduced into a single curve, if self-similarity is exhibited by the velocity profile. Fig. 4 shows a comparison of the radial profile obtained from the present numerical analysis with the experimental results of Singh et al. (2003) and developed empirical relation by Becker et al. (1963). The comparison clearly shows that the results from the numerical analysis matches well with the previous works.

Figure 5 shows the validation of the present CFD k-ε model with the experimental results available from Singh et al. (2003). The entrainment ratio $Q_{suction}/Q_{inlet}$ as shown in this validation graph, was varied as a function of jet location for a circular jet. It can be seen from the plot the present CFD results matched well with the existing experimental results.

5.2 Analysis of Pipe Configuration 'A'

5.2.1 Influence of Domain Size on Air Ingress Into Pipe

The size of computational domain may affect the air entrainment into the pipe. So, to get a particular value of the domain, computations were performed using three different domain sizes as shown in the Figure 6. For each case, the pipe diameter was fixed at 0.2 m and pipe length fixed at 1 m while the diameter of the computational domain was varied from 4 to 8 times the pipe diameter and its length was varied from 2 to 4 times the pipe length. All these parameters are mentioned in the plot.

Figure 4. Comparison of the radial profile of axial velocity obtained from the present CFD model with the empirical correlation developed by Becker et al. (1963)

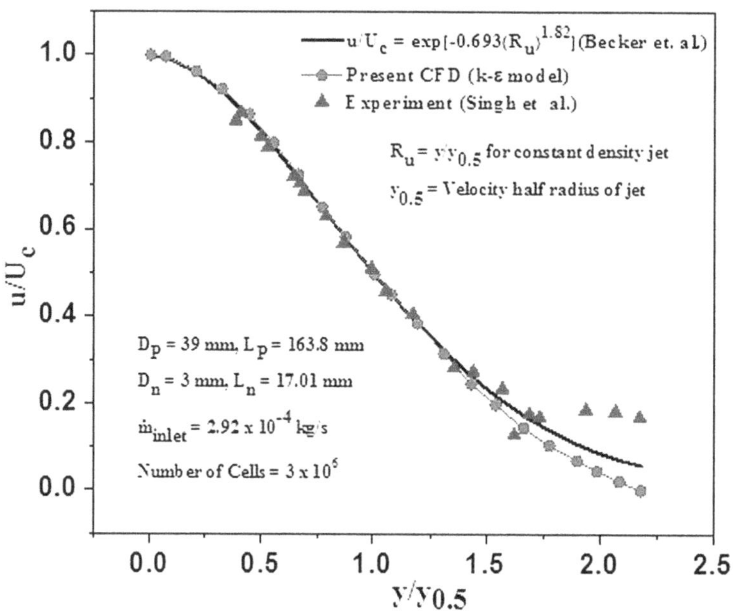

Figure 5. Entrainment ratio variation with jet location

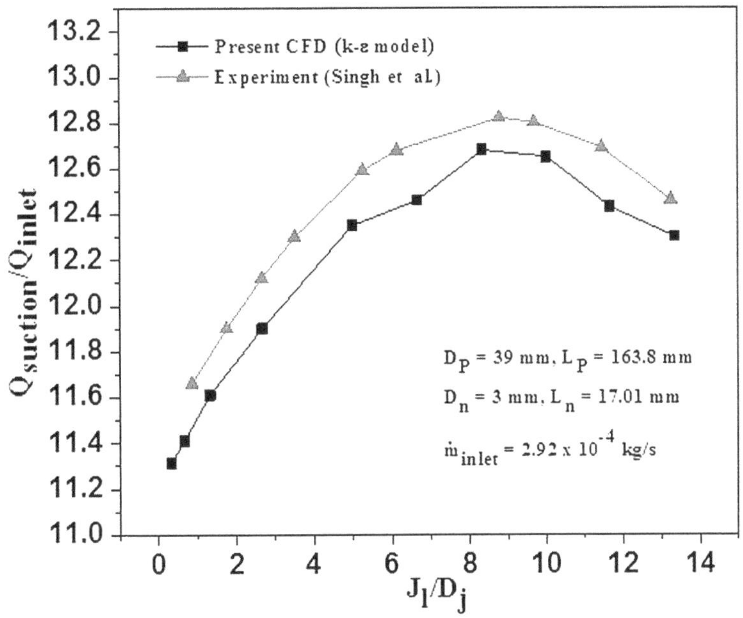

Figure 6. Influence of domain size on suction rate

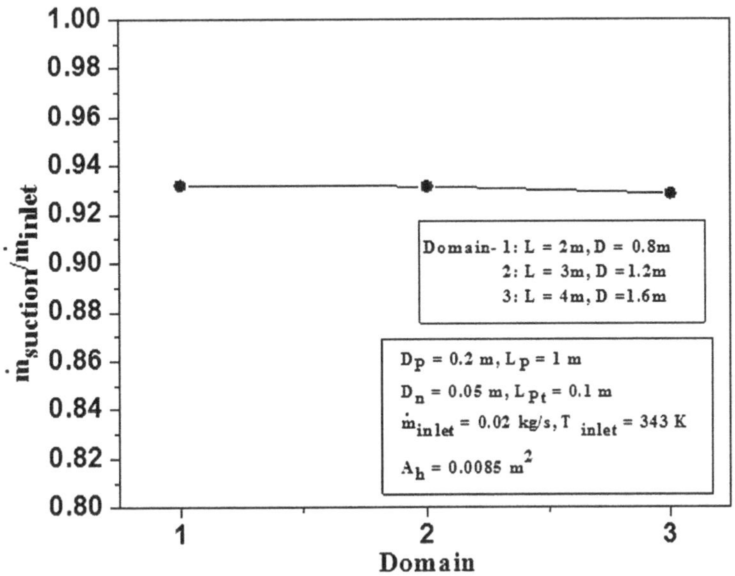

Figure 6 clearly shows that the domain size had no appreciable effect on mass ingress of air. So the computational domain was selected as 2 m x 0.8 m (L_{CD} x D_{CD}) for the pipe of size 1 m x 0.2 m (L_p x D_p) which is a minimum in size. Pressure outlet boundary condition was used on all the faces of the computational domain. The top surface of the computational domain admitted the air mass to stream out of the pipe, while other surfaces of the domain were used to admit the surrounding air to come into the mixing pipe from unlike locations on the pipe. Pressure outlet boundary was the only choice for the boundary condition for these types of cases.

5.2.2 Influence of Number of Cells on Air Ingress Into the Pipe

The tests conducted in section 1.1 helped in determining the particular dimensions of the computational domain. After this point, the numerical investigation was started by taking the diameter (D_p) and length (L_p) of the pipe as 0.2 m and 1 m. The diameter (D_{CD}) and length (L_{CD}) of the surrounding cylindrical computational domain as found in section 1.1 was 1.2 m and 3 m. The nozzle diameter (D_n) was taken as 0.05 m whereas protrusion was kept at 0.1 m. Twelve circular louvers per row (D_h = 0.03 m, A_h = 0.0085 m^2) were cut around the periphery of the mixing pipe. The effect of cell number on air entrainment was studied with single row of louver. The number of rows of louver was then varied after getting an optimum grid interval size and its effect has been shown later in the study. The distance between each row was maintained at 0.5 m. The nozzle flow rate of was maintained at 0.02 kg/s. The number of cells was varied from 1 x 10^5 to 2 x 10^6. The variation of air suction rate with respect to the number of cells can be visualized from Figure 7. It can be seen that the suction rate almost became constant from 6 x 10^5 numbers of cells. So, the mesh which gave 6 x 10^5 numbers of cells was adopted for performing the further computations.

An unstructured tetrahedral meshing was used for the computational domain, pipe, nozzle and the louvers. Figure 8 shows the meshing of pipe with louvers cut around the periphery.

Figure 7. Influence of number of cells on air suction rate

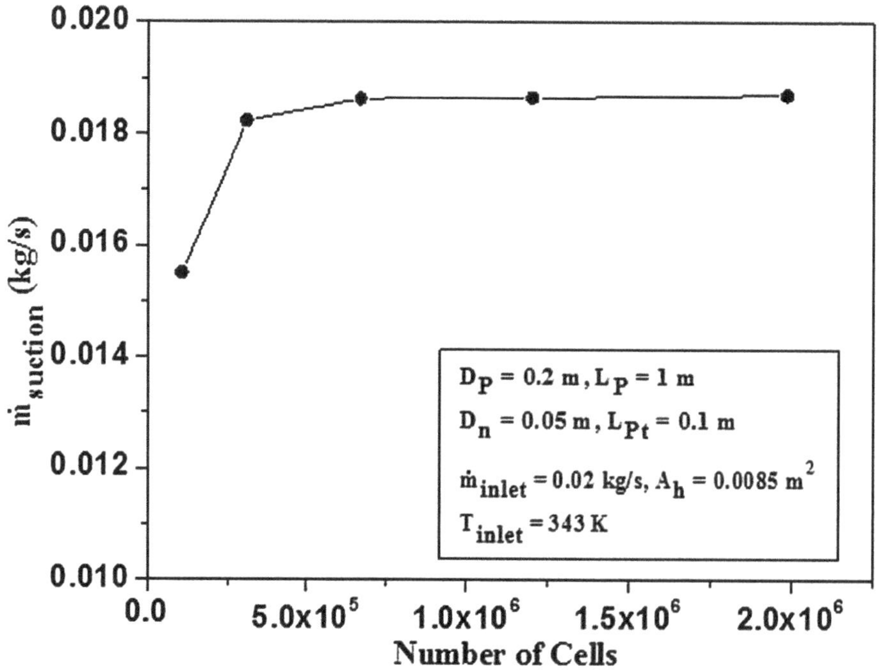

Figure 8. Cutaway view of louvered pipe (top) showing the unstructured tetrahedral meshing

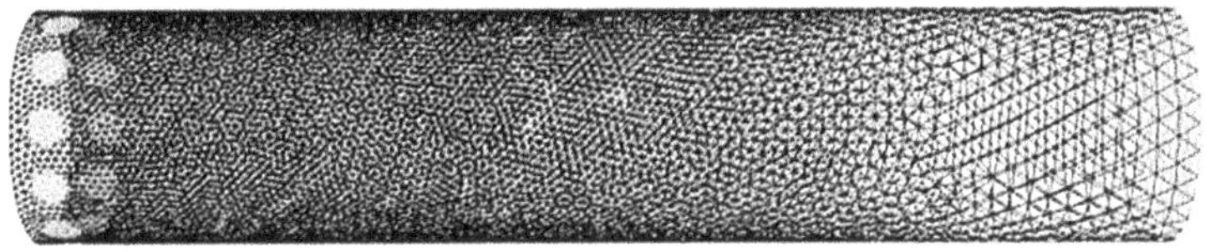

5.2.3 Influence of Louvers Opening Area on Air Ingress into the Pipe

After obtaining a particular value for the domain and cell size, the next job was to study the effects of various parameters on the air suction rate. This section was devoted to show that how air suction rate is influenced by louver opening area. Here, Figure 9 shows an instant of change in air suction rate taking into consideration the change in the louver opening area. For carrying out the simulations, the rows of louvers were increased up to 10 and thus having the louver opening area as 0.0085 m^2, 0.017 m^2, 0.0255 m^2, 0.034 m^2, 0.0425 m^2, 0.051 m^2, 0.0595 m^2, 0.068 m^2, 0.076 m^2 and 0.085 m^2. The nozzle flow rate was kept unvarying in nature at 0.02 kg/s. The louver opening area was enhanced by making addition in rows of louvers placed proportionally around the periphery. The length between each row was held at 0.5 m. The distance between the first row of louvers and the bottom of the pipe was kept as 0.3 m. The nozzle diameter, nozzle protruding length, pipe diameter and pipe length were kept constant at 0.05 m, 0.1 m, 0.2 m and 1 m. The diameter and length of the computational domain was fixed at 1.2 m and 3 m. An analysis from Figure 9 clearly shows that air suction rate was raised by increasing the louvers opening area, which eventually decreased the pipe outlet temperature. The air entrainment was increased by 1.8 times as the louvers opening were increased by two times and when the opening area

Figure 9. Influence of louver opening area on air suction rate

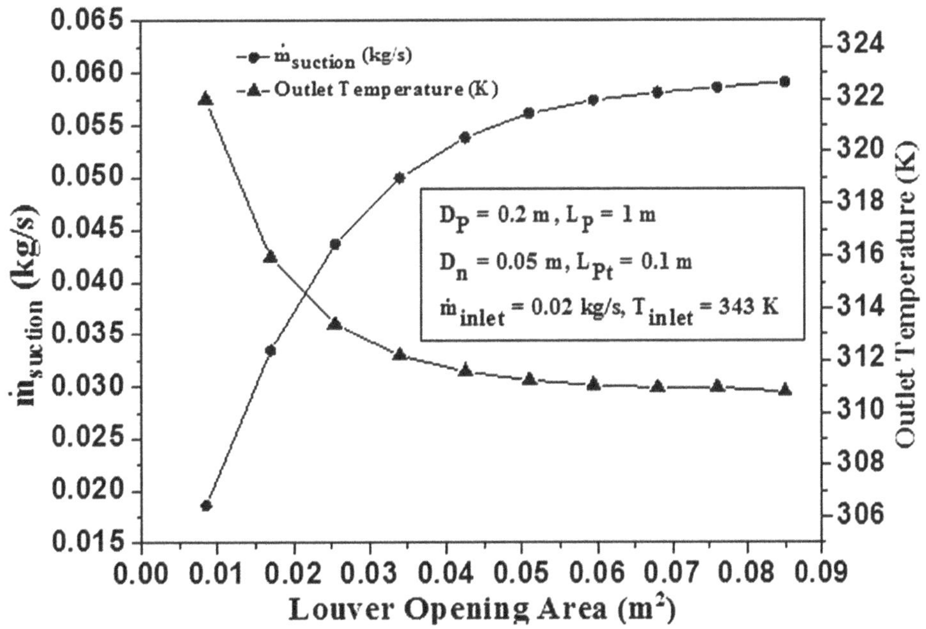

was increased by six times, the mass ingress of air increased by three times. It may be visualized from Figure 9 that by enhancing the louver opening area beyond 0.051 m^2 had no measurable effect on the increase of air suction rate and so the decrease in pipe outlet temperature also became unappreciable. This gives a clear indication that the louvers located close to the nozzle towards the bottom of the pipe gives higher entrainment rate.

It became obvious that the distance of the louver rows from the pipe bottom goes on increasing as more number of louver stacks is added. Therefore, if the louver rows are positioned at the bottom half of the pipe, i.e., close to the nozzle, more effective air suction effect would be observed. When visualizing the velocity vector plot (Figure 10), the entrainment process taking place in the pipe can be observed. Figure 10 shows the air suction from atmosphere into the pipe. Air suction from the surrounding atmosphere into the pipe through the louvers can well be observed from the figure. The hot air streams (temperature about 70-80 °C) being released from the nozzle, mixes with the entrained atmospheric air (temperature about 27 °C) inside the pipe, and then released back to the atmosphere from the top surface of the pipe.

5.2.4 Influence of Number of Nozzles on Air Ingress Into the Pipe

Figure 11 shows how the air suction rate was determined by the number of nozzles. A total of 4 cases were considered for simulation by taking 1, 2, 4 and 6 nozzles respectively. For the case of single nozzle, it was placed at the centre of the entrance face of the pipe. When the number of nozzles was increased to 2, 4 and 6, they were then placed at a constant pitch circle diameter (D_{pd}) of 0.1 m as shown in Figure 11. For simulation of single nozzle, its diameter was taken as 0.05 m. For simulating cases of 2,4 and 6 nozzles, diameters of 0.035 m, 0.025 m and 0.02 m were considered respectively so that a constant mass flow rate and a total cross section area can be maintained. For observing the effect of number of nozzles on air suction rate, parameters like mass flow rate, louver opening area, protruding length of nozzle, pipe

Figure 10. Velocity plot showing air suction into louvered pipe in case of 4 rows of louvers

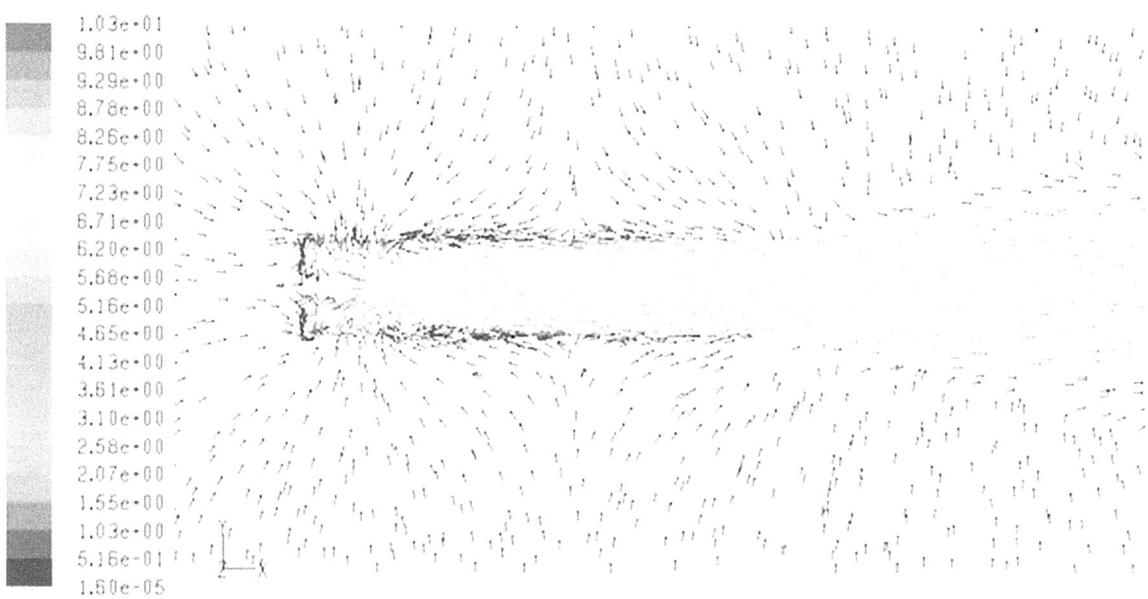

For a more accurate representation see the electronic version.

Figure 11. Influence of number of nozzles on air suction

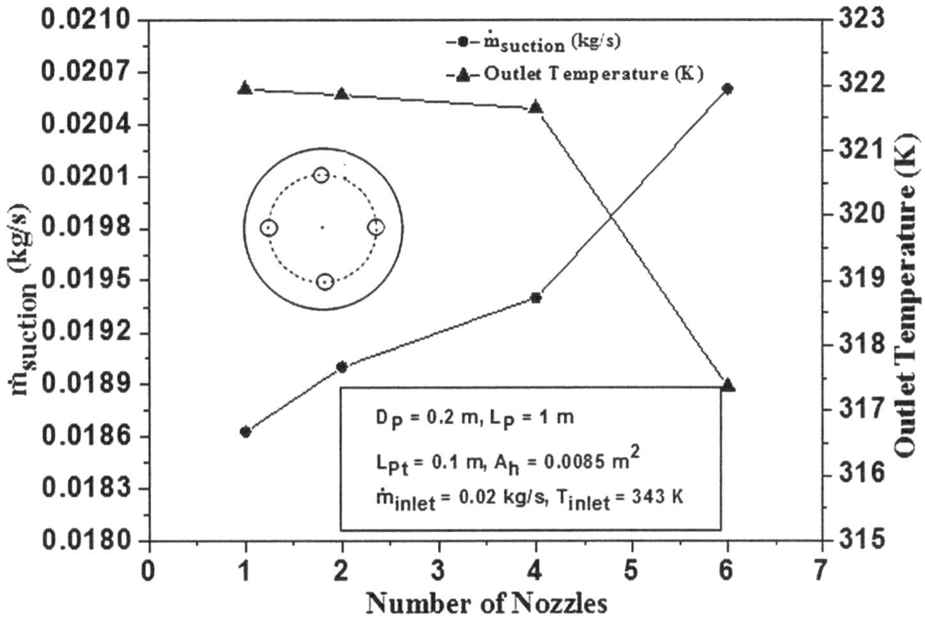

diameter and length were kept constant 0.02 kg/s, 0.0085 m^2, 0.1 m, 0.2 m and 1 m respectively. It can be seen from the plot 11 that the change in air suction rate was little when the number of nozzles was increased up to 4. But, for the case where 6 number of nozzles were used, an enhancement of 1.1 times in the entrainment rate was observed which sharply decreased the pipe outlet temperature.

When the nozzle number was increased it produced a consistent and a uniform low pressure inside the pipe which gave rise to a higher suction rate and thus, better mixing of the hot fluid with cold was observed. As a results of which the outlet temperature was decreased considerably and can be visualised from Figure 11.

5.2.5 Influence of Nozzle Protrusion on Air Ingress Into the Pipe

Figure 12 shows mass ingress of air as a function of nozzle protrusion. To observe the nozzle protrusion effects, parameters like the nozzle fluid inlet temperature, nozzle diameter, and mass flow rate, louvers opening area, pipe diameter and length were kept constant as can be seen from the plot. The air suction rate first increased as the protrusion increased from 0 to 0.75 m, then decreased with the increase in protrusion. So, the optimum protrusion for the nozzle was found to be at 0.75 m because at this point the air suction rate was at maximum and the outlet temperature of the pipe was at minimum. The air enters into the pipe due to low pressure inside the pipe and it had to cross a pressure hill to again come to the surrounding pressure at the expenditure of its own kinetic energy. Placing the nozzle tip towards the entrance of the pipe expenses more kinetic energy to move the air from the pipe to the surroundings, whereas placing of the nozzle tip away from the entrance, created a low pressure at a location far away from the louvers owing to their presence at the entrance of the pipe leading to a fall in the air suction. So from this analysis it was determined that the air suction rate was highest at an optimum distance of 0.75

Figure 12. Influence of nozzle protrusion on air suction rate

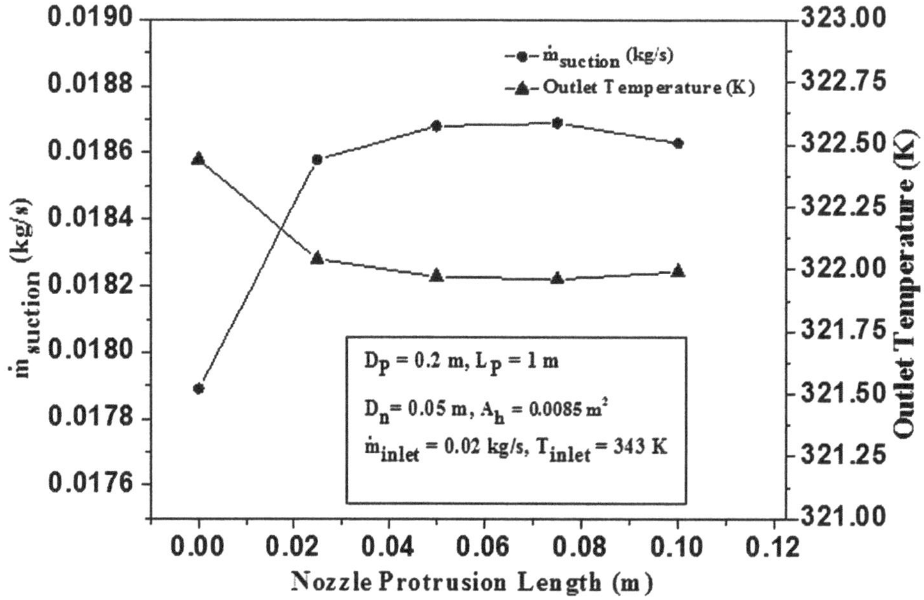

m from the entrance of the pipe. It can be concluded from this experiment that for maximum air suction through the louvers in the pipe, an optimum protrusion length of 0.075 m for the nozzle was found.

5.2.6 Influence of Change in Pipe Diameter on Air Ingress Into the Pipe

Figure13 shows the influence of changing the pipe diameter on air suction rate and pipe outlet temperature. The parameters that were kept constant were the size of the computational domain, louvers opening area, pipe length, mass flow rate, nozzle diameter, protrusion length and nozzle fluid inlet temperature as shown in the figure. When increasing the pipe diameter from 0.15 m to 0.2 m, the air suction rate was found to decrease. Because of this decrease in air suction rate, the temperature at the outlet of the pipe was thus increased. When the diameter pipe was increased, the louvers present at the periphery of the pipe moved far away from the centre of the pipe and so, the low pressure created at the centre of the pipe became ineffective. So it can be concluded from the investigation that the diameter pipe should be low to entrain maximum amount of air.

5.2.7 Influence of Nozzle Inclination Angle on Ingress of Air Into the Pipe

Figure 14 shows how the air suction rate was influenced by the nozzle inclination angle. The pipe diameter and length, opening area, nozzle flow rate and nozzle protrusion length and nozzle inlet temperature were kept constant. To see the effect of nozzle inclination angle, 4 number of nozzles were used for the computations. The angle made by the nozzle axis with the vertical was considered as the nozzle inclination angle as could be seen from Figure 14. The nozzles were inclined at an angle of 5°, 10° and 15° towards the pipe axis and 5° away from the pipe axis. The negative sign as seen in the figure

Figure 13. Influence of change of ratio of pipe diameter to nozzle diameter on air suction rate as a function of pipe outlet temperature

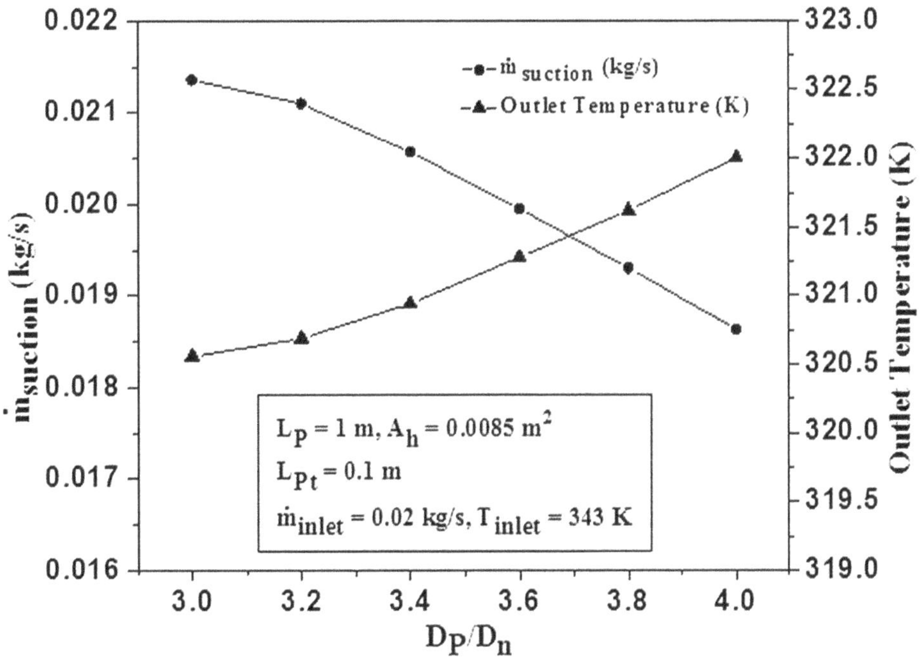

refers to the nozzle inclination towards the pipe axis and similarly the positive sign refers to the nozzle inclination away from the pipe axis.

It can be seen from the Figure 14 that the air suction rate was maximum when the nozzles were parallel to the horizontal axis, i.e., the nozzles had 0° of inclination, and at zero inclination maximum cooling of the hot air could be achieved. When the nozzles are tilted towards the pipe axis, the jets strike against each other due to which they don't expand properly. This improper expansion of nozzle jets leads to insufficient generation of low pressure inside the pipe which then reduces the air suction rate and hence unable to bring down the temperature of the hot air.

When the nozzles are tilted away from the pipe axis, there becomes a chance for the nozzle jet to escape through the louvers which again leads to insufficient generation of low pressure inside the pipe and as a result, low air suction rate with high pipe outlet temperature was observed.

5.2.8 Influence of Nozzle Pitch Circle Diameter on Air Ingress Into the Pipe

Figure 15 gives an idea on how the air suction rate was influenced when there was a change in the pitch circle of nozzles. For this analysis, the parameters like the louver opening area, pipe diameter and length, nozzle diameter and protrusion length, mass flow rate and the inlet temperature of the nozzle fluid nozzle fluid were kept constant. The number of nozzles used for this purpose was taken as 4. The pitch circle diameters taken were 0.05 m, 0.075 m, 0.1 m, 0.125 m and 0.15 m. With the increase in pitch circle diameter, the air suction rate increased but after certain point there was a sharp decrease as can be seen from the figure. Maximum suction effect was found at the pitch circle diameter of 0.1 m.

Figure 14. Variation of air suction rate and exit temperature at different nozzle inclination angle

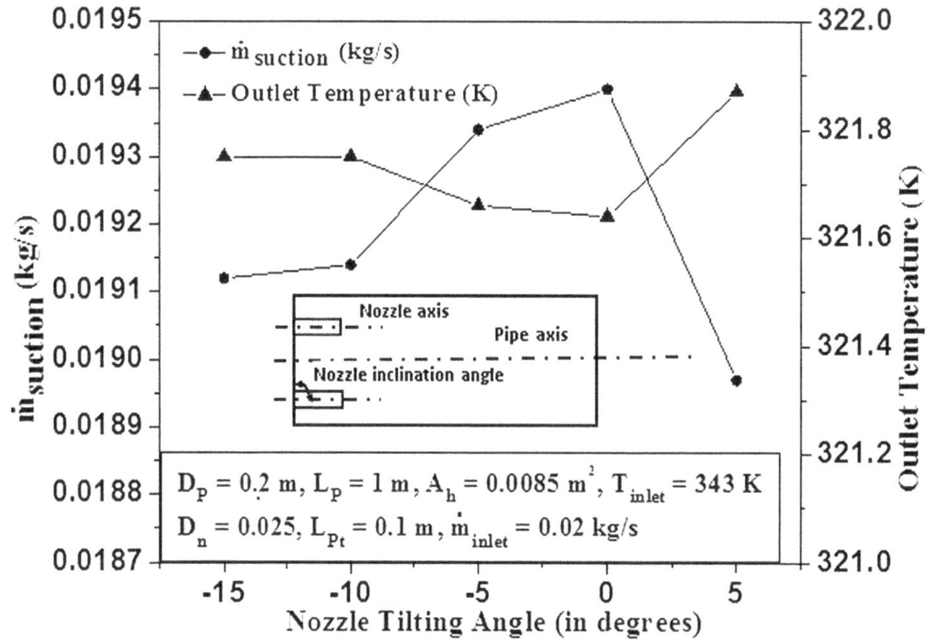

Figure 15. Variation of air suction rate and exit temperature at different nozzle pitch circle diameter

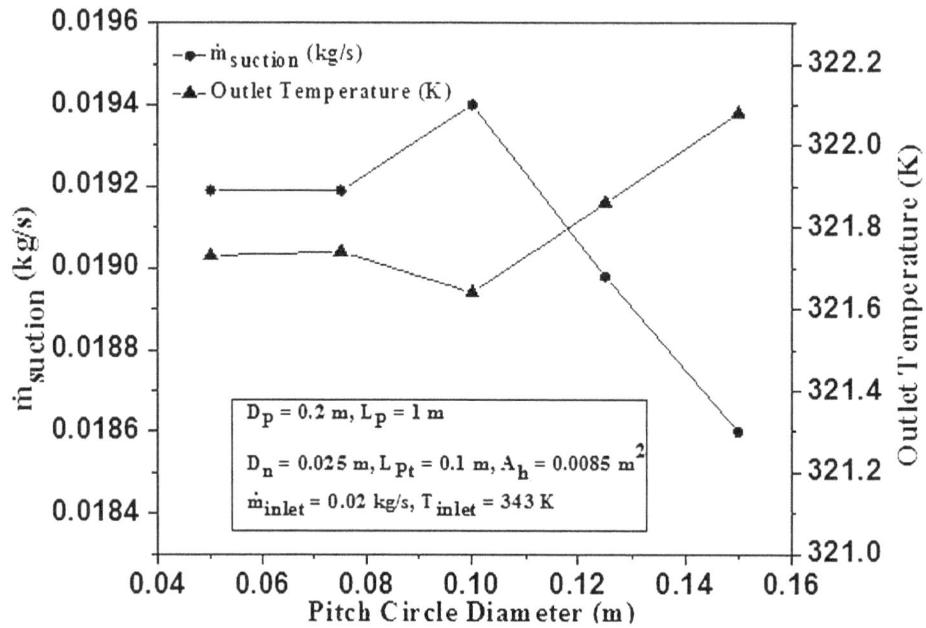

When the nozzles were located close to the centre, i.e., far from the louvers, the high velocity jets from the nozzles were unable to create low pressure inside the pipe for the atmospheric air to be sucked in through the louvers. When the nozzles were placed too close to the louvers, the viscous effect became more prominent due to which the jets were unable to create enough low pressure inside the pipe leading to sharp fall in air suction rate.

5.3 Effect of Pipe Configuration 'B' On Air Suction Rate and Pipe Outlet Temperature

5.3.1 Influence of Total Opening Area on Air Ingress Into the Pipe

Figure 16 shows the influence of increasing total opening area on air ingress into the pipe. To carry out the simulations, the pipe's entrance face area was kept constant at 0.03142 m^2, while, by symmetrically placing and adding of rows of louvers around the pipe periphery, the louver opening area was made to increase from 0.0085 m^2 to 0.068 m^2. The nozzle flow rate was kept constant at 0.02 kg/s. The distance between each row was maintained at 0.5 m. The parameters that were kept constant are the nozzle's diameter and its protrusion length, pipe's diameter, its length, and its entrance face as can be seen from the figure. From Figure 16, it can be clearly perceived that air suction rate was enhanced when the total opening area was increased. This phenomenon led to a decrease in the pipe outlet temperature. When the total opening area was increased by 2.5 times, the air suction rate is increased by 1.05 times, which decreased the pipe outlet temperature. But after a certain total opening area close to 0.09 m^2, the variation in pipe outlet temperature was unappreciable.

Figure 16. Variation of air suction rate with respect to total opening area

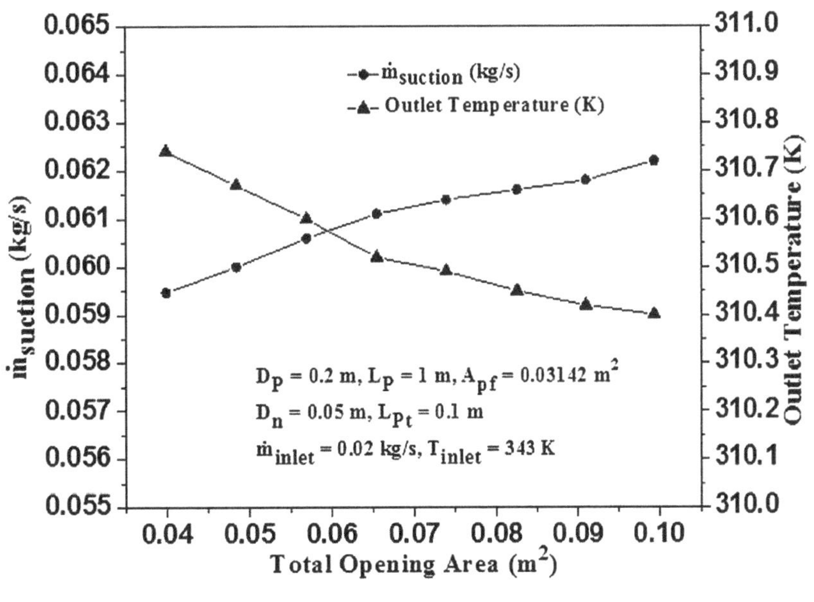

Atmospheric pressure was imposed on the exit as well as entrance faces of the pipe, so the effectiveness of the louvers located nearby the top would be less. When the louvers were located close to the nozzle, better suction effect was observed. The hot air streams (temperature about 70-80 °C) released from the nozzle were mixed inside the pipe with a cooler atmospheric air (temperature about 27 °C), and finally came out from the pipe's top surface.

5.3.2 Influence of Number of Nozzles on Air Ingress Rate Into the Pipe

The influence of number of nozzle on air ingress rate can be observed from Figure 17. The simulation was carried out for 1, 2, 4 and 6 nozzles by keeping the mass flow rate, louvers opening area, nozzle protrusion, pipe diameter and length constant as seen in the figure. The single nozzle was located or placed at the centre of the pipe's entrance face while for the rest of the cases, the nozzles were symmetrically placed at a pitch circle diameter of 0.1 m. The diameter of the single nozzle was taken as 0.05 m. Total cross section area of the nozzle was kept constant by using the nozzle diameters as 0.035 m, 0.025 m and 0.02 m for number 2, 4 and 6 nozzles.

There was slight change in air suction rate by increasing the nozzle number up to 4. But increasing the number of nozzles to 6 brought an enhancement of 1.1 times in the entrainment rate which led to a sharp decrease in pipe outlet temperature. With the increase of number of nozzles, the creation of low pressure inside the pipe became more uniform, thus gaining more suction rate. Moreover, with such placement of the 6 nozzles, a better mixing of the hot and cold fluid took inside the pipe, leading to a drastic fall in the pipe outlet temperature.

Figure 17. Influence of number of nozzles on mass ingress of air

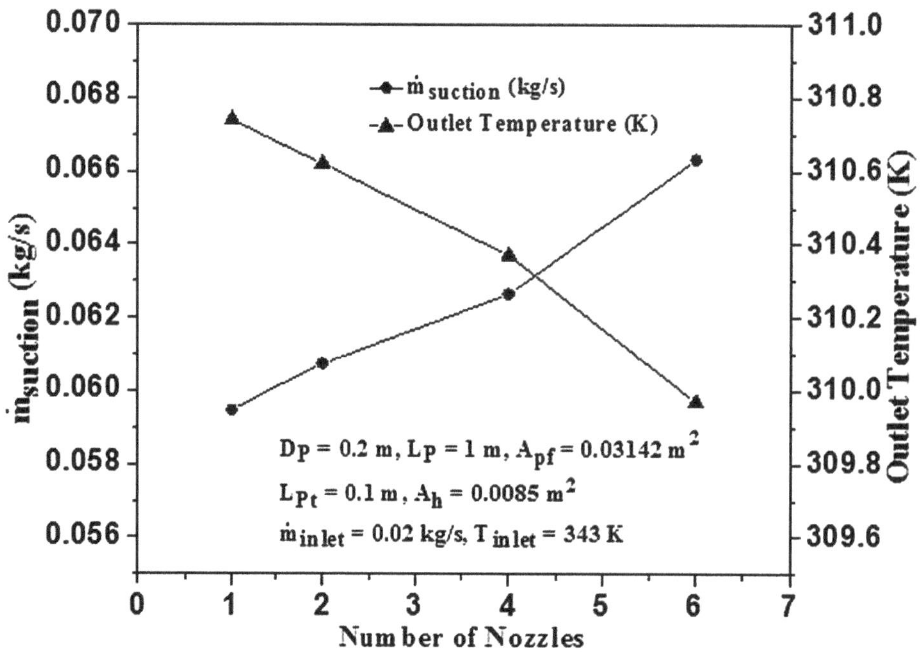

5.3.3 Effect of Nozzle Protrusion on Air Ingress Into the Pipe

Figure 18 shows the effect of nozzle protrusionon mass ingress of air into the pipe. The pipe diameter and length, nozzle diameter, mass flow rate, louver opening area and the inlet temperature of nozzle fluid along with the area of the entrance surface of the pipe were the constant parametersas shown in the figure. The protrusion of nozzle is done with reference to this nozzle location. The positive sign in the figure indicates the protrusion of nozzle inside the pipe while the negative sign indicates the moving of nozzle away from the pipe along the same axis. It can be visualized from the plot, as the nozzle moves away from the entrance of pipe, increased the entrainment ratio due to the exposure of the high velocity jet creates a low pressure close to the entrance of the pipe. However, moving the nozzle further away expanded the jet almost up to the size of the pipe causing spillage and hence, reducing the air suction rate. On the other hand, moving the nozzle into the pipe continuously decreased the air suction rate. This observance thus showed that an optimum nozzle exit location exists for any jet configuration and mixing tube diameter where a peak entrainment value is obtained. So, the maximum air suction was obtained when the placement of nozzle 0.075 m outside the entrance of mixing pipe.

5.3.4 Influence of Pipe Diameter on Air Ingress Into the Pipe

Figure 19 shows the variation of air suction rate with respect to the ratio of pipe diameter and nozzle diameter. For the simulation of the present case, only the pipe diameter was varied and the rest of the parameters were kept constant which can be seen from the figure. The pipe diameter was varied from 0.15 m to 0.2 m. For a given mass flow rate through the nozzle, the amount of air sucked in increases

Figure 18. Effect of nozzle protrusion on air suction rate

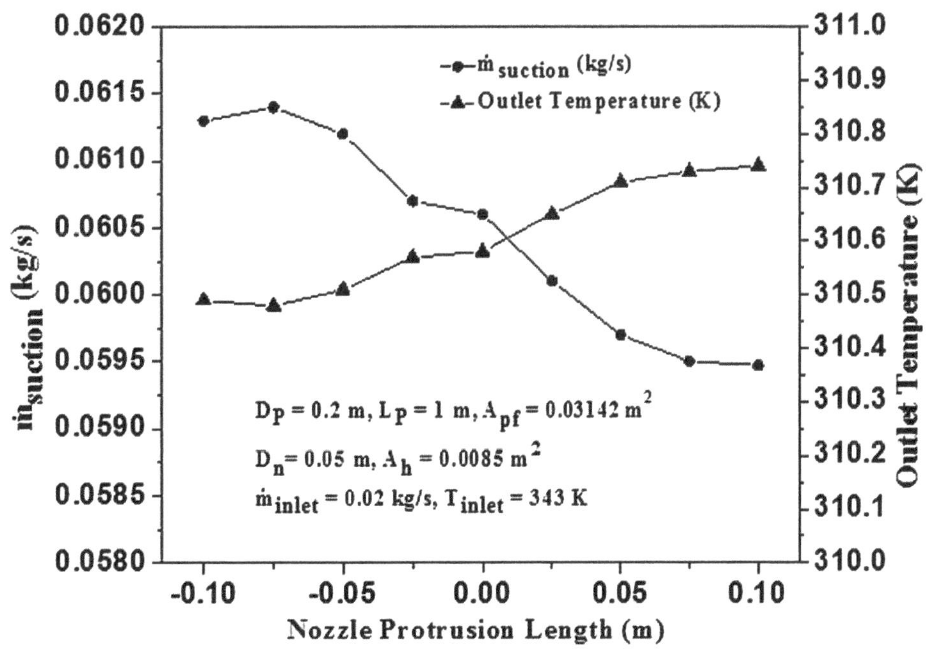

Figure 19. Variation of air suction rate with respect to the ratio of pipe diameter and nozzle diameter

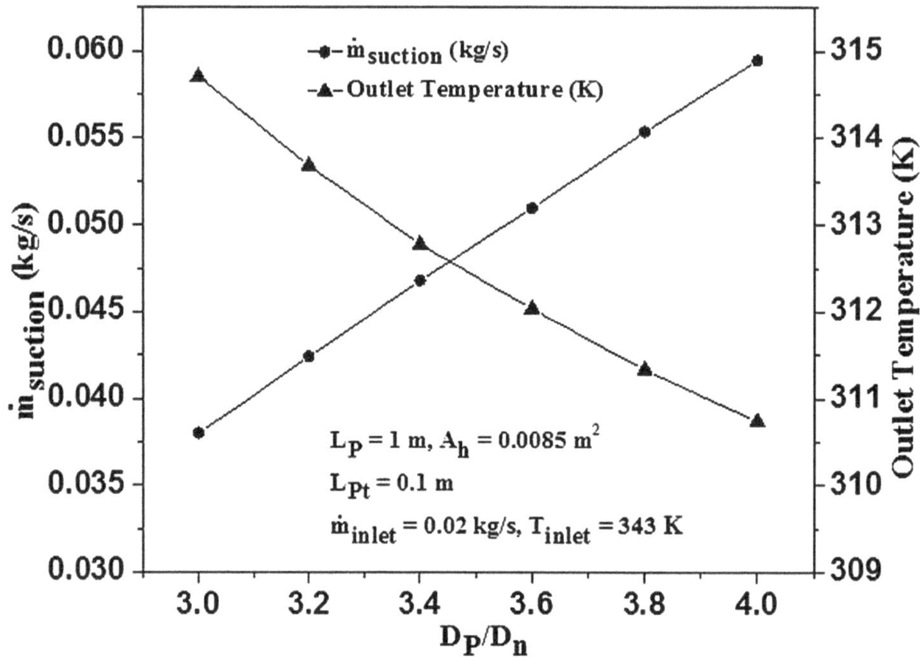

with the increase in pipe diameter. Since, the entrance face of the pipe is kept open to the atmosphere, more air was sucked in due to the increase in its area, thus reducing the pipe outlet temperature.

5.3.5 Effect of Nozzle Pitch Circle Diameter on Air Ingress Into the Pipe

Figure 20 shows the mass ingress of air as a function pitch circle diameter. The analysis was carried out by changing the pitch circle diameter where as other parameters are kept constant that can be visualised form plot. 4 nozzles were used for the analysis by placing them at different pitch circle diameters of 0.05 m, 0.075 m, 0.1 m, 0.125 m and 0.15 m. It was found from the analysis that increasing the pitch circle diameter, the air suction rate was increased but decreased sharply after certain point. Maximum suction effect was found at the pitch circle diameter of 0.1 m with minimum the pipe outlet temperature.

When the air stream was released from the nozzles, a low pressure was created inside the pipe which enabled the atmospheric air at a higher pressure to move into the pipe. Placing the nozzles either close to or far away from the centre does not facilitate this phenomenon. In these cases, thus viscous effect played role and so the air suction was sharply reduced.

5.4 Agreement and Disagreement With Other Similar Investigation

There have been many ideas of making the exhaust gas less harmful to the environment. Cooling of the exhaust gases became one of them. Although, this method was only applied to naval ships and not to the public vehicles as stated in the literature. Both the works have different applications as well as operating conditions. So, it will be difficult for the present work's result to be compared to those presented in the literature as they have different flow conditions and applications. Still, if a comparison is made, it can

Figure 20. Effect of changing nozzle pitch circle diameter on air suction rate

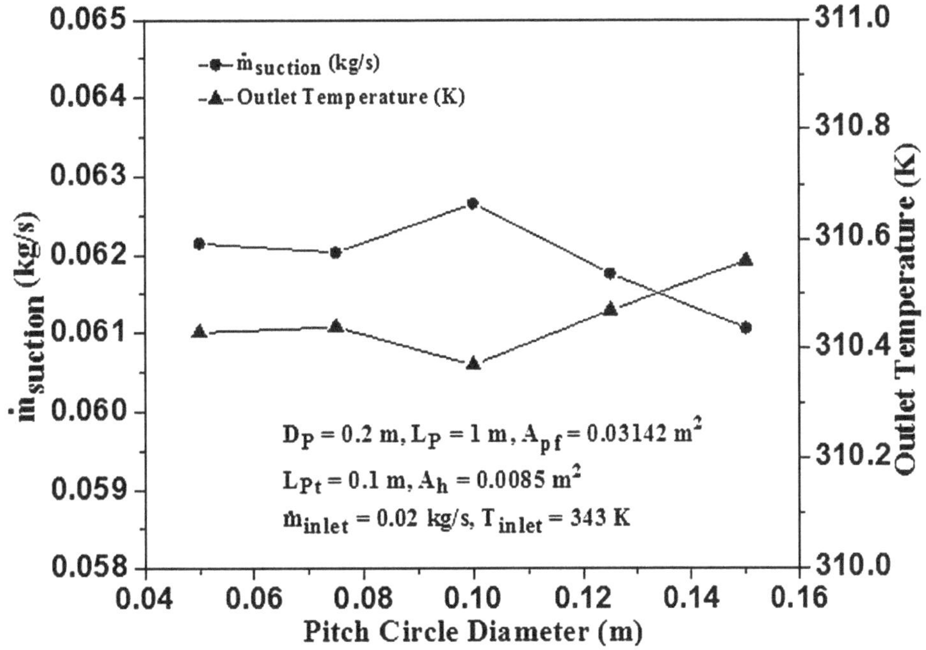

be said that for pipe configuration A, under an operating temperature of 353 K, the increase in the ratio of pipe diameter to nozzle diameter (for the case under study) increased the suction rate by 66.67% as compared to an increase in suction rate by 130% when increasing the ratio of funnel diameter to the nozzle diameter (Mishra and Dash, 2010a) where the operating temperature was 573 K.

6. CONLCUSION

The numerical analysis was carried out by varying the various parameters affecting the air suction rate into the cylindrical louvered pipe for both the pipe configurations. The conservation equations of mass, momentum and energy with two equation based k-ε turbulence model were solved using ANSYS FLUENT 14.0. After the analysis of the results obtained from the computations of various variables on air suction rate in both the pipe configurations, the following conclusions were drawn:

1. The air ingress rate improved by enhancing in opening area (louver opening area for Configuration A and total opening area for Configuration B) which thereby decreased the pipe outlet temperature considerably. The CFD analysis indicated that the louvers made close to the entrance half of the pipe and closer to the nozzle gave maximum suction effect. The louver opening proved to be a dominant factor affecting the air suction rate.
2. Increasing number of nozzles to 6, increased the air ingress rate owing to better mixing of hot and cold fluids which led to considerable decrease in pipe outlet temperature as compared to single nozzle. The increase in air suction rate by using 6 nozzles was about 1.11 times for both the pipe configurations but with pipe Configuration B, the pipe outlet temperature obtained was much lower compared to Configuration A.

3. Considering Configuration B of the pipe, the protrusion of nozzle into the pipe had an adverse effect on the air suction rate but moving the nozzle away from the pipe increased the air suction rate up to a certain distance. But in case of pipe Configuration A, the protrusion of nozzle into the pipe first increased and then decreased the air suction rate. When considering Configuration A of the pipe, the nozzle protrusion of 0.075 m into the pipe was found to be optimum while for Configuration B, placing the nozzle at a distance of about 0.075 m away from the pipe's entrance was found to be optimum for maximum air suction rate.
4. When the nozzle flow rate and opening area were kept constant, increasing the pipe diameter in Configuration A adversely affected its air suction rate whereas for the same increment in pipe diameter, the air suction rate got enhanced for Configuration B.
5. From the CFD analysis it was found that, tilting the nozzle either away from or towards the pipe axis decreased the air suction rate. Maximum air suction was obtained when the nozzle axis was parallel to the pipe axis.
6. When the nozzles were placed either too close or too far from the pipe centre, a sharp decrease in the air suction rate was observed. It was clear from the CFD analysis the optimum pitch circle diameter was found to be 0.1 m for both pipe configuration for maximum air ingress into the pipe.

REFERENCES

Barik, A. K., Dash, S. K., & Guha, A. (2014b). New correlations for prediction of air entrainment into an Infrared suppression (IRS) device. *Applied Ocean Research*, *47*, 303–312. doi:10.1016/j.apor.2014.06.007

Barik, A. K., Dash, S. K., & Guha, A. (2015). Experimental and numerical investigation of air entrainment into an infrared suppression device. *Applied Thermal Engineering*, *75*, 33–44. doi:10.1016/j.applthermaleng.2014.05.042

Barik, A. K., Dash, S. K., Patro, P., & Mohapatra, S. (2014a). Experimental and numerical investigation of air suction into a louvered funnel. *Applied Ocean Research*, *48*, 176–185. doi:10.1016/j.apor.2014.08.009

Becker, H. A., Hottel, H. C., & Williams, G. C. (1963). Mixing and flow in ducted turbulent jets. In *9th International Symposium on Combustion*. Pittsburg, PA: The Combustion Institute. doi:10.1016/B978-1-4832-2759-7.50009-1

Han, D., & Mungal, M. G. (2001). Direct measurement of entrainment in reacting/nonreacting turbulent jets. *Combustion and Flame*, *124*(3), 370–386. doi:10.1016/S0010-2180(00)00211-X

Hill, B. J. (1972). Measurement of local entrainment rate in the initial region of axisymmetric turbulent air jets. *Journal of Fluid Mechanics*, *51*(04), 773–779. doi:10.1017/S0022112072001351

Jie, L. (2011). *Environmental Effects of Vehicle Exhausts, Global and Local Effects – A Comparison between Gasoline and Diesel* (Published Master thesis of Applied Environmental Science). Halmstad University.

Markatos, N. C. (1986). The mathematical modelling of turbulent flows. *Applied Mathematical Modelling*, *10*(3), 190–220. doi:10.1016/0307-904X(86)90045-4

Mishra, D. P., & Dash, S. K. (2010a). Numerical investigation of air suction through louvers of a funnel due to high velocity air jet. *Computers & Fluids, 39*(9), 1597–1608. doi:10.1016/j.compfluid.2010.05.012

Mishra, D. P., & Dash, S. K. (2010b). Prediction of entrance length and mass suction rate for cylindrical sucking funnel. *International Journal for Numerical Methods in Fluids, 63*, 681–700.

Mishra, D. P., & Dash, S. K. (2011). Maximum air suction into a louvered funnel through optimum design. *J Ship Res, 55*, 1–11.

Mishra, D. P., Dash, S. K., & Kishan, P. A. (2010c). Isothermal jet suction through the lateral openings of cylindrical funnel. *J. Ship Res., 54*(4), 268–280.

Mishra, D. P., Samantaray, M. K., & Dash, S. K. (2014). Maximum air entrainment into a mixing pipe through optimum design. *Ships and Offshore Structures, 9*(6), 605–618. doi:10.1080/17445302.2014.881246

Pritchard, R., Guy, J. J., & Conner, N. E. (1977). *Industrial Gas Utilization*. New Providence, NJ: Bowker.

Ricou, F. P., & Spalding, D. B. (1961). Measurements of entrainment by axisymmetrical turbulent jets. *Journal of Fluid Mechanics, 11*(01), 21–32. doi:10.1017/S0022112061000834

Singh, G. S., Sundarajan, T., & Bhaskaran, K. A. (2003). Mixing and entrainment characteristics of circular and non-circular confined jets. *Journal of Fluids Engineering, 125*(5), 835–842. doi:10.1115/1.1595676

Singh, G. S., Sundarajan, T., & Shet, U. S. P. (1999). Entrainment and mixing studies for a variable density confined jet. *Numer. Heat Transfer. Part A, 35*, 205–223.

APPENDIX: NOMENCLATURE

A_h: Louvers opening area
$C_{1\varepsilon}$: Constant of value 1.44
$C_{2\varepsilon}$: Constant of value 1.92
D_{CD}: Diameter of computational domain
D_p: Diameter of pipe
D_n: Diameter of nozzle
D_h: Diameter of circular louver
D_{pd}: Pitch circle diameter
D_k: $\frac{\partial}{\partial x_j}\left[\left(\mu + \frac{\mu_t}{\sigma_k}\right)\frac{\partial k}{\partial x_j}\right]$
D_ε: $\frac{\partial}{\partial x_j}\left[\left(\mu + \frac{\mu_t}{\sigma_\varepsilon}\right)\frac{\partial \varepsilon}{\partial x_j}\right]$
L_{CD}: Length of computational domain
L_P: Length of pipe
L_{Pt}: Protruding length of nozzle
$\dot{m}_{inlet}$: Mass flow rate through the nozzle
$\dot{m}_{suction}$: Air suction rate through louvers
k: Turbulent kinetic energy
P: Pressure
Pr: υ/α
Pr_t: Turbulent Prandtl number
Re_n: Reynolds number based on nozzle, $\rho V_n D_n/\mu$
T: Temperature
U: Velocity
V_n: Velocity at nozzle exit

Greek Symbols

υ: Kinematic Viscosity
ρ: Density
μ: Shear viscosity
μ_t: Turbulent viscosity
ε: Dissipation rate
σ_k: Turbulent Prandtl number for k
σ_ε: Turbulent Prandtl number for ε
ϕ: Scalar variable either k or ε
α: Thermal diffusivity

Subscripts

n: Nozzle
∞: Free stream

Section 2

Development of New or Alternative Material for an Existing Component/Product

Chapter 7
Lower Body Orthotic Calipers With Composite Braces

Nisha Kumari
Birla Institute of Technology, India

Kaushik Kumar
Birla Institute of Technology, India

ABSTRACT

Composite based materials are finding application in a large number of research and engineering spectrum due to its better mechanical properties (strength and stiffness), inherent surface finish, easiness in fabrication and installation and corrosion resistant. They are very strong and firm, yet very light in weight due to which lower weight-to-volume ratio can be achieved and stiffness to weight is 1.5 times greater than the non-ferrous materials like Aluminum. The work is undertaken in two parts. First and foremost being modeling and virtual estimation of mechanical properties using CREO and ANSYS for currently used aluminum based calipers and fabrication of the composites and testing of the same. A comparison is performed between the virtual and experimental results and also the effectiveness of composite based calipers over Aluminum ones is studied. Here two polymeric based composites are proposed for fabrication which are thermoplast and thermoset based composites respectively. The braces are modeled using a solid modeling Software, CREO and the same is tested using ANSYS.

INTRODUCTION

Disability related to locomotion has affected many individuals and one such locomotor disability is poliomyelitis which is spread due to polio virus. Due to polio virus they suffer from different joint problems, sitting cross-legged, walking and running. Few decades ago, a brace or caliper came into existence which is used to support the feeble joints which require support for any kind of movement in different and transverse positions. Calipers or braces also known as Orthotic device are appliances which are being referred by the orthopedist considering amputees problem (Figure 1). Calipers are categorized into lower body extremity, upper body extremity and spinal. Calipers for lower body extremity restore weight bearing capabilities to the affected leg and hence allow the patients to walk and move from one place

DOI: 10.4018/978-1-5225-3401-3.ch007

to another without any risk of falling and contracture of joints. It also prevents further disfigurement of the joints and makes walking more effective by correct positioning of the braces/ calipers. But not every different type of brace/calipers suites the affected handicapped survivor. Calipers are designed by keeping in mind the problems being faced by the handicapped patients. Patients who suffer from knee pain are advised to have Knee supporting braces. Those who are suffering from ankle and knee related injuries they are given Ankle-Knee-Orthotic calipers also known as (AKO) and KAFO also known as Knee-Ankle-Foot-Orthosis are worn by the patients having weak knee, ankle and foot so that the braces given to them provide strength to their joints and they are able to move and walk without any issues. Underbracing and overbracing can be done considering the patient's height, weight, occupation and their problems related to the joints. Calipers are made as lighter as possible for the handicapped patients and are made available to them at the nearest hospitals and rehabilitation centers.

Lower extremity or lower body here is meant by the part which stops functioning properly after the muscular dysfunction. Due to muscular dysfunction paralysis of one side of body part or paralysis of both sides, poliomyelitis occurs and then requirement of braces becomes essential to lead a normal life. For lower part of the body foot orthotics are available, ankle braces and knee braces are designed to provide support the joints which are not able to function and are weakened. All the calipers/Braces designed for amputees consist of an adjustable lock which helps them in standing and sitting by adjusting it. Leather straps are present at calf and knee to make it more efficient for the affected survivors. For the fabrication of orthotic calipers researchers chose material which is lighter in weight, cost-effective, environment friendly and provides higher strength, stiffness and lower weight to volume ratio.

Figure 1. Commercially available orthotic braces/calipers

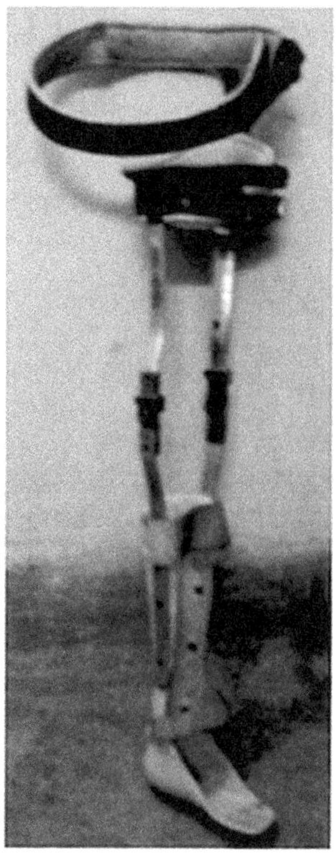

BACKGROUND

In olden days calipers were made up of wooden pieces. Two wooden pieces were joined with leg placed in between and then joined with some kind of strap or belt so that they are comfortable in walking. Wood is one of the most commercially used organic materials but one of the major disadvantages with calipers made up of wood was its hygroscopic property. It absorbs moisture very easily due to which it was not a better option for polio affected survivors to use wooden based calipers. In rainy season those wooden calipers used to become wet and because of which it was impossible for the amputees to wear them and move from one place to another. Foot Orthotics was also developed for the handicapped patients who use to suffer from problems related to foot only (Sethi et al., 1978, Ferber et al., 2005). SACH known as Solid Ankle Cushion Heel type foot also came into picture to provide the polio survivors with shoes which are extra cushioned from inside with the layers of sponge rubber intact into it (Sethi, 1990). It helped the patients in restricting their unwanted movement but presence of wooden keel does not allowed degree of freedom of movement if amputees try to sit cross-legged on floor due to the strain transmitted to the socket and the pressure felt on the strap. Development of Muller Solid Foot replaced the SACH foot but resilience and rigidity which is a vital factor in ankle foot mechanism was missing in the Muller Solid Foot.Then the researchers/scientists thought of designing calipers made up of ferrous metal like Iron due to their splendid properties. Ferrous metals (Iron) were decided as material for calipers due to its tensile strength, firmness. But due to its magnetic property, rusting property, corrosion resistant and higher weight it was found that it was not helpful for the polio affected patients. Polio affected survivors were in need of those materials which are lower in weight; cost-effective provides lower weight-to-volume ratio and higher strength and stiffness so that they are able to bear the weight of the individuals affected. As age increases they are more reluctant to fatigue problem due to which they start gaining weight so to bear that weight the calipers should also be strong and firm enough so that they help the polio patients in their movement (Genet et al., 2010, Moreno et al., 2006, Giovacchini et al., 2015; Bregman, 2011).

Nowadays calipers are being fabricated and manufactured considering the polio affected patient's condition, their age, their workplace, height and every other things which is required to provide them with best calipers (Berwick., 1971, Rakib et al., 2015). It is very important for the researchers/scientists to keep in mind the cost, ease of usage, affordability, higher strength, firmness while designing calipers for the polio affected patients(Kastner et al., 2010).So, Presently Aluminum based calipers are used by the handicapped patients but weight of the orthotic calipers is still an issue which should be worked on. The handicapped patients are in need of such material which is lighter in weight, provides maximum strength and stiffness when compared to other materials, should be long-lasting, cost-effective and should have lower weight-to-volume ratio. One such material is composite material which has developed as an emerging material in the field of engineering and research. In this work the authors have tried composite materials for braces/calipers for polio affected survivors. Composite materials are those materials which are being extensively used due to its properties (Etcheverry and Barbosa., 2012, Bader., 2002, Yadav et al., 1993, El-Tayeb and Mostafa., 1996).

Composite materials are made up by combination of two or more macro-constituents (Kröling et al., 2014, Liew et al., 2015, Ashik et al., 2015). Macro-constituents are of matrix and reinforcement type. To make a composite material there is always requirement of a matrix and a reinforcement having different physical and chemical properties and the material formed will have splendid properties such as high strength, stiffness, and lighter in weight (Ku et al., 2011, Powell., 1994, Ravikumar et al., 2017) totally different and better than the individual components combined (Mitra, 2014). The macro-constituents are not dissolved into each other. Matrix provides the shape and size to the composite and when load is applied the matrix shares its load with reinforcement. Reinforcement acts like load bearing material as it supports the matrix and transfer the load among different types of reinforcement. So, when matrix and reinforcement are combined together they lead to the formation of composite materials. Reinforcement sometimes are abrasive in nature due to which erosion takes place but presence of matrix leads to separation of the reinforcement so erosion does not take place. Reinforcements can be present in the form of flakes, fibers, particles, whiskers, layers, etc. To classify a material as composite material both the matrix as well as reinforcement should be present in reasonable proportion greater than 5 percent. The final material formed will always have properties different from the individual macro-constituents. If 60 percent of matrix is being used for fabrication of the composites then 40 percent reinforcement should be mixed with it. Matrix can be of polymer, ceramic and metal matrix composites whereas reinforcement depending upon the shape can be of continuous, short fibers, particles, whiskers. Composite materials find number of applications in the fields of aerospace, automobile, sports, energy etc. (Arpitha et al., 2014, Jagannatha and Harish., 2015, Kinsella et al., 2000, Deogonda et al., 2013, Subash and Pillai., 2015, Ansar et al. 2011, Liu et al., 2012, Yu et al., 2016). They are friendly to nature and environment and can be decomposed easily (Gibson, 2015, Suresh et al., 2003). In our work we have primarily focused on polymeric matrix composites

Polymeric matrix composites also known as PMC as matrix isin the form of polymers and reinforcement is in the form of fibers and finds use in the bearing applications (Sayed et al.,1995). Polymers can be of different type such as thermoplastic and thermoset based polymeric matrix. The properties of polymeric based matrix are ascertained by the properties of the fiber and the resin, geometry and orientation of the fibers in the composite and the fiber volume fraction (Mantiaand Morreale., 2011, Shivakumar and Guggari., 2011, Ala-Kleme et al., 2006). Plastic materials/polymers which are present in the raw form can be molded and shaped with the help of processing techniques. Plastic comprises of thermoset and thermoplast. Material that cures in a reversible manner and when heated above glass transition temperature it softens, when heated the material which becomes hard are said to be thermoplastic materials. There are individual chains which are very weak as they are linear in nature. Examples of thermoplastics are polyethylene, polycarbonate, polystyrene and polypropylene (Otheguy et al., 2009). Thermoset materials after curing in an irreversible manner convert pallets into tangible product it again can be converted into pallets by heating it above a transition temperature. In thermoset mass of polymers is very strong as cross-linking chains are present. A thermoset material is hard, brittle in nature and does not soften with heating. Thermoset materials consist of epoxies, polyesters, phenolic, Bakelite, etc.

WORK UNDERTAKEN

The work was conducted in two parts. Firstly, 3D parametric software CREO was used for modeling and virtual approximation of mechanical properties such as stress, strain and total deformation using ANSYS. Modeling is done so that the results of the material which is used nowadays and proposed materials (thermoset and thermoplast based polymeric composite) can be validated. To carry out modeling of presently used Aluminum Based calipers in CREO a small rectangular piece with dimension of 100 mm gauge length was taken, width was taken as 20.22 mm and thickness of the piece was 6.03 mm. Only a small part of the rectangular piece with above dimensions was taken to perform the experimental part and with the same dimensions 3D model was created using CREO parametric and then further analysis and simulation process was done in ANSYS.

After the preparation of 3D model in CREO Parametric it was imported in ANSYS Workbench for simulation. For analysis process the static structural analysis was performed on the member both with axial as well as transverse loading with Aluminium and polymeric based composites as materials as orthotic caliper are subjected to these loads while in operation. The calculation of the equivalent Von-mises stress, strain and total deformation form the yardstick for the selection of composite material which will provide better results like higher strength, stiffness and lower weight to volume ration when compared to presently used Aluminum based calipers.

Secondly fabrication of the proposed composite (Polymeric based composites) was done. Thermoplast (Nylon with Glass)and thermoset (Carbon with Epoxy) was chosen for preparation of the composite material. These samples were subjected to various tests and results were compared with virtual testing results.

MATERIALS AND COMPOSITE FABRICATION

Epoxy

Low solidifying curing epoxy resin (standard Epoxy AW 106) is very easily cured at any temperature ranging from 5^0 C to 150^0C; they chemically belong to the 'epoxide' family. Epoxy resin and corresponding hardener (HV 953U) were mixed in a quantitative relation (as explicitly stated by the manufacturer) for preparation of the thermoset matrix. High strength and mechanical properties such as adhesiveness and good insulation properties can be achieved with the use of Epoxy resin.

Carbon

Carbon in the form of particulate (about 70µ) was used as filler. The properties of carbon filler like high stiffness, firmness, low weight-to-volume ration, high reliability, better chemical resistance, heat tolerance and low coefficient of thermal expansion (Cao et al., 2017) makes it usable in construction and building components, products in industrial, military and sports etc. (Meredith et al., 2015).

Nylon

For thermoplast based composite Nylon was chosen as a matrix material. Nylon having base of polymeric composites exposes greater number of better physical and mechanical properties that provides better rigidity, higher strength, better stability due to which it finds application in the textile industry and many other industrial applications.

Glass

Glass in the form of particulate size (about 70µ) was used as the material which provides strength to manufacture thermoplast composites. Glass being moldable filler can be used with Nylon to manufacture structural elements by injection or compression molding. It has got properties which provide better strength to weight ratio and good thermal insulation when mixed with matrix material like Nylon.

Conditioning

Nylon has power to absorb so it cannot be dried effectively so heating beforehand is required. Preheating of the Nylon would be accomplished in dehumidifier chamber(Figure 2) at specified temperature and duration of 2-3 hours.

Composite Fabrication for Thermoset Based Composite

For preparation of thermoset based Composites, Samples were created by mixing carbon fillers (30%) having density (1.15 gm/cc) with low curing epoxy resin having density (1.4 gm/cc). If the composite is 30% by volume fiber then Epoxy should be 70% by volume matrix, so they are mixed in the ratio of (1:2.35). The temperature was set as 160^0 Centigrade. Plates of 99*88 millimeter with thickness of 4.5 millimeters were fabricated. Thickness of the composite potently depends upon the count of consistence. Vacuum pressure was used throughout the heating stage to remove air bubble and wetness if present and also a pressure of five metric tons was applied. Totally different plates were factory-made out of those 6-7 different samples with size of 99*88*4.5 mm^3 were investigated. When hardening of the specimen was done the product made was taken out from the carver press (Figure 3) and was investigated.

Composite Fabrication for Thermoplast Based Composite

For fabrication of thermoplast based composite Nylon was taken as the matrix material having density of 1.14 gm/cc and 30% glass filler having density of 2.5 gm/cc. The mixing ratio for the matrix and the

Figure 2. Dehumidifying chamber
Source: http://www.igi-global.com/chapter/estimation-mechanical-tribological-properties-epoxy/156904(Roy et al. (2016))

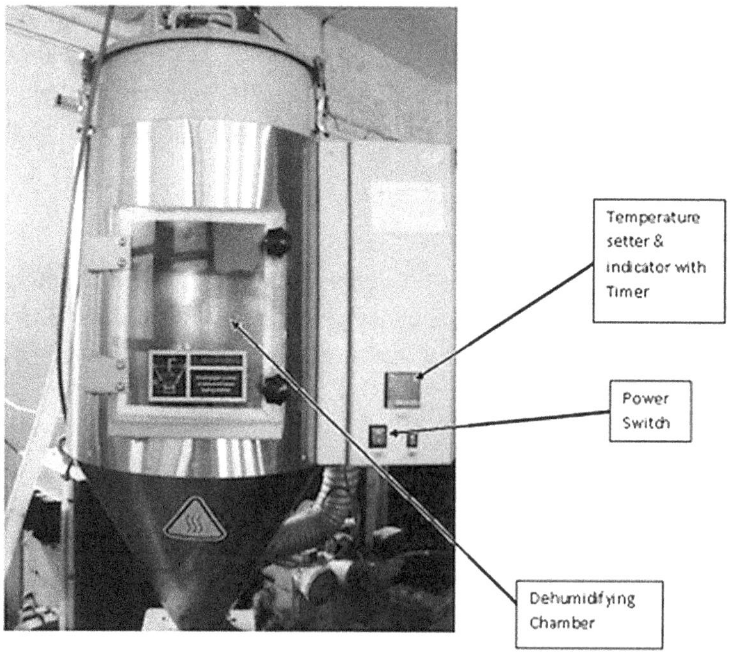

Figure 3. Carver press

reinforcement for thermoplast based composite was 1:2.35.In order to obtain homogeneous mixing of Nylon and glass, Haake single screw extruder (Figure 4) was utilized.

The strands thus formed were chopped into granular form and fed into an Injection molding machine(Figure 5). Palettes were put inside the hopper. A screw present in the molding machine carries the palettes and melts the palettes within the barrel. Screw speed was set according to cooling cycle for particular run. Normally screw rpm setting for the nylon ranges from 60-100 revolutions per minute. Screw starts rotating first and then after few seconds the mold starts rotating. Melted polymer is pro-

Figure 4. Haake single screw extruder

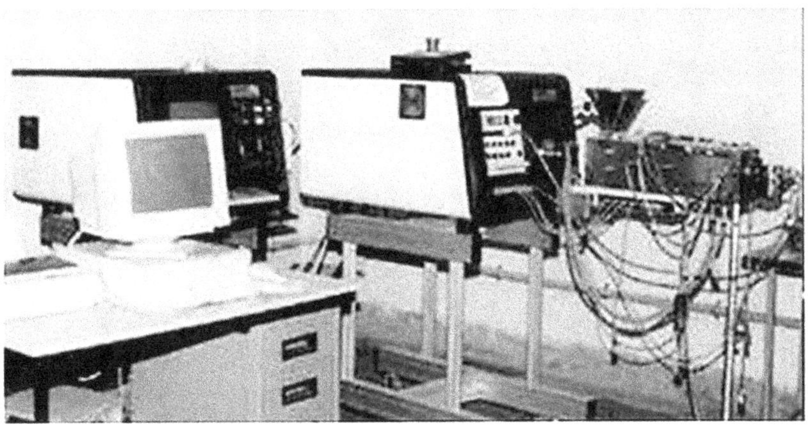

duced into the closed mold; the part is then removed and cooling is done inside the mold itself. After few minutes the mold is opened and the part produced is taken out or removed from the mold and the cycle repeats. The final product is taken for the testing.

Sample for Testing

The samples fabricated were cut down to the exact sizes (gauge length 100 mm, width 20.22 mm and thickness 6.03 mm) and then testing was performed. One sample each of the three types is shown in Figure 6.

TESTING

Density and Void Content

The density of polymer composites depends upon the spatial property of the composite. It is defined as the ratio of mass per unit volume of a substance. The theoretical densities of composite material would be calculated by using weight additive principle equation and Mettler Toledo electronic balance (Figure 7) which is a simple immersion method is used for experimental determination of actual density of the composite. For experimental calculation of density and void content number of observation taken should be more than 3 so that an average value would be taken for considering the best void content of the composite. The difference would indicate the %age void. Void content, in the fabricated composites, is an important physical property which can significantly affect the physical and mechanical properties of composite material. Hence approximation of void content/density is important to measure the quality of composites.

Figure 5. Injection moulding machine

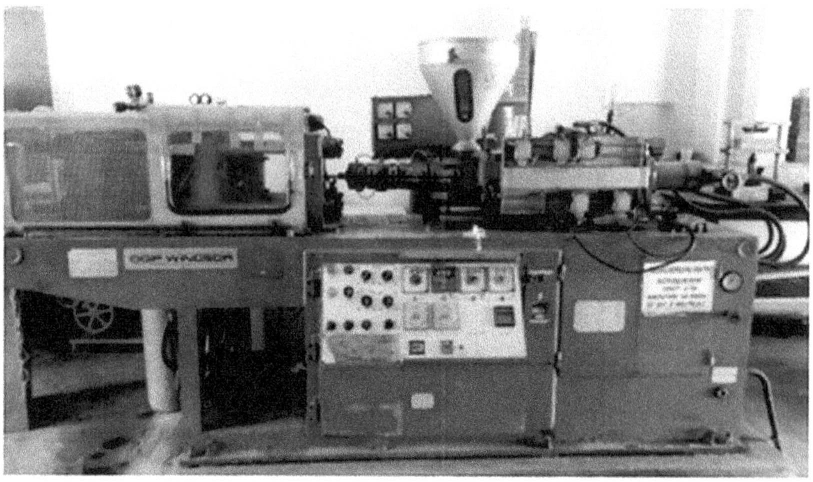

Figure 6. Samples for testing
A: Aluminium Alloy for an existing calliper; B: Composite with Nylon and glass filler; C: Composite with Epoxy and carbon filler.

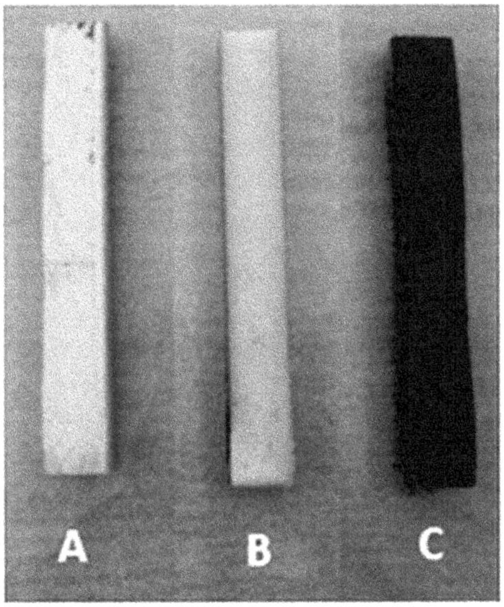

Figure 7. Mettler Toledo electronic balance

Tensile and Flexural Tests

Universal testing machine (UTM) is used to examine the tensile and flexural behavior (Tensile strength, tensile modulus, flexural strength and flexural modulus) of the composite materials. The mechanical tests are carried out using Instron universal testing machine (Instron Ltd, UK) (Figure 8). The tensile tests are conducted according to ASTM D 3039. The specimens for mechanical tests are of the shape of a rectangular bar of size 100 X 12 X 8 mm. The flexural or 3 point bending tests are undertaken as per ASTM D 790 under the same condition as that of the tensile tests.

RESULTS AND DISCUSSION

Modeling

In order to have a similarity in virtual and actual experimentation results, rectangular sample was modeled in CREO with 100*20.22*6.03 mm^3 size. Figure 9 shows the modeled sample exported to ANSYS workbench for simulation.

Simulation

ANSYS Workbench was used for simulation purpose and then equivalent von-mises stress, strain and total deformation was calculated for both axial as well as transverse loading. The boundary condition for the simulation part was edited while choosing the new material. Boundary condition such as density, Young's modulus and Tensile Yield strength of the material was provided. After putting the boundary conditions geometry has to be imported and mesh size should be considered as coarse as given in the ANSYS Workbench. The maximum deflection for axial loading was found out to be 0.50613 m and the maximum deflection for transverse loading was found out as 0.0044239 m. The equivalent von-mises

Figure 8. Instron Universal Testing Machine: (a) tensile; (b) flexural

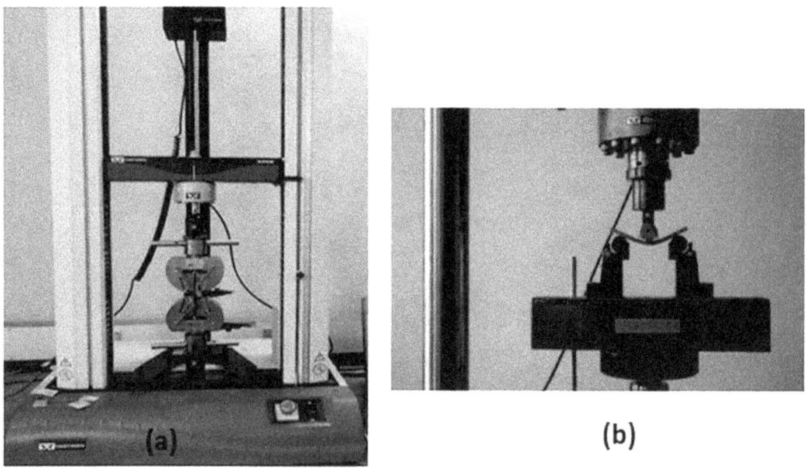

Figure 9. 3D model of sample imported to ANSYS

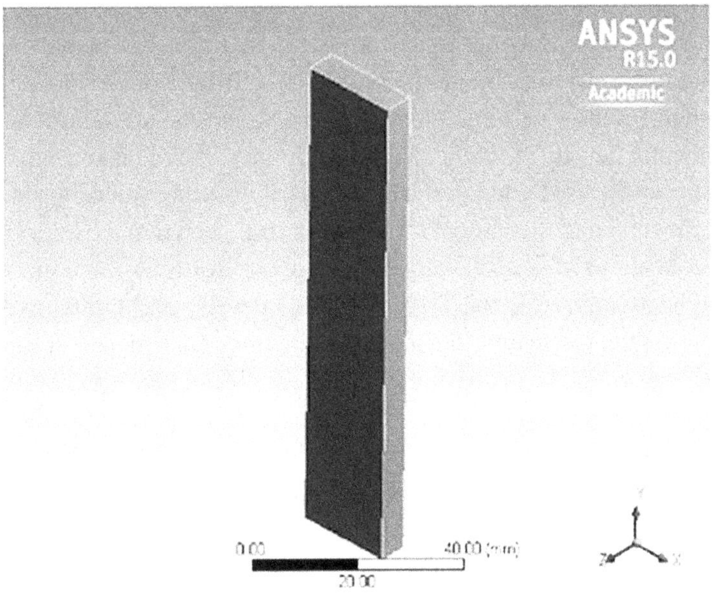

stress for axial and transverse loading was 430.03 MPa as maximum and 194.13 was found out to be minimum value for stress as clearly shown in Figure 10 and 2.31e9 Pa respectively. The maximum strain under both loading conditions was observed to be 6.0568e-003 mm/mm and 3.2872e-002 mm/mm. A sample picture can be viewed in Figure 10 showing equivalent von-mises stress for axial. All the other results are tabulated along with the experimental results to observe the validity of the outputs of the work.

Figure 10. Equivalent (von-Mises) stress under axial loading

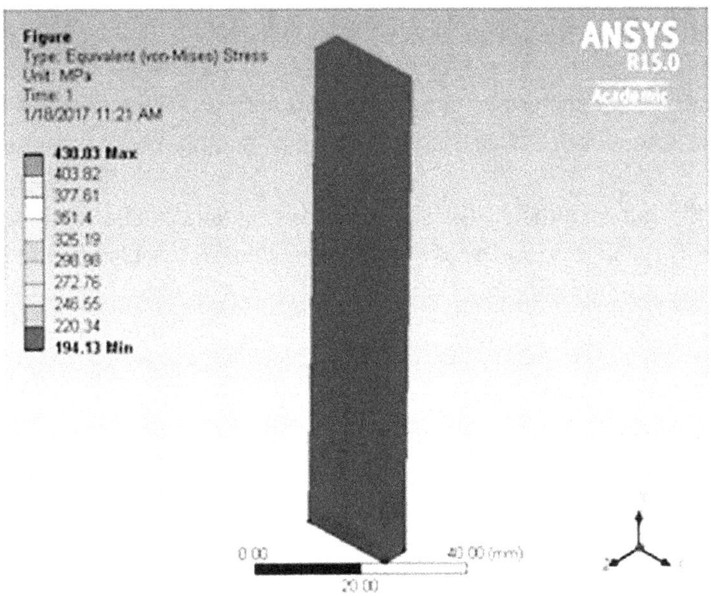

Density and Void Content

Density and void content is incredibly necessary physical properties of chemically compounded polymeric based composite materials to estimate the prominent attribute of composites. The experiment is being carried on Mettler Toledo electronic balance by victimization straightforward water immersion technique and experimental density is found. Weight additive principle is used to find out the theoretical density of the composites. Void content for the primarily based composites is examined by taking into account the experimental and theoretical information. It is seen that the experimental data is lesser than the theoretical showing existence of voids at the surface forming the common boundary within the composites. It can be assured from the Table 1 that the density of Aluminum alloy (presently used orthotic calipers) is more than the nylon (thermplast) and epoxy (thermoset) composites. Void content must be less than 5 percent which is an appropriate level for all industrial and household products. The void present in the composites are due to the entrapment of air bubbles during mixing and fabrication of composites. Thermoplast based composites have less void content when compared to thermoset based composite because of the fact that thermoplast based composite was injection molded and shaped (processed at usually higher temperature than compression moulding) whereas thermoset based composite were compression molded entrapping the moisture.

Tensile and Flexural Tests

The tensile testwas carried out in accordance to ASTM D 638 where the specimens are unit placed within the grips and are unit force till failure. The tensile tests are performed with the best possible load capacity of 1 kN with 50 mm gauge length and cross head speed of 2.5 mm / min. The information regarding the data of tensile modulus and tensile strength are entered on a computer being joined in close association with machine setup. The flexural or 3 point bending tests was executed as per ASTM D 790 where the test was started by enforcing the load on the specimen at the specified rate till the breaking of the specimen. The tests are carried out with a span length of 50 mm and cross head speed of 2.5 mm / min and the flexural stress, flexural strain and lateral deformation was recorded from the tests. To carry out the mechanical tests minimum of three samples for each composition and the average of the results are reported. The plots of the various values are depicted in the graphs. Figure 11 shows the tensile strength values for each of the materials along with their virtual values whereas Figure 12 illustrates variation in maximum strain for the same. Figure 13 reveals maximum deformation undergone in axial loading condition.

The difference in the values of the virtual and experimental data can be attributed to the fact that the fabricated composites contains void as already found using density tests but the distribution of the same

Table 1. Density and void content of composites and Aluminum alloy

Material of Construction	Density (g/cc)		Void Fraction (%)
	Experimental	Theoretical	
Aluminium	2.705	--	--
Thermoset Based Composite	1.088	1.144	4.89
Thermoplast Based Composite	1.319	1.378	4.28

Figure 11. Comparison of Von misc stresses under axial loading with 5% error bars

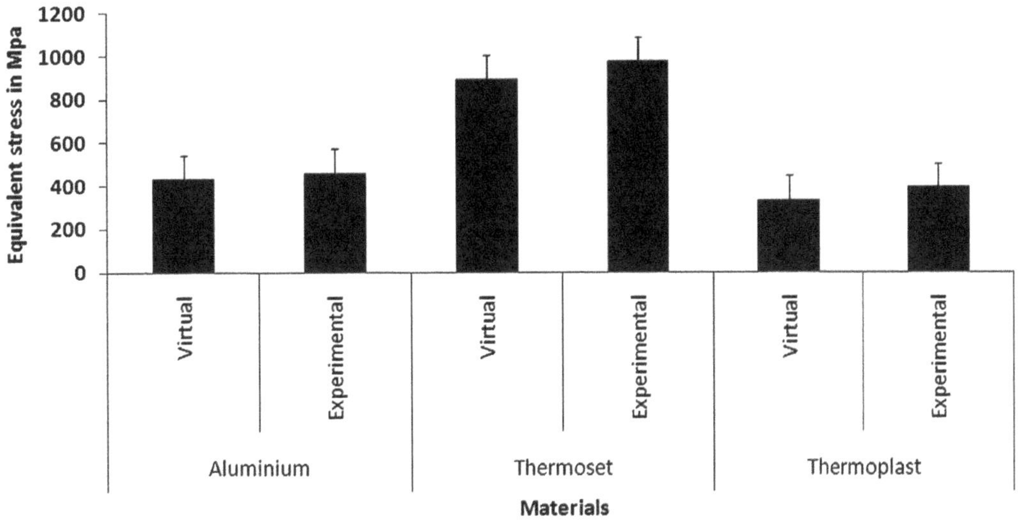

Figure 12. Comparison of strain under axial loading with 5% error bars

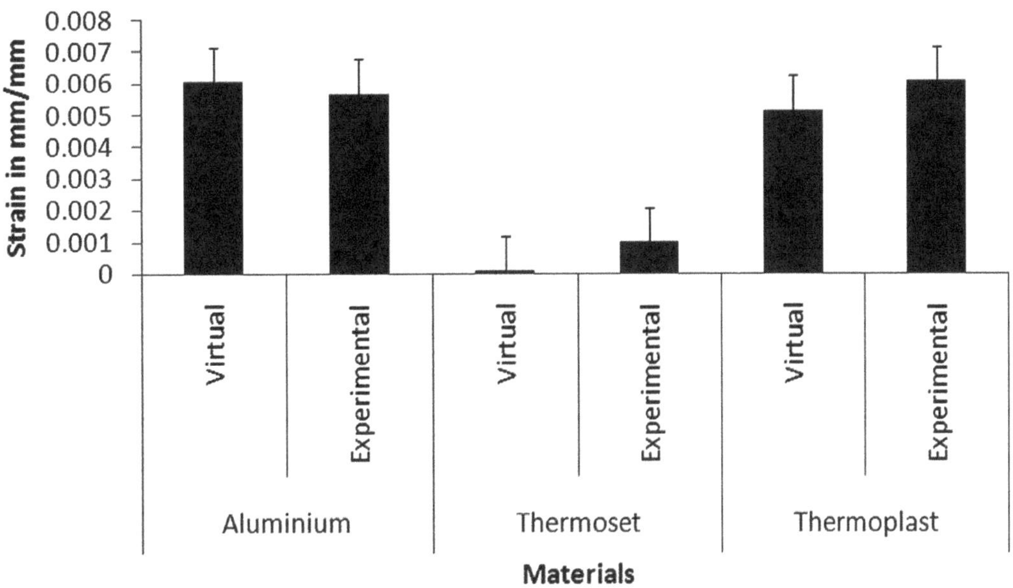

could not be determined. Hence, if the voids are not uniformly distributed throughout the product but gets clustered at a place, the results of the experimental data would get affected.

Figure 14 shows flexural strength data whereas Figure 15 indicates the flexural strain. The lateral deformation can be viewed in Figure 16. From all the graphs it is clear that the mechanical strength of thermoplast based composite (Nylon + Glass) is of smallest quantity and for thermoset based composite (Epoxy + Carbon) values are more eminent than both Aluminium alloy and thermoplast based composite. The Equivalent Von-mises strain and Total Deformation is least for thermoset based composite. This can

Figure 13. Comparison of maximum deformation under axial loading with 5% error bars

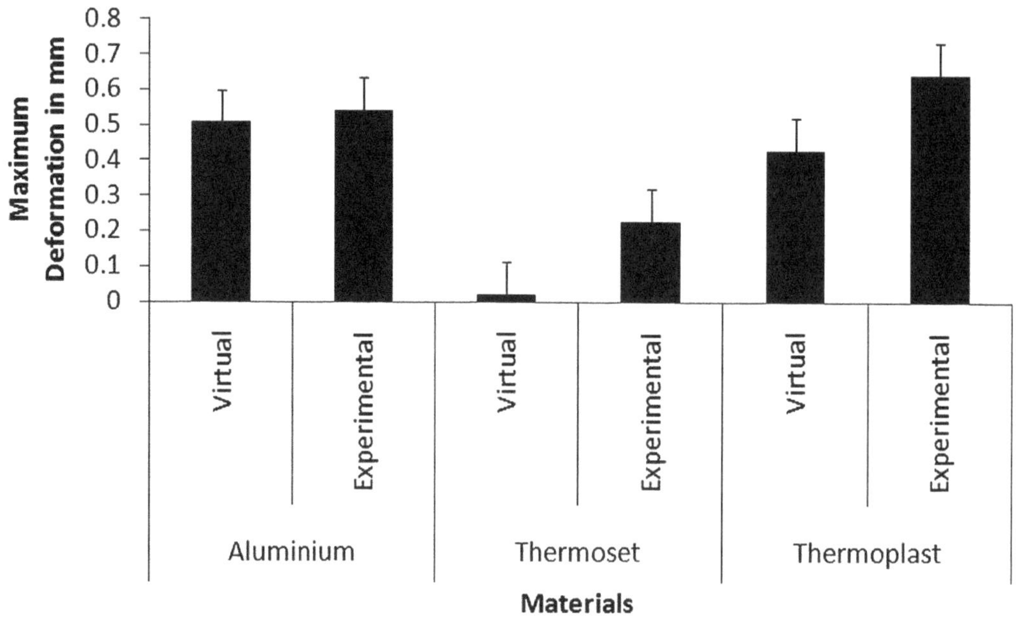

Figure 14. Comparison of Von misc stresses under lateral loading with 5% error bars

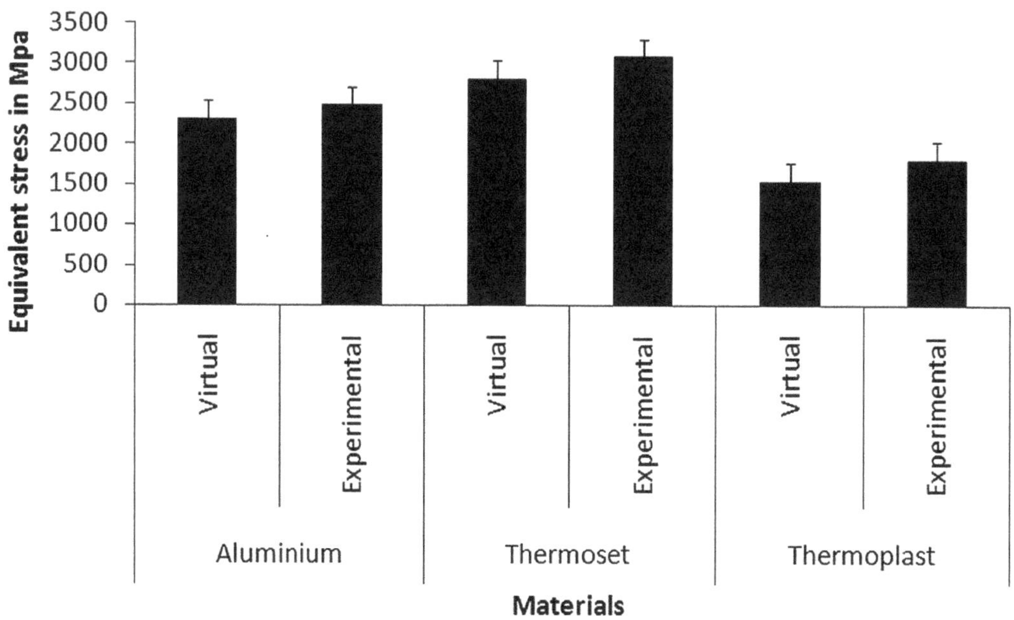

Figure 15. Comparison of strain under lateral loading with 5% error bars

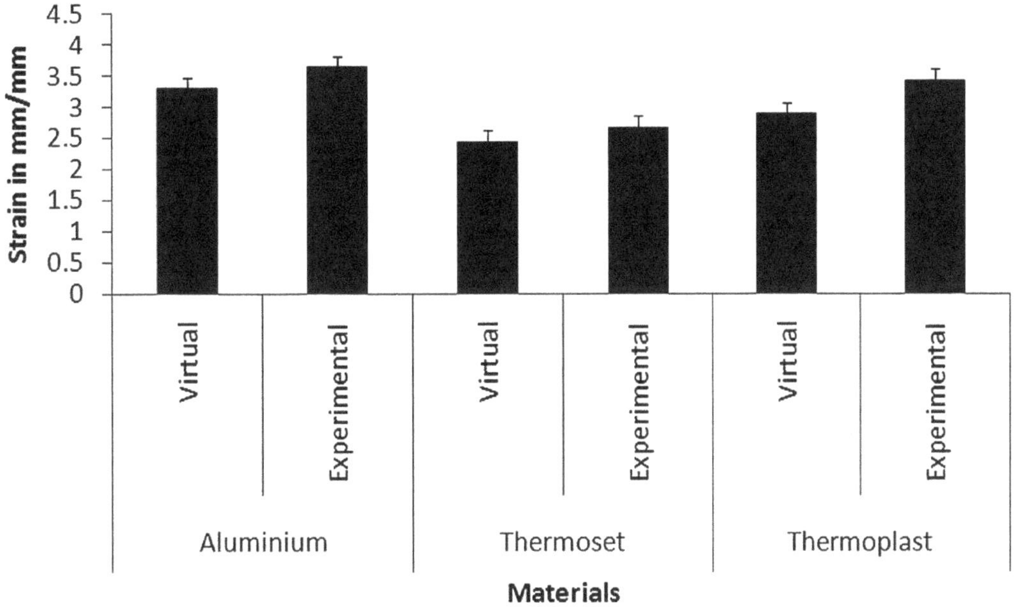

Figure 16. Comparison of maximum deformation under lateral loading with 5% error bars

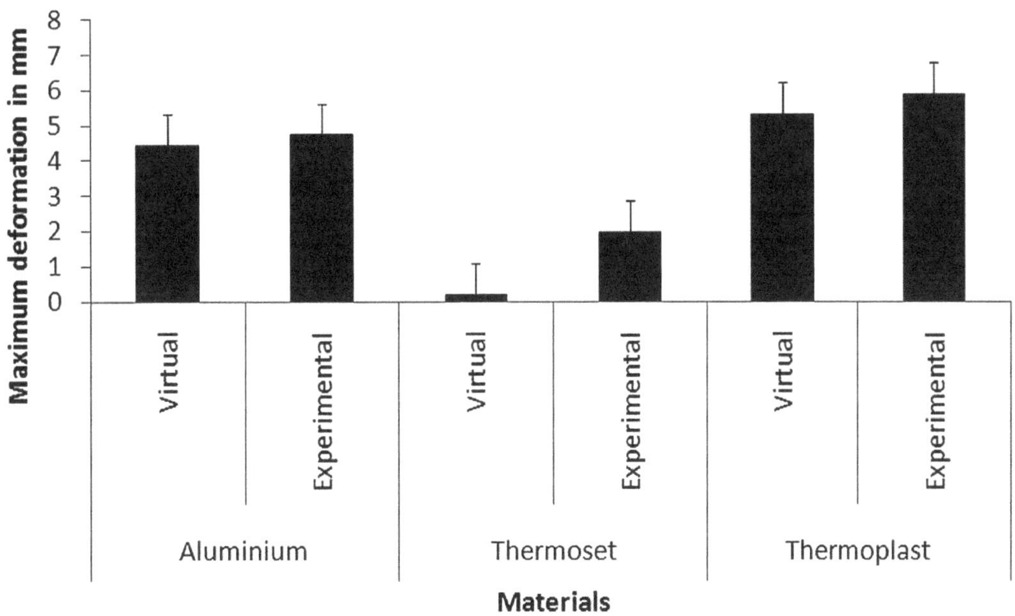

be imputed to the high strength of the carbon reinforcement in epoxy. This shows that the Aluminium alloy made calipers can be fancied using thermoset composite. Moreover the similarity of the results between the virtual and experimental data confirms the validity of the experimentation.

As discussed above and depicted in the figures the maximum deformation and strain also varies in the similar trend as in case of Equivalent Von-mises stress both under axial and lateral loading condition.

CONCLUSION

An alternative material, if chosen, should provide results better than that of the presently used material (Aluminum alloy)at a lower cost for acceptability by the manufacturer and users of the lower body orthotic calipers. The literature review suggests that one of such material can be polymeric based composite which apart from giving better strength would also provide lower weight to volume ratio. Hence, would decrease the self weight of the calipers. This would surely increase the ease of mobility of the polio infected patient. The carried out experimentation suggested that the proposed thermoset based material can be epoxy based composite reinforced with carbon cloth. The results indicated that it should provides better strength and stiffness and lower weight to volume ration when compared with the results of presently used Aluminum based calipers. In order to validate the accuracy of the experimental results, the test pieces were subjected to virtual testing using ANSYS and the variation found was minimal. In order to compare the performance of the suggested composite, a thermoplast composite was also undertaken. But it was found to be of weaker mechanical properties when compared to aluminum alloy and thermoset composite.

FUTURE WORK

Since these supports undergo wear and tear and are always subjected to adverse climatic conditions hence Tribological testing and dynamic mechanical testing should also be performed. A complete lower body orthotic caliper is required to be fabricated and to be clinically tested under actual usage before replacing the aluminum alloy made calipers with epoxy – carbon composite ones.

ACKNOWLEDGMENT

The authors sincerely acknowledge the comments and suggestions of the reviewers that have been instrumental for improving and upgrading the paper in its final form.

REFERENCES

Ala-Kleme, S., Kivikytö-Reponen, P., Liimatainen, J., Hellman, J., & Hannula, S. P. (2006). Abrasive wear properties of metal matrix composites produced by hot isostatic pressing. *Proc. Estonian Acad. Sci. Eng*, *12*(4), 445–454.

Ansar, M., Xinwei, W., & Chouwei, Z. (2011). Modeling strategies of 3D woven composites: A review. *Composite Structures*, *93*(8), 1947–1963. doi:10.1016/j.compstruct.2011.03.010

Arpitha, G. R., Sanjay, M. R., Naik, L., & Yogesha, B. (2014). Mechanical Properties of Epoxy Based Hybrid Composites Reinforced with Sisal/SIC/Glass Fibers. *International Journal of Engineering Research and General Science*, *2*(5), 398–405.

Ashik, K. P., & Sharma, R. S. (2015). A Review on Mechanical Properties of Natural Fiber Reinforced Hybrid Polymer Composites. *Journal of Minerals & Materials Characterization & Engineering*, *3*(05), 420–426. doi:10.4236/jmmce.2015.35044

Bader, M. G. (2002). Selection of composite materials and manufacturing routes for cost-effective performance. *Composites. Part A, Applied Science and Manufacturing*, *33*(7), 913–934. doi:10.1016/S1359-835X(02)00044-1

Berwick, A. (1971). The Short Leg Brace For Ankle Foot Stability In Adults. *The Australian Journal of Physiotherapy*, *17*(3), 81–84. doi:10.1016/S0004-9514(14)61110-7 PMID:25025930

Bregman, D. J. J., van der Krogt, M. M., De Groot, V., Harlaar, J., Wisse, M., & Collins, S. H. (2011). The effect of ankle foot orthosis stiffness on the energy cost of walking: A simulation study. *Clinical Biomechanics (Bristol, Avon)*, *26*(9), 955–961. doi:10.1016/j.clinbiomech.2011.05.007 PMID:21723012

Cao, L., Bai, Z., Huang, J., OuYang, H., & Wu, J. (2017). Ablation properties of Cf/C-SiC MoSi2 Composites: Effects of Hydrothermal penetration temperature. *Journal of Alloys and Compounds*, *703*, 45–55. doi:10.1016/j.jallcom.2017.01.316

Deogonda, P., & Chalwa, V. N. (2013). Mechanical Property of Glass Fiber Reinforcement Epoxy Composites. *International Journal of Scientific Engineering and Research.*, *1*(4), 6–9.

El-Sayed, A. A., El-Sherbiny, M. G., Abo-El-Ezz, A. S., & Aggag, G. A. (1995). Friction and wear properties of polymeric composite materials for bearing applications. *Wear*, *184*(1), 45–53. doi:10.1016/0043-1648(94)06546-2

El-Tayeb, N. S. M., & Mostafa, I. M. (1996). The effect of laminate orientations on friction and wear mechanisms of glass reinforced polyester composite. *Wear*, *195*(1-2), 186–191. doi:10.1016/0043-1648(95)06849-X

Etcheverry, M., & Barbosa, E. (2012). Glass Fiber Reinforced Polypropylene Mechanical Properties Enhancement by Adhesion Improvement. *Materials (Basel)*, *5*(12), 1084–1113. doi:10.3390/ma5061084 PMID:28817025

Ferber, R., Davis, I. M., & Williams, D. S. III. (2005). Effect of foot orthotics on rear foot and tibia joint coupling patterns and variability. *Journal of Biomechanics*, *38*(3), 477–483. doi:10.1016/j.jbiomech.2004.04.019 PMID:15652545

Gene^t, F., Schnitzler, A., Mathieu, S., Autret, K., Théfenne, L., Dizien, O., & Maldjian, A. (2010). Orthotic devices and gait in polio patients. *Annals of Physical and Rehabilitation Medicine*, *53*(1), 51–59. doi:10.1016/j.rehab.2009.11.005 PMID:20022835

Gibson, R. F. (2010). A review of recent research on mechanics of multifunctional composite materials and structures. *Composite Structures*, *92*(12), 2793–2810. doi:10.1016/j.compstruct.2010.05.003

Giovacchini, F., Vannetti, F., Fantozzi, M., & Cempinia, M. (2015). A light-weight active orthosis for hip movement assistance. *Robotics and Autonomous Systems*, *73*, 123–134. doi:10.1016/j.robot.2014.08.015

Jagannatha & Harish. (2015). Mechanical properties of carbon/glass fiber reinforced epoxy hybrid polymer composites. *International Journal of Mechanical Engineering and Robotics Research*, *4*(2), 131 – 137.

Kastner, J., Plank, B., Salaberger, D., & Sekelja, J. (2010). Defect and Porosity Determination of Fibre Reinforced Polymers by X-ray Computed Tomography. *2nd International Symposium on NDT in Aerospace 2010*, 1 – 12.

Kinsella. (2000). Innovative Automotive Applications Based on ZenTron® High Strength Glass Fiber. *Proc.* 45th*Science of Advanced Materials and Process Engineering Series, 45*, 1979 – 1994.

Kröling, H., Fleckenstein, J., Nubbo, N., Endres, A., Miletzky, F., & Schabel, S. (2014). Non-Woven and Paper Based Epoxy Composites. *Science and Technology DAS PAPIER*, *6*(7), 2 – 5.

Ku, H., Wang, H., Pattarachaiyakoop, N., & Trada, M. (2011). A review on the tensile properties of natural fiber reinforced polymer composites. *Composites. Part B, Engineering*, *42*(4), 856–873. doi:10.1016/j.compositesb.2011.01.010

La Mantia, F. P., & Morreale, M. (2011). Green composites: A brief review. *Composites. Part A, Applied Science and Manufacturing*, *42*(6), 579–588. doi:10.1016/j.compositesa.2011.01.017

Liew, K. M., Lei, Z. X., & Zhang, L. W. (2015). Mechanical analysis of functionally graded carbon nanotube reinforced composites: A review. *Composite Structures*, *120*, 90–97. doi:10.1016/j.compstruct.2014.09.041

Liu, D., Tang, Y., & Cong, W. L. (2012). A review of mechanical drilling for composite laminates. *Composite Structures*, *94*(4), 1265–1279. doi:10.1016/j.compstruct.2011.11.024

Meredith, J., Bilson, E., & Powe, R. (2015). A performance versus cost analysis of prepreg carbon fibre epoxy energy absorption structures. Composite Structures, 124, 206–213.

Mitra, B. C. (2014, May). Environment Friendly Composite Materials: Biocomposites and Green Composites. *Defence Science Journal*, *64*(3), 244–261. doi:10.14429/dsj.64.7323

Moreno, J. C., Rocon de Lima, E., & Andr'es, F. (2006). Design and implementation of an inertial measurement unit for control of artificial limbs: Application on leg orthoses. *Sensors and Actuators. B, Chemical*, *118*(1-2), 333–337. doi:10.1016/j.snb.2006.04.039

Otheguy, M. E., Gibson, A. G., Robinson, M., Findon, E., Cripps, B., & Ochoa Mendoz, A. (2009). Recycling of end of life thermoplastic composite boats.*Plastics. Rubber and Composites:Macromolecular Engineering*, *38*(9-10), 406–411. doi:10.1179/146580109X12540995045642

Powell, R. M. (1994). Utilization of Natural Fibers in Plastic Composites, *Lignocellulosic. Plastic Composites*, *4*(27), 23–51.

Rakib, M. I., Choudhury, I. A., Hussain, S., & Osman, N. A. A. (2015). Design and biomechanical performance analysis of a user-friendly orthotic device. *Materials & Design*, *65*, 716–725. doi:10.1016/j.matdes.2014.09.075

Ravikumar, K., Kiran, K., & Sreebalaji, V. S. (2017). Characterization of mechanical properties of aluminium/tungsten carbide composites. *Measurement*, *102*, 142–149. doi:10.1016/j.measurement.2017.01.045

Roy, S., Davim, J. P., Bhowmik, S., & Kumar, K. (2016). Estimation of mechanical and tribological properties of epoxy based green composites. In D. Verma, S. Jain, X. Zhang, & P. C. Gope (Eds.), *Green Approaches to Biocomposite Materials Science and Engineering* (pp. 96–124). Hershey, PA: Engineering Science Reference. doi:10.4018/978-1-5225-0424-5.ch005

Sethi, P. K. (1990). Designing Aids for Physically Handicapped in Developing Countries. *Indian Journal of Physical Medicine and Rehabilitation*, *3*, 1–10.

Sethi, P. K., Udawat, M. P., Kasliwal, S. C., & Chandra, R. (1978). Vulcanized rubber foot for lower limb amputees. *Prosthetics and Orthotics International*, *2*, 125–136. PMID:733459

Shivakumar, S., & Guggari, G. S. (2011). Literature review of fiber reinforced polymer composites. *International Journal of Advances in Engineering and Technology*, *1*(5), 218–226.

Subash, T., & Pillai, S. N. (2015). Bast fibers reinforced green composites for aircraft indoor structures applications [review]. *Journal of Chemical and Pharmaceutical Sciences*, *7*, 305–307.

Suresh, K. P., Niranjan, H. B., Martin Jebaraj, P., & Chowdiah, M. P. (2003). Tensile and wear properties of aluminum composites. *Wear*, *255*(1-6), 638–642. doi:10.1016/S0043-1648(03)00292-8

Yadav, S., Mathur, M. K., Dr, S. C., & Kasliwa, S. C. (1993). Saltiel's Design Modification: Short leg brace for knee extensor weakness. *Indian Journal of Physical Medicine and Rehabilitation.*, *6*, 24–29.

Yu, G.-C., Wu, L., Feng, L., & Yang, W. (2016). Thermal and mechanical properties of carbon fiber polymer-matrix composites with a 3D thermal conductive pathway. *Composite Structures*, *149*, 213–219. doi:10.1016/j.compstruct.2016.04.010

Chapter 8
Improvement of Mechanical Properties and Morphological Studies of Friction Stir Processed Composites:
A Short Review

Sanjay Sharma
I. K. Gujral Punjab Technical University, India & Graphics Era Hill University, India

Gurmeet Singh Cheema
Punjab Technical University, India

Sahib Sartaj Singh
Punjabi University, India

Deepak Verma
Graphic Era Hill University, India

ABSTRACT

Friction-stir processing (FSP) is a property enhancement technique which, not only removes the defects of initial casting process, but also improves the microstructure of the metals and metal matrix composites (MMCs). The process is based on frictional heating which results in a considerable dynamic plastic deformation within the metals. FSP can be specifically applied to develop fine-grained microstructures throughout the thickness of metal surface, to impart super plasticity and ensure homogeneous distribution of reinforced particles, if any. This chapter is a dedicated effort to consolidate the latest developments contributed by different researchers in last few years. The work covers various components and parameters, selected and used, for FSP to obtain specific desired results. Also, it includes past researches to exhibit various changes in mechanical properties with a keen focus on morphological study (by scanning electron microscopy) of these MMCs. In the last, a brief discussion on application and future scope of FSP processed MMC materials, is presented.

DOI: 10.4018/978-1-5225-3401-3.ch008

INTRODUCTION

Friction stir processing (FSP) is a technique applied to fabricate or modify the microstructural features of surface composites (Mishra RS et al.2003). It is serving as a common metallurgical tool for refinement of grains and homogeneous alloying of reinforcing elements, which establishes and governs the desired properties in the base metal, via dynamic recrystallization. Friction Stir Processing works on the principle of friction stir welding, which was developed by "The Welding Institute" in UK (Thomas WM et al. 1991). After inception, its successful adaption and wider acceptability through its applications to many areas, it is further modified to Friction Stir Processing (FSP) (Mishra RS et al. 2000, Ma ZY 2002), which has also attracted attention. It is considered to be a ''green'' technique, due to its efficient energy utilization, environment friendliness, versatility of use and without emission of harmful gases and noise. Also, FSP technique does not alter the shape and size of the processed materials. Presently it is viewed as a promising technology for property enhancement of the metals and composites, homogeneous reinforcements. A complete manufacturing technology is connected to these on large scale, because of its wide acceptability and growing recognition. It is used for enhancing the strength of the metals, along with ductility and corrosion resistance. It also contributes in enhancement of resistance to fatigue and formability, in addition to other properties. The resulted fine grain structures impart super plasticity to the materials. Figure 1 demonstrates the various micro structural features and friction stir process attributes. [Mishra RS et al.2014].

FRICTION STIR PROCESSING CONCEPT

To fabricate surface composites (SC), many techniques have been reported so far, but FSP has been identified as one of the most promising techniques. FSP is developed my R S Mishra (Mishra RS et al 1999), which works on the same principle, as that of friction stir welding (FSW). Conventional liquid phase processing techniques suffers from the limitation like, formation of intermediate detrimental phase

Figure 1. Evolution of microstructural features
Adapted from (Mishra RS et.al.2003).

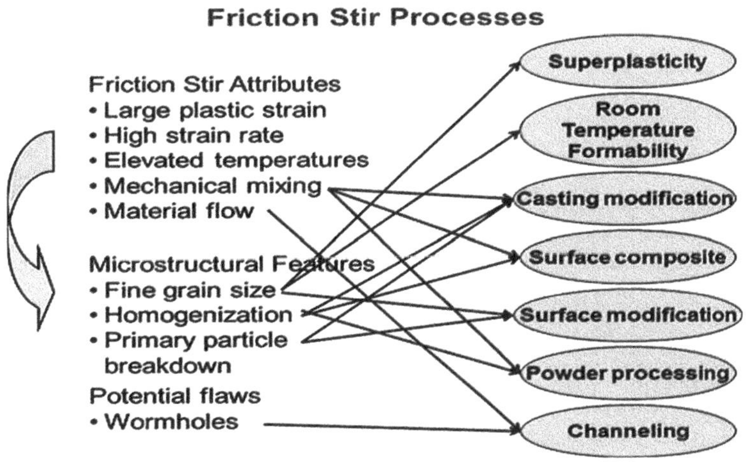

and porosities due to elevated temperature and various processing conditions, and requires critical control of process parameters to get control on the solidified structure (Mishra RS et al 2001).

The main advantages of Friction Stir processing over other methods i.e. it has one step process for depth control of processing thickness in the material and only pin length is governing the feature. It can be easily repeatable and ensures meticulous grain refinement of the area processed, along with homogenization. Friction stir processing is free from any fumes, smoke, noise and solvents for degreasing and cleaning, which makes it an environmental friendly technique of metal processing. Figure 2 shows the process of friction stirring of a metal surface.

The process involves a specially designed cylindrical rotating tool with a large diameter shoulder and a small diameter protruded pin at the bottom of the tool shoulder. When it is descended to the work piece material, firstly the pin comes in contact and due to frictional heat it softens the metal below and gets penetrated inside the metal. After the insertion up to penetration depth, i.e. till the complete length of the pin, a small diameter column area around the pin is plasticized. Next when shoulder contacts the work piece, due to tits larger diameter, contact area increases and hence the friction, which results in more amount of heat generation. This total heat plasticizes a larger cylindrical metal column around the inserted pin of diameter equal to the shoulder diameter. After full insertion of the pin, the tool is transverse to a predefined direction. During its lateral movement, tool has two sides, leading side and trailing side. The plasticized material is stirred from leading side to trailing side.

In addition to above, the heat is generated due to localized plastic deformation of the work piece material bulk, which contributes significantly to the heat already present in the processing zone due to frictional interaction. The amount of heat generated due to plastic deformation energy of the work material is added to the already existing heat of friction, and this total sum varies between 80% and 100% of the total input (Hodowany J. et al. 2000, Kapoor R et al.1998). The total plastic deformation energy developed in the work material can be utilized in two ways, the first part fraction is stored in the microstructure, and second part fraction is converted to heat. The combination of tool rotation and translation results in the stirring movement of material from leading side to trailing side. The flow of material around the tool and evolution of microstructure are the two main inherent features of FSP, resulting noticeable

Figure 2. Schematic illustration of FSP
Adapted from (RS Mishra et.al.2003).

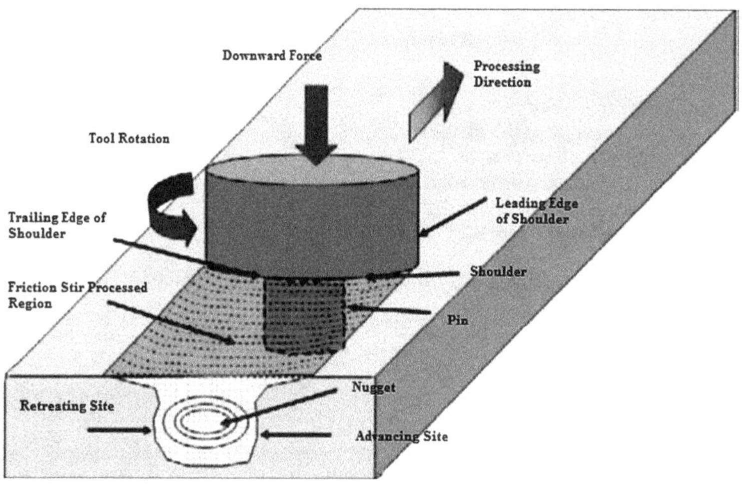

micro structural changes in the processing zone. In this process, Tool rotation and tool translation motion combination serves as a governing factor of overall characteristics induced in the work metal or composite. Frictional stirring of the work piece material modifies the coarse grain microstructures into fine grains due to plastic deformation and faster cooling rates (Zhang Y. et al. 2008). The evolution of microstructure in the process is a unique property of FSP which results in dynamical recrystallization of the micro structure with enormous numbers of high angle grain boundaries (Kapoor R. et al. 2010).

Inherent casting defects are eliminated locally by the process of Friction Stirring, and dramatically refine the grain structure of the processed zone. This improves properties of strength, ductility, formability etc (RS Mishra et.al.2003, Saito, N. et al. 2001, Mahoney et al. 2003, Santella M L et al. 2005). Microstructural refinement of the material reflects in the enhancement of the mechanical properties particularly densification and homogeneity. This refinement of processed zone is governed by the optimizing tool design, controlling different FSP parameters, and cooling / heating of the work piece. Friction stir processing has been successfully applied to different metals and alloys including aluminium alloys in improving mechanical properties (Dash Wood et al. 2001, Rhodes CG et al. 1997, Karthikeyam, L et al. 2011, Kumar N et al. 2011, Afsaneh Dorri et al. 2015). In case of fabrication of Metal Matrix composites (MMC) reinforced by different particulates, can be achieved by (a) Solid State processing; (b) Liquid State processing; and (c) Vapour deposition (Shamanian et al. 2014). FSP serves as a solid state reinforcement techniques, avoiding almost all the thermal defects as temperature in the process is usually 70% to 90% of the melting temperature of the metal (Darras B M et al. 2007). In case of FSW, which is similar to FSP, the joining is achieved at about 70 to 80% of the melting temperature of the base material (Sorensen C D et al. 2007). Thermodynamic impacts on to the metals and composites are less than the processes of liquid state.

In addition to grain refinement, reinforcement of specific additives can also be done, to get desired surface composite layers. The stirring action is generally used to homogeneously distribute the material particles uniformly across the processing zone. This method includes pouring of the reinforcement additive powder into a groove or hole, created by different tool, in the processing zone. Figure 3 represents the steps of creating and filling the groove, in which a pin less tool is, used which covers the groove top after powder filling and prevents scattering of the powder during FSP technique (Besharati M K et al. 2014). After the filling of the groves, FSP is carried out throughout the metal surface, which ensures the proper homogeneous distribution of the reinforced particulates in the material surface.

Figure 3. Surface composite layer fabrication via FSP
Adapted from (RS Mishra et.al.2000).

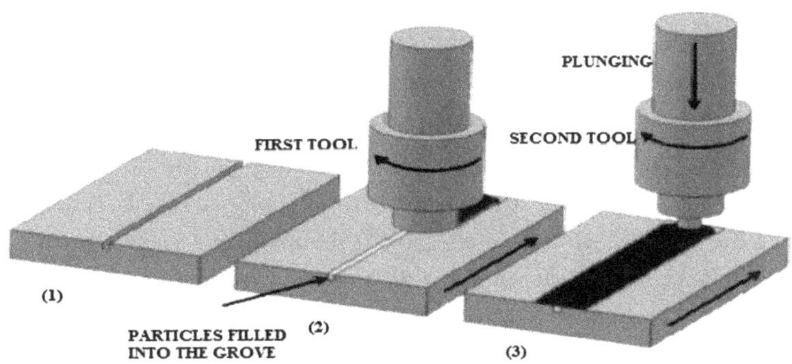

COMPONENTS OF FRICTION STIR PROCESSING

Friction stirring is generally referred as a thermomechanical deformation process. The components of FSP can be categorized as a) Geometrical component related to tool and b) Proper selection of processing parameters.

A proper selection of optimum tool design and adequate tool material is necessary for a quality friction stir processed material for desired application. All friction stir tools contain some specific features specifically designed for a particular function.

TOOL GEOMETRY AND TOOL MATERIALS

The tool of FSP is similar to that of FSW, which has a shank, to clamp in rotating chuck or collect, a shoulder, of larger diameter, which causes the friction to generate heat, a pin, which is inserted into the metal surface and stirred to obtain refinement of grains and homogeneous distribution of reinforcements. Tool geometry affects the traverse force, torque, heat generation rate and gets affected by the thermomechanical environment experienced by the tool. There are two types of parameters in the FSP, firstly tool parameters, such as shoulder diameter, shoulder surface angle, pin geometry including its shape and size, tilt angle of tool and the nature of tool surfaces and secondly the process parameters, such as rotation speed, transverse speed, spindle speed, feed rate, plunge depth and axial pressure applied. (Mishra RS et al.2005, Lorrain O. et al. 2010, Kawanami S. et al. 2010, Sun N. et al. 2009, Kumar K et al. 2008, Hattingh D.G. et al. 2008, Fujii H. et al. 2006, Fujii H. Cui L. et al. 2006, Vijay S. J. et al. 2010, Badarinarayan H. et al. 2009, Sorensen C.D. et al. 2009, Colegrove P. A. et al. 2004, Nandan R. et al. 2008),

Considering process parameters such as spindle speed, feed rate, plunge depth etc are fixed, the heat generated depends on:

1. Surface contact between work piece and the tool
2. Flow of semi molten material under the tool shoulder
3. Axial pressure of shoulder on the work piece material (De Giorgi M. et al. 2009)

Tool material plays an important role in FSP, as it is exposed to very harsh working environment. Also, tool wear after extensive use, which is a predominant factor, results in more power consumption to process the same material. Thus proper selection of the adequate tool material is required, based on specific properties of the under process material. Rai et al. presented an overview of various aspects such as the selection of the tool material, geometry and load bearing ability, mechanisms of tool degradation and economics in context with FSW of high melting temperature materials. They submitted that feasibility of the FSW process for joining high melting temperature materials highly depends on proper tool material selection and tool design. These factors greatly influence the performance of the tool, the overall process cost and the quality of weld joint produced. Many scholars have reported different tool materials for friction stirring of different metals and composites, such as illustrated below:

Tool steel is one of the most commonly used tool materials in the welding/processing of aluminum copper and magnesium alloys and can weld up to 50 mm in these materials (Akos M. et al. 2013). In case of tool steels, AISI H13, which is a hot worked, air hardened and chromium–molybdenum steel, has been the most accepted as tool material (Feng A.H. et al. 2007, Najafi M. et al. 2008). Steels are easily available and have good machinability and thermal fatigue resistance.

Polycrystalline cubic boron nitride (pcBN) was specifically developed for machining of cast irons, different tool steels and latest super alloys (Mishra R. S. et al. 2007, Sorenson C.D. et al. 2004), but presently it is a well accepted as a friction stir tool material. This is because of its inherent properties such as sustainability at higher temperatures and high mechanical performances.

Nickel and cobalt based super alloys were developed for making components of aircraft engine and offer considerably good creep value, extensive high strength, and property of resistance to corrosion, makes it an important tool material for FSW and FSP.

Ceramic particle reinforced metal matrix composites have also been used as tool materials, but due to their brittle structure, they get fractured during initial plunging phase of the tool into the work material. Ceramics also have been also used as FSW and FSP tool materials; however, it is too brittle and normally fractures during the plunging phase (Sorenson C.D. et al. 2004, Mishra R.S. et al. 2014).

ELEVATED TEMPERATURE STRENGTH AND STABILITY

In case of FSW and FSP, the heat generation also depends on the tool geometry and different properties of tool and work piece materials, such as thermal conductivity, stability at elevated temperatures, reactivity, wear resistance etc., along with the processing parameters. The thermal expansion coefficient is also included as it thermal stresses may affect the tool (Rai R. et al. 2011). According to Ramachandran et al., in case of friction stirring of high melting temperature materials, the stirring tool is exposed to very severe thermomechanical environment and degradation of the tool of predominant nature is observed, particularly during the initial plunge stage. In case of FSW, it has been observed that about 70–80% of the temperatures of melting temperature of the base material are sufficient to make a successful joint. Even during FSW of steels, the peak temperature reported at the weld zone is around 1200 °C (Lorrain O. et al. 2010). The base material (BM) under processing, when subjected to FSP tool, it has three different zones, as established by many scholars. They are termed as thermo-mechanically affected zone (TMAZ), heat-affected zone (HAZ), and Stir zone (SZ) are present in the area under welding or processing, in case of FSW and FSP respectively (as shown in Figure 4). The central zone i.e. around the rotating pin and under the shoulder diameter is referred as SZ. It has the highest of the temperature generation being main area under process during FSW and FSP.

In TMAZ, which is the name given to the area in between SZ and HAZ, the generated microstructure is characterized by a bimodal population mix of fine recrystallized grains and highly deformed but coarse grains. As this zone experiences partial dynamic recrystallization because of the less heat than that of SZ, this causes the development of some ultra-fine grains distributed among the coarse and elongated grains. Lastly in HAZ, i.e. the area in between the TMAZ and undisturbed material, the grains are heated but not deformed due to lack of adequate temperature needed for deformation to take place (Chai F. et al. 2015). Hence the processing zone is subjected to tremendous heat during the process and ideally a FSW or FSP tool has to sustain in the processing zone with all its geometrical properties and unaltered shape.

In FSP, when tool comes in contact with the work material, it experiences sufficient compressive loads and shear force. High temperatures are generated due to mutual rubbing action. For the long processing lengths Creep (and creep fatigue) phenomenon is considered, however it has been observed that poor creep resistance would generally change the dimensions of the tool during processing. Tool materials that obtain their strengths from precipitates, work hardening, or transformation hardening have described maximum-use temperatures. However it has been noticed that tools used above the maximum use temperatures will, in time, shows a decrease in mechanical properties. The tool must sustain its specific properties and remain stable without fracture or distortion in the process. The tool should have good strength and dimensional stability during the time of its use.

RESISTANCE TO WEAR

The FSW technical handbook (Welding Automation SE-695 81 LAXA) outlined that the resistance to wear is very critical because no traces of the tool material should be left in the joint. Tool wear affects not only the tool life but also the weld and process characteristics. It has been observed that the considerable tool wear changes the shape of the tool diminishing the tool features. This alteration of shape, ultimately changes the quality of microstructures in processing zone, including grain refinement and increasing the probability of defects and undesired grain structures. In friction stirring processing, tool wear can be noticed by different wear mechanisms such as by adhesive, abrasive, or chemical. The actual wear mechanism can be observed in the interaction zone of tool material, tool parameters and work material. It is observed that, wear at low tool rotation speed is caused by adhesive wear, while wear at high tool rotation speed is caused by abrasive wear (Sorenson C.D. et al. 2004).

FSW and FSP tools made from conventional materials have considerable endurance in case of processing materials having less strength than their own materials. But they exhibit wear in case of dealing with harder materials such as metal matrix composites and different steels (Bist Ashish et al. 2016). It has been observed that the amount of wear of a tool in a specific process varies inversely with the hardness ratio H, while processing composites. The hardness ration H, is a ration of hardness of tool material (Ht) to the hardness of reinforcement (Hr). There will be no tool wear in case of H >1, i.e. Ht>Hr, as the hardness of tool material will be more than that of the reinforcement. But in case of H < 1, Ht<Hr, i.e. hardness of reinforcement material is more than that of the tool material, the tool wear is inevitable and can be compensated by increasing the hardness of tool material (Prater T.et al. 2013).

It has been observed as the design features are found altered, such as, erosion of the probe features, probe length and change in shape etc. This alters the desired stirring of material and results in altered flow of material in the processing zone. Stirring action is inherent feature of FSP and results in homogeneous mixing of the reinforcement and generation of "optimal microstructure" i.e. finer grain structure in the processing zone (Pirondi A et al. 2008). The mechanism of tool wear depends on the interaction conditions of the materials, i.e. materials participating in the process, tool geometry and processing parameters. Abrasive wear is found predominant in the case of Friction stir processing and welding. (Collier M, et al. 2003, Mishra R S et al. 2005)

The amount of tool wear depends on factors, such as 1) the ratio of tool material hardness in comparison with hardness of the reinforcement material in case of the composites, where reinforcement materials are added to provide adequate strength to the base material, 2) percent of reinforcement particulate, and 3) processing parameters chosen for the process such as rotation speed of tool, transverse speed of tool and axial load applied. Prater et al. in their study elaborated the two factors in details i.e. relation in between tool material and tool wear (Figure 4) and relationship between percentage reinforcement with tool wear (Figure 5).

Figure 4. Tool material and tool wear relation
Adapted from (Prater T.et al. 2013).

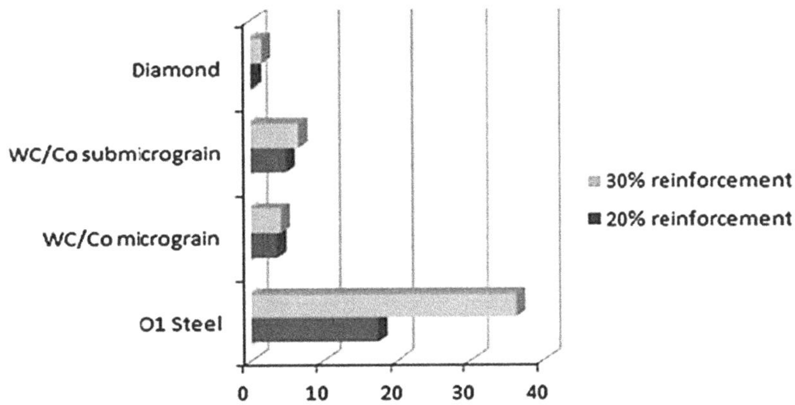

Plot of % wear vs. tool material. Each bar represents the % wear recorded for the corresponding tool material after a sequence of three successive welds at 1000 RPM/3 IPM in an Al 359 MMC with either 20 or 30% SiC reinforcement. Values for diamond are extrapolated

Figure 5. Reinforcement percentage and tool wear relation
Adapted from (Prater T. et al. 2013).

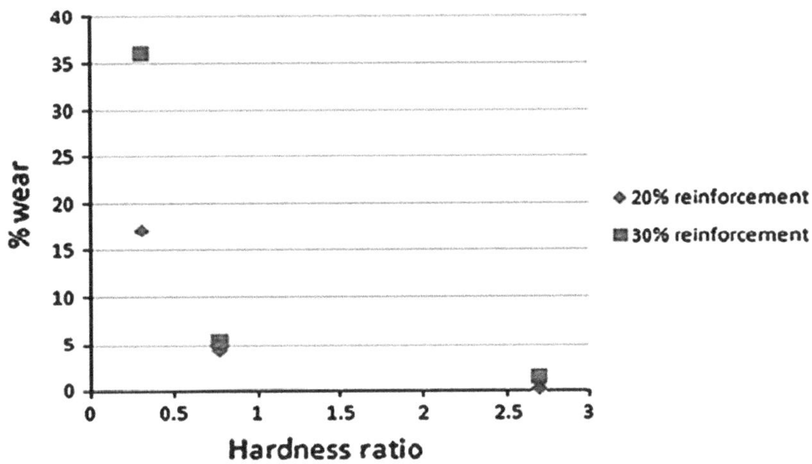

Plot of hardness ratio vs. percent cumulative wear

FRACTURE TOUGHNESS OF TOOLS

The fracture toughness is a remarkable property of the tool material, which plays an important role during the plunge and dwell actions of the tool. It has been observed that when the tool touches the work piece then generation of the local stresses and strains takes place, which are sufficient to break a tool, even when mitigation methods are used. It is generally noticed that the plunge and dwell periods produces the most damage to a tool (Lienert T et al.2003). The spindle run out of friction stir machine (lateral movement during spindle rotation) should also be considered while selecting a tool material. The use of high fracture toughness tools is recommended for FSP, so as to overcome the run out problem and premature tool failure (Sorenson C.D. et al. 2004).

REACTIVITY OF TOOLS

Reactivity is the most important aspect in friction stir processing technique. The tool materials should not react with the work piece material, which would ultimately results in change of surface properties of the tool. On the other hand titanium is found to be reactive at elevated temperatures; so, the reaction of titanium with the tool material results in change in the tool properties and affects the quality of the processed zone. Environmental reactions such as oxidation of the tool and production of toxic substances, can also affect the tool wear resistance. Park et al. examined FSW of ferritic, duplex and austenitic steels with a pcBN tool and reported the presence of higher nitrogen content in the SZ, while processing austenitic stainless steels, rather than ferritic and duplex stainless steel welds. They reported a reaction of Boron, librated from pcBN tool and reacted with chromium in austenitic steels to form borides, which results in pitting of welds and makes it susceptible to corrosion. The above establishes a need of careful selection of tool material is preferred to avoid the reactivity of the tool with the work piece material. Work piece material can also show some environmental reactions, as it also subjected to elevated temperatures without a protective shielding. As example in case of titanium alloys processing, a covering non reactive gas is needed to prevent work piece from oxidation. (Sorenson C.D. et al. 2004)

PARAMETERS SELECTION FOR FRICTION STIR PROCESSING TECHNIQUE

The processing parameters are the governing factors of the extent of plastic deformation, which ultimately are the main cause of frictional heat generation and flow of plastically molten material around the non consumable pinned tool shoulder, in the processing zone. The geometry of the tool is a leading factor in concern with material flow in the processing zone. The FSP tool consists of the two main parts such as pin and shoulder. The geometrical features such as shape of the pin and its length, pattern of the surface on tool shoulder and diameter of the tool, have a considerable impact on the flow of material in the processing zone. The consideration of exerted forces and acting torques on to the tool are important for several reasons. Firstly, as larger the value of torque, greater will be the power requirement for FSW & Processing (Atharifar H. et al.2009). Secondly, increased load on to the tool results in higher tool deformation and wear, incurring more will the processing cost due to frequent tool replacements. Thirdly, tool wear will ultimately result in contaminated welds, deteriorated joint properties and improper distribution of reinforcement particles in the metal processed.

TOOL ROTATION AND TRANSVERSE SPEEDS

These process parameters are directly responsible for heat generation, and induction of plastic deformation phenomenon into the material under processing. Tool rotation and traverse speeds are treated as governing factors and individually or in combination have a considerable impact on the properties of the processed zone. A well known and proved concept categorizes the processing of materials, in two different categories, as: "Cold" and "Hot". In case of Cold processing, the ratio of rotating and traverse speed is below 3 and in Hot processing, this ratio exceed 6. Though there's no scientific basis for this border line, it is noticeable that an increase in this ratio, i.e. towards the values more than 3, makes the TMAZ and a very fine structure with a larger area is observed. Whereas, under "Cold" conditions i.e. value of the ratio is 3 or less, the TMAZ is not well defined as generated heat is inadequate to initiate and promote grain refinement due to recrystallization. It is evident that, increasing the rotational speed of the tool, plastic deformation is more exquisite and larger TMAZ is produced, due to larger amount of heat generation. This leads to more homogeneous mixing and grain refinement in the TMAZ. Transverse speed is related to movement of the tool and affects the exposure time of the tool to the generated heat and viscosity of the material. Lower transverse speeds results in larger exposure times and finer grain structure evolution can be expected.

TOOL AXIAL FORCE

This parameter is responsible for the rubbing of shoulder against the material surface, under process, and contributes in heat generation. Initially pin comes in contact with the work material followed by the shoulder during start of the process, the forces acting on the tool varies considerably due to shear and work hardening due to axial compression (Figure 6). This results in generation of heat and softening of the material. After the plunge, and a little dwell time, when tool is traversed towards the rest of the work piece material, the forces acting on the tool tends to stabilize at a value which is a lower value than the peak force values at the plunge state (Atharifar H.et al. 2009, Sorensen C.D. et al. 2007). Therefore the tools in FSW and FSP, are under severe stress during the initial plunge state than while in transition state of the tool followed by the plunge. Tools of brittle materials such as ceramics, carbides and pcBN are more likely to fail during the initial stage of plunge rather than transition. Phenomenon of preheating and dwelling after insertion in the work piece, are adopted to reduce the tool stresses during the initial plunge.

PIN PROFILE AND TILT ANGLE

In the process of FSW and FSP, a non-consumable friction stirring tool with a shoulder and a profiled pin in inserted into the work material and moved along the center line of the material or weld, which results in generation of frictional heat an causes deformation, needed to produce a desired surface feature or a solid state weld (Farrokhi H et al. 2013). Both the tool parameters and process parameters are to be controlled to result finer grain structures in the processing zone of the material through dynamic recrystallization in FSP and to enhance the mechanical properties of the FSW joints. Tool parameters include the design and geometrical features of the tool and process parameters include traverse speed, rotational speed and axial force.

Figure 6. Forces acting on the tool

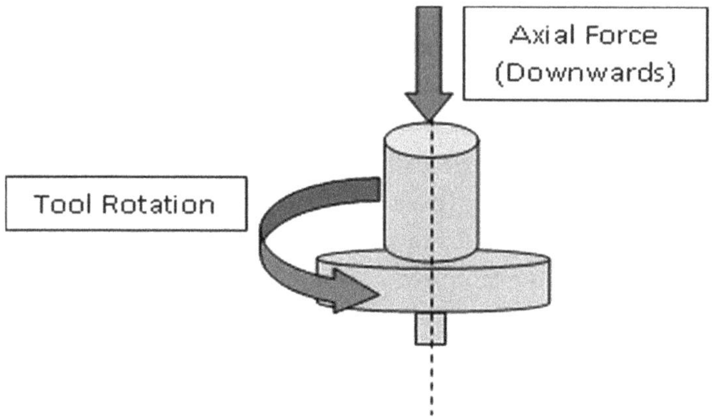

The affecting parameters such as tool geometry and process parameters are studied in detail by many researchers, with their impact on the mechanical and microstructural properties of the FSP and FSW. Different types of pin geometry and pin profiles have been suggested with their comparative results on different materials (Figure 7).

Elangovanet. al. have recorded the impact of five different tool pin profiles on the formation of friction stir processing zone in AA2219 aluminum alloy. It has been observed and recorded that a square pin profiled tool with transverse speed of 0.76 mm/s renders welding joints of fine grain structure with improved tensile strength, and enhanced hardness. Also, Farrokhi H et al. have supported the above statement that welding joints processed by tools of a square or a triangular pin profiles, exhibit improved ultimate strength as compared to the joints welded with cylindrical pinned tools. Reshad Seighalani et al. examined the threaded cylindrical pin profile and demonstrated that better mechanical properties are resulted in the welding joints as compared to the joints welded by using a simple cylinder pin profile.

Figure 7. Different types of pin geometry and pin profiles
Adapted from (Farrokhi H. et al. 2013).

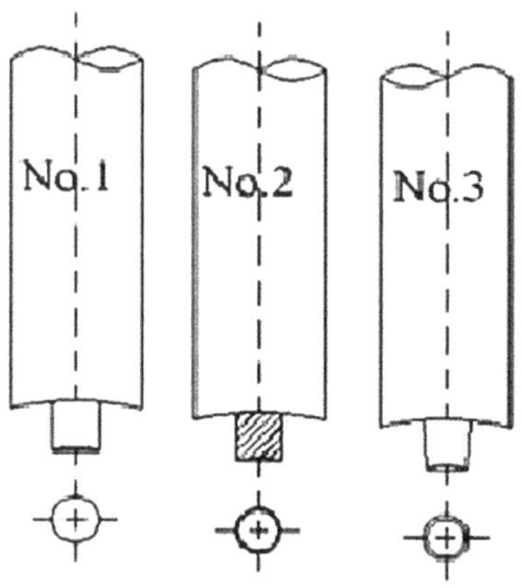

Schematic illustration of the different tool pin designs: (No. 1) simple cylindrical, (No. 2) screw threaded cylindrical, and (No. 3) taper.

Chowdhury et al. have studied magnesium alloys and recorded the impact of thread orientation of pinned tool on properties of AZ31B magnesium, specially fatigue and tensile properties. They confirmed that fatigue strength and tool life improves in the joints, by altering thread orientation, in case of FSW.

Tilt angle is referred as a considerable factor in FSW and specially in FSP. It is the angle between the tool axis and the work piece surface. This results in homogeneous distribution of the mix in the material, packing of blow holes (if any), improvement in surface finish and evolution of fine grain structures. Several authors and researchers have documented and agreed that heat generation and plastic deformation are responsible for flow of material, which results in good consolidation. To ensure better performances a tilt angle of 2-4° is used in practice. (Miranda R.M.et al. 2013).

FRICTION STIR PROCESSING OF COMPOSITE

Surface Modification of Composites

In last few years, various surface-modification techniques, such as high-energy electron beam irradiation (Choo S.H. et al. 1999), plasma spraying, high-energy laser melt treatment (Hu C.et al. 1997), cast sinter (Wang Y.S. et al. 2001), and casting (Attia A N. et al. 2001), have been developed and utilized for fabrication of surface of MMCs. Laser melt treatment or laser processing, is the most utilized technique among all and most widely used for surface modification of metals and MMCs. Laser melt treatment is a process which uses a laser beam for melting the surface of the substrate and material to be deposited. It may be either carbide powder of silicon or titanium, or a combination of both or a binding material such as cobalt, aluminum, or nickel. Along with the laser beam, coating material is either preplaced for melting or injected through a nozzle simultaneously. While using injection method, the material to be deposited, is carried through a nozzle by a carrier inert gas in powder form and propelled to the surface to be treated, and gets glued in the pool of laser melted surface. Nearly every available surface processing technique can be termed as a liquid phase processing at elevated temperatures. It makes it difficult to eliminate reactions in between the metal matrix and reinforcement material, which results in inevitable formation of some pernicious phases.

LOCALIZED SURFACE MODIFICATION

To obtain desired combination of properties i.e. a gradient of properties within a monolithic structure, the technique of surface modification can be a very powerful tool. The potential exists in enhancing design possibilities using MMC surfaces. For all such approaches, a particularly sustainable nature of particles is required.

In case of aluminum alloys, micro structural features such as prior-particle boundaries, micro structural inhomogeneity, and size of primary intermetallic particles resulting porosity, are very important, as these limit their full potential. These micro structural features particularly impede the ductility in very high-strength aluminum alloys (Spowart J.E. et al. 2003, Berbon P.B. et al. 2001).

MECHANICAL PROPERTIES OF FRICTION STIR PROCESSING COMPOSITES

Morisada et al. 2006 used FSP technique for the reinforcement of Multi-walled carbon nanotubes (MW-CNTs) into a magnesium alloy (AZ31). The most important parameter such as speed of the tool plays a major role in the dispersal of the Multi-walled carbon nanotubes in the magnesium alloy. It has been noticed that the grain size of the MWCNTs reinforced AZ31 surface composites was lesser than that of the unreinforced friction stir processed AZ31. The inclusion of the MWCNTs found to be efficacious for the development of the composites which consists fine matrix grains. Figure 8 shows the penetration marks of the micro hardness test by using 200 g load. The optical micrographs with the same magnification are obtained in order to compare the size of the penetration marks. It has been found that the average value of micro hardness of the unreinforced AZ31 was 41Hv shown in Figure 8(a). However, the penetration mark for the MWCNTs reinforced AZ31 surface composite was lesser as compared to the parent or unreinforced AZ31 Figure 8(b). The hardness value of 78 Hv is found to be the maximum micro hardness value of the composites. Also the maximum micro hardness of the unreinforced FSP processed samples was 55Hv.

Lee CJ et al. 2006 performed the micro hardness testing, Hv, just at the central cross-sectional zones of the Friction Stir processed samples and are shown grey regions. Also, from Figure 9, the gray regions represent the location of the groove. In this groove, without undertaken the FSP into account, the silica powders were introduced into AZ61. It has been noticed the hardness value of the unreinforced AZ61 alloy increases and becomes double. After performing FSP on the specimens, the distribution of Hv in the FSP nugget region is found to be less, because the pin stirring efficiently in to the sample and uniformly distributed the nano-silica powders. The AZ61 alloy without reinforcing silica powder gives and increases the Hv from 60 and up to 72. It is also found that the value of the yield stress of the FSP composites was increases to 214 MPa in case of the one dimensional sample and for two dimensional samples the value increases to 225 MPa, in contrast to 140 MPa of the parent AZ61 billet and 147 MPa of the unreinforced Friction Stir processed AZ61 alloy.

Figure 8. Prints of indentation of material: (a) as-received Mg AZ31; (b) surface composite of MWCNTs/AZ31 at 1500 rpm and 50 mm/min of transverse speed
Adapted from (Karthikeyam L et al. 2011).

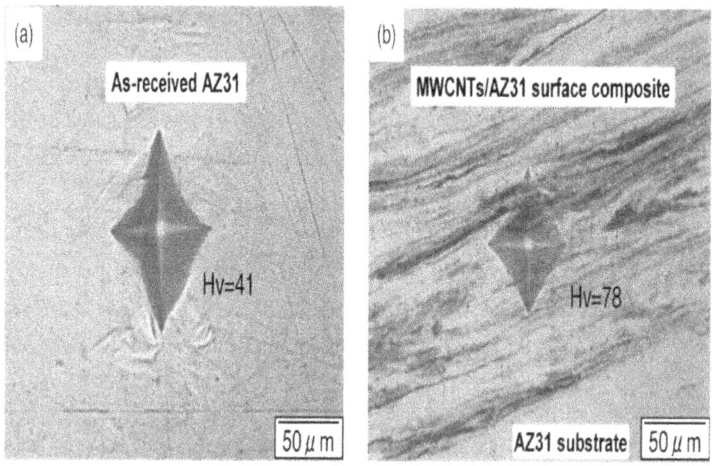

Figure 9. Typical variation in micro hardness (Hv) distributions of Mg AZ61 (without SiO2) and composites
Adapted from (Karthikeyam L et al. 2011).

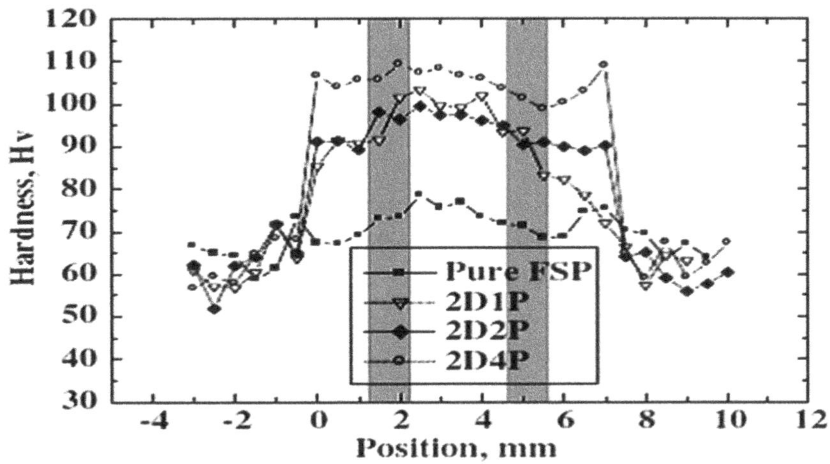

Chang et al. 2006 fabricated composite by using ZrO2 and SiO2 particles at different vol% such as 20 vol% and 10 vol % respectively in Mg-AZ31. Figure 10 shows the values of Vickers hardness readings. The specimen with approx. 20 vol% ZrO_2 particles reinforced composite gives approx. double value of the hardness testing as compared to the parent AZ31 alloy. After Friction Stir Processing, the distribution of Hv in the FSP nugget zone is examined to be comparatively low and showed that the stirring action of the pin has effectively and uniformly distributed the nano-ZrO_2 particles just after one more pass.

Figure 10. Typical variations in micro hardness (Hv) distribution in FSP AZ31 composites and the FSP AZ31 (without reinforcement)
Adapted from (Afsaneh Dorri et al. 2015).

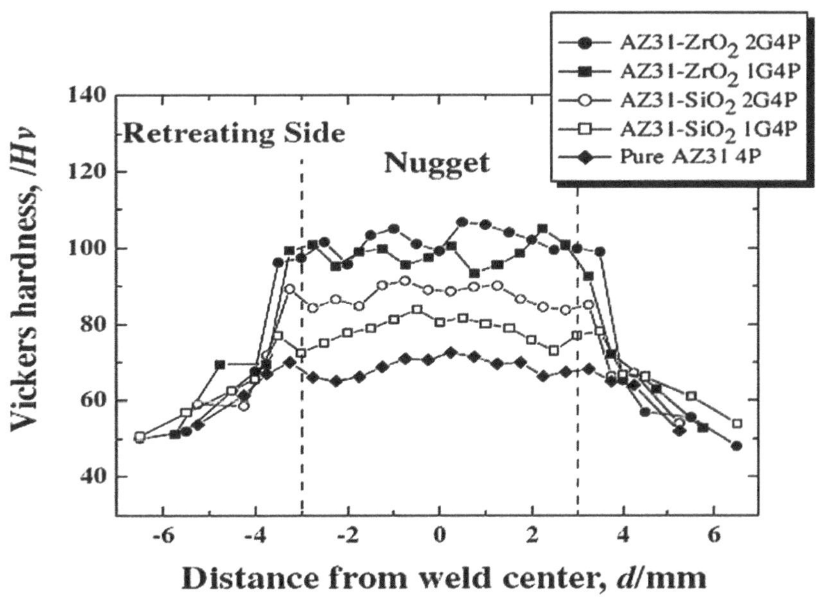

Chang et al. 2007 studied and showed the hardness measurement results of composites on the transverse cross-sectional plane in Figure 12. Also from Figure 11 it has been seen that the hardness value is gradually improved by increasing Al and Zn contents. The mean hardness value gives 350 Hv in the $Mg_{50}Al_5Zn_{45}$ and $Mg_{37.5}Al_{25}Zn_{37.5}$ systems after 3 FSP passes without using cooling system. With the increase in the Friction Stir Processing passes, the reduction in the difference of Hv values found to be more uniform just inside the nugget zone, which also resulted in uniform microstructure. The hardness value of $Mg_{37.5}Al_{25}Zn_{37.5}$ after 10 FSP passes gives lower variation.

Shamanian et al. 2014 developed a Cu–Ti C composite and reported the impact of FSP on the mechanical properties of the composite. Figure 12 shows the micro hardness values of the samples of Friction Stir processed material. The mean hardness value of pure copper is found to be 65 HV. The hardness of the SZ was more than that of BM just because of a harder phase reinforcements of the TiC particles. This happens due to the development of a composite layer which ultimately reduced the grain size of the matrix in the SZ, as observed from the SEM micrograph. As per the well established Hall–Petch relationship, a higher hardness is resulted due to decreased grain size. Also, as per the Orowan mechanism, TiC particles generally act as a barrier in the dislocation movement results in the increase in micro hardness. The wear test results are shown in Figure 13. The weight losses of composite samples are lower as compared to the pure copper.

Shafiei-Zarghaniet al. 2009 reinforced nano-sized Al_2O_3 into AA6082 aluminum alloy by using Friction Stir Processing and developed composite surface layer (particulate). The same can be seen in Figure 14 (a) The size of the Al_2O_3 particles used was about 50 nm. Figure 14 (b) shows the good bonding between the composite surface and the aluminum alloy. Mechanical properties such as wear resistance and micro hardness were performed. Figure 14 (c), shows comparison of wear rate of the nano composite layered surface produced after four passes and as received alloy material. A noticeable improvement in the nano-composite surface was observed due to improved wear resistance as compared to the parent Al alloy.

Figure 11. Typical variation of micro hardness (Hv) along the transverse cross-sectional plane of the Mg70 Al5 Zn25, Mg50 Al5 Zn45 and Mg37.5 Al25 Zn37.5 alloys after FSP passes
Adapted from (Shamanian et al. 2014).

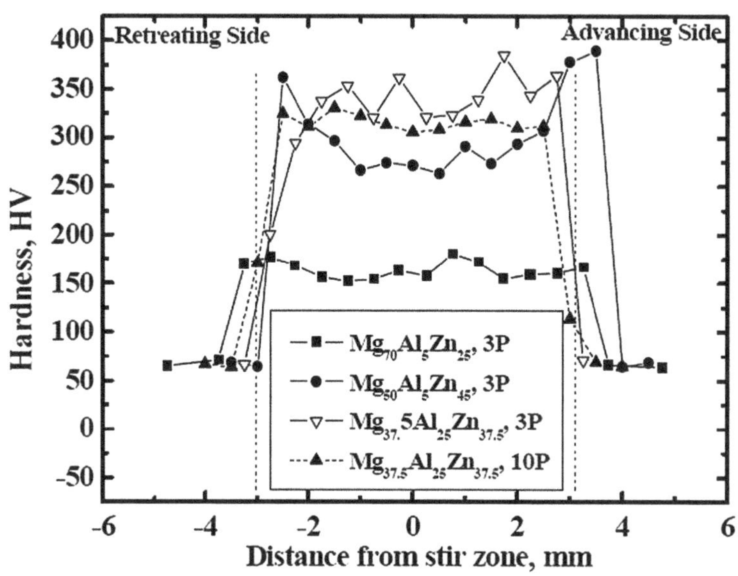

Figure 12. Micro hardness profile of the friction stir processed material (a) at surface (b) in cross-section
Adapted from (Shamanian et al. 2014).

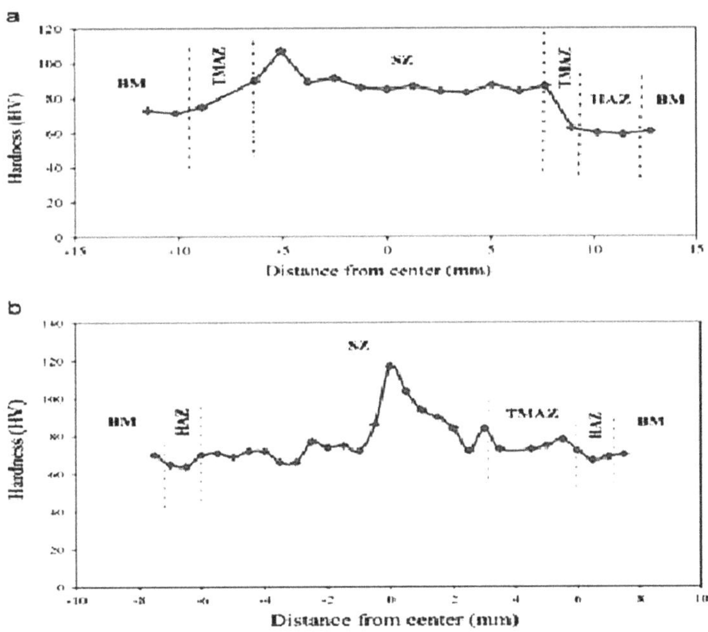

Figure 13. Wear test results of the friction stir processed material
Adapted from (Shamanian et al. 2014).

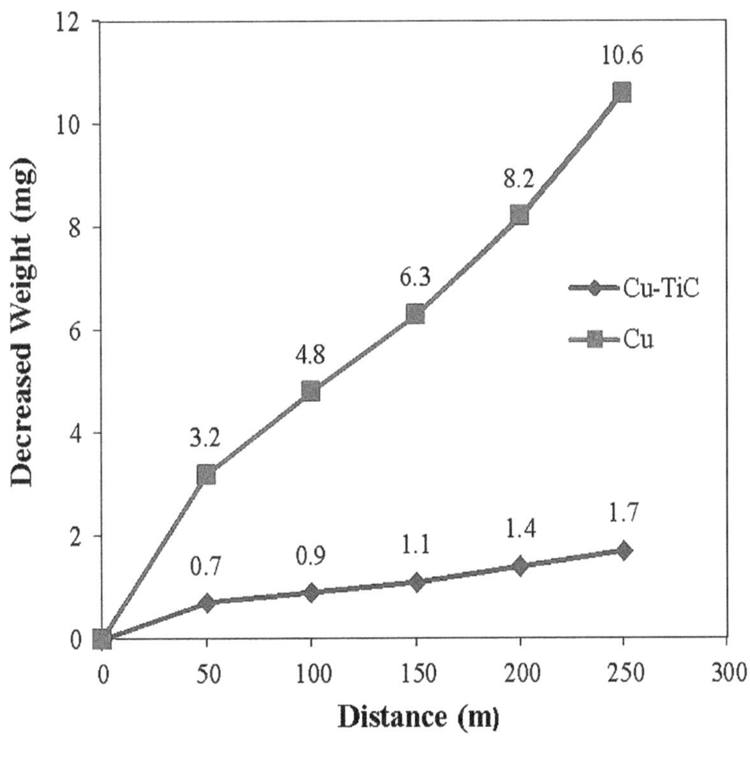

Figure 14. Microstructure and wear property of alumina particle reinforced A6082 alloy: (a) optical micrograph showing the top and bottom layers of composite; (b) SEM image showing the well bonded interface between the two layers; (c) comparison of wear rate of the as-received alloy and reinforced material surface nano composite layer, produced by four FSP passes
Adapted from (Besharati M K et al. 2014).

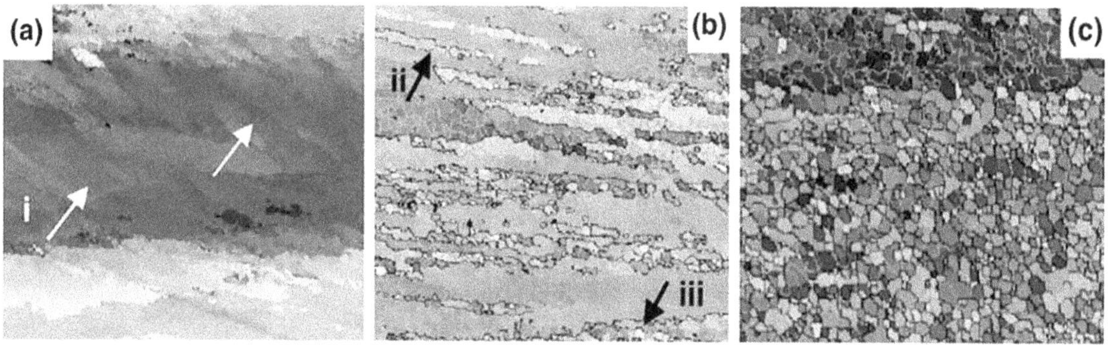

*For a more accurate representation see the electronic version.

Sathiskumar et al. 2013 applied FSP to fabricate composite by reinforcing the boron carbide (B4C) in particulate form into the copper. The effect of various FSP parameters such as processing speed, tool rotational speed and groove width on the microstructure and micro hardness of the composites was examined. In this research a groove was made on the 6mm thick copper plates and reinforced with B4C particles. Friction stir processing was undertaken by using five different processing speeds, tool rotational speeds and groove widths.

Asadi et al. 2010 used FSP technique for the development of SiC-reinforced AZ91 composite. In this research authors used AZ91 as a matrix material and SiC powders of 5 μm size as reinforcing particles for the fabrication of AZ91/SiC surface composite layer. The plate of 5mm thickness was used. In this research a square pin steel tool of 5 mm diameter and length of about 2.5 mm with 15 mm shoulder diameter was used. The various speeds such as tool rotational and linear speeds were used. It was observed that by increasing the linear speed and decreasing the rotational speed ultimately results a decrease in the grain size.

Shafiei-Zarghani et al. used FSP technique for developing a surface nano composite layer of Al/Al_2O_3 and studied the mechanical properties composite in accordance with its microstructures. They used a commercially extruded 6082 Al bar of thickness 7 mm, as substrate and a powder of Al_2O_3 of 50 nm diameter, as reinforcement. FSP tool of H-13 steel with a pin of 5mm diameter and 4mm of length is used to process the composite. The processing parameters were kept as rotation speed of 1000 rpm and transverse speed of 135mm/ min. Grove method was adopted to reinforce the particulate in the metal with a grove of 1 mm width and 4mm depth.

Lee et al. 2006 used AZ61 magnesium alloy and reinforced nano scale $SiO_2$of dimension of 20 nm into it. Friction stir processing method was used to reinforce 5-10 vol % nano particles into the alloy. It is seen that the dispersion of the SiO_2 particles was uniform just after performing four FSP passes. It is also observed that the grain size of the FSP composites was refined to 0.8 μm, as compared with the FSP AZ61 alloy. A reaction between SiO_2 particles and the Mg alloy was observed during FSP technique. The products of reactions include Mg_2Si and MgO with size range from 5 to 200 nm, which is found to be very fine and perfect for the mechanical properties. The hardness and strength at room temperature of the AZ61Mg composite with nano-particles were improved, as compared with the AZ61 cast billet.

Asadi et al. 2011 performed an experiment on magnesium-base nano composite by using FSP technique to confirm the number of FSP passes and effects of particle types. They used AZ91 as-cast magnesium alloy material with grain size of 150 µm. The particles used for reinforcement were SiC and Al_2O_3 with mean diameter of 30 nm and 99.98% purity. The reinforcing particles were introduced into a groove of size 0.8 mm×1.2 mm made on the AZ91 as-cast plate. The various tool speeds such as tool rotational speed and linear speeds of around 900 rpm and 63 mm/min were used. During FSP, it was noticed that accumulation of alumina particles were seen at various points in the matrix and uniformly distributed in the stir zone, on the other hand, SiC particles didn't sticked and the distribution of SiC particles in the Mg matrix was not uniform.

MORPHOLOGICAL STUDIES OF FRICTION STIR PROCESSING COMPOSITES

Shamanian et al. 2014 fabricated a Copper based composite by reinforcing TiC through friction stir processing technique. Figure 15 shows the optical micrographs of the different Friction Stir processed materials. Fig. showed four zones such as thermo mechanically affected zone (TMAZ), stir zone (SZ), base metal (BM) and heat affected zone (HAZ). From Fig.BM Figure 15(a) showed equiaxed grains of size 150 µm. Due to the high thermal conductivity of Cu (copper) and evolution of heat during FSP technique, resulted in the visibility of heat affected zone(HAZ) Figure 15(b) (Liu, FC et al. 2008). The interface between BM and TMAZ can be seen in Figure 15(c) However Figure 15(d) showed the interface

Figure 15. The microstructures of the friction stir processed material: (a) BM (equiaxed grains of size 150 µm); (b) BM–TMAZ interface(visibility of heat affected zone HAZ); (c) interface between BM and TMAZ; (d) interface between TMAZ and SZ; (e) powder distribution in SZ
Adapted from (Shamanian et al. 2014).

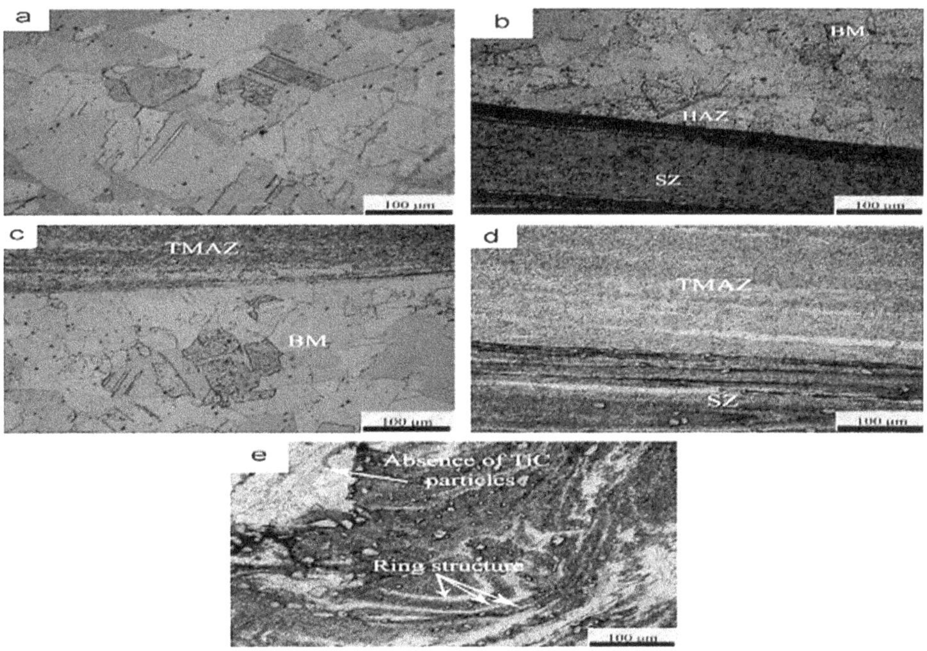

between TMAZ and SZ. The DRX (dynamic recrystallization) mechanism in SZ gives a microstructure, which consists of much finer grains as compared to the BM (Mishra R. S. et al. 2005). Figure 15(e) represents SZ with the dispersion of TiC particles in the Cu matrix. The stir zone showed a ring pattern because of the tool rotation rate and transverse speed (Asadi P.et al. 2011).

Sathis kumar et al. 2013 selected the tool rotation parameter by increasing the speed from 800 rpm to 1200 rpm and then studied the variation of microstructure on the same speed (varied speed) and presented it in Figure 16. The tool rotational speed ultimately affects and increased the surface area of the composite calculated from an image analyzing software and found to be 27 mm^2 at 800 rpm to 39 mm^2 at 1200 rpm. The effect of rotational speed of tool on the microstructure of Cu/B_4C composite surface is shown in Figure 17 (a, b, c, d & e), which shows the dispersion of B_4C particles in the Cu matrix. It is also observed from the Optical photomicrograph of Cu/B4C surface composite that the dispersion is not uniform at speed of 800 rpm because of the presence of B4C particles at some regions. Then at 900 rpm, 1000 rpm and 1100 rpm Figure 17 (a, b, c &d), The agglomerations were disappeared by increasing the tool rotational speed. The optical micrograph (shown in Figure 17 e) of surface composite fabricated at 1200 rpm shows excellent distribution of B4C particles.

Chang et al. 2007 studied the microstructure of the friction stir processed composite, after single-pass and recorded more uniform dispersion of the nano particles within the central cross-sectional area of the FSZ, as shown in Figure 18 (a). It is also observed that the clustered particle size is in the range of 0.1–2 mm, which is much more than that of the individual nano-particle size (approx. 20 nm). From 18 (b), some inhomogeneous areas of the particles can be seen just after single pass of FSP.

SUMMARY AND FUTURE SCOPE OF FRICTION STIR PROCESSING COMPOSITES

Presented review article is a brief introduction to a avenue which is widely adopted and needs to be focused. Latest developments in processes, its tools, geometrical features of tool, process parameters

Figure 16. Macrostructure of friction stir zone at tool rotational speed: (a) 800 rpm; (b) 900 rpm; (c) 1000 rpm; (d) 1100 rpm; and (e) 1200 rpm
Adapted from (Sathis kumar et al. 2013).

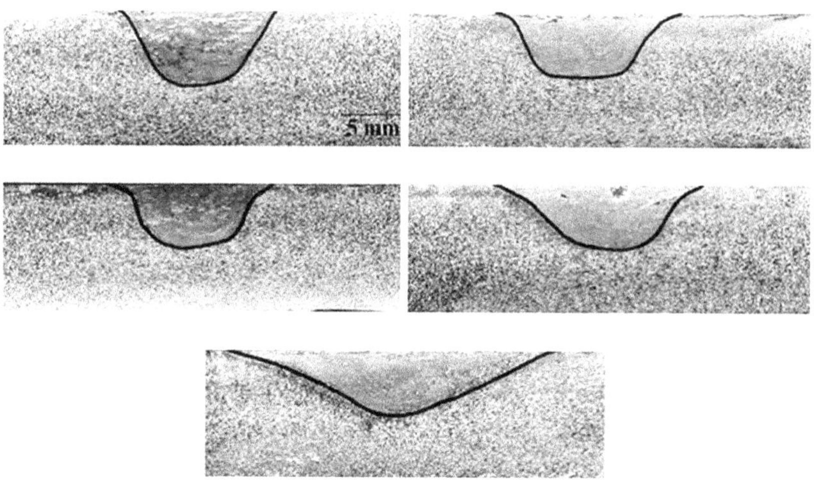

Figure 17. Optical photomicrograph of Cu/B4C surface composite at tool rotational speed: (a) 800 rpm; (b) 900 rpm; (c) 1000 rpm; (d) 1100 rpm; and (e) 1200 rpm
Adapted from (Sathis kumar et al. 2013).

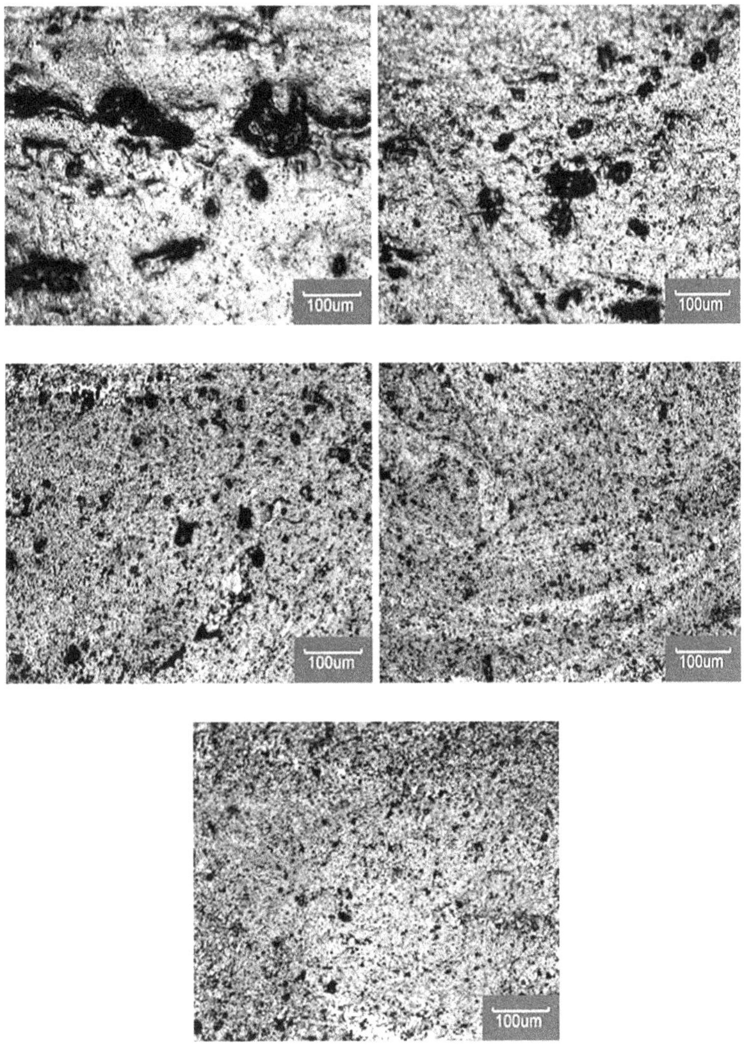

and their impacts growth of microstructure in the processing zone, generation of heat and its impacts and morphology of the studies are addressed.

The shape and geometry of the tool in FSP is an important factor to enhance the properties of the surface and the welds. The tool designing is still in a state of exploration and under study, hence are generally proprietary to individual researchers. The information available is too less in the open literature, however a few researchers have published optimum and progressive designs. Many tool materials have been examined by scholars, and observed that the choice of the tool material depends on the properties of base metal, reinforcement used and process parameters decided. The processing parameters such as tool rotation, traverse speed, axial force, and tool tilt angle are important features of the processing and must be optimized before use. Tool wear is an inevitable phenomenon and due focus is required to study and develop a standard design of tool for a specific purpose with minimum wear criteria.

Figure 18. SEM/SEI images of the AZ31/10 vol%ZrO2 FSP composite showing: (a) relatively homogeneous dispersion; and (b) local in homogenization of the nano-particle clusters within the stirred zone after one-pass FSP
Adapted from (Chang et al. 2007).

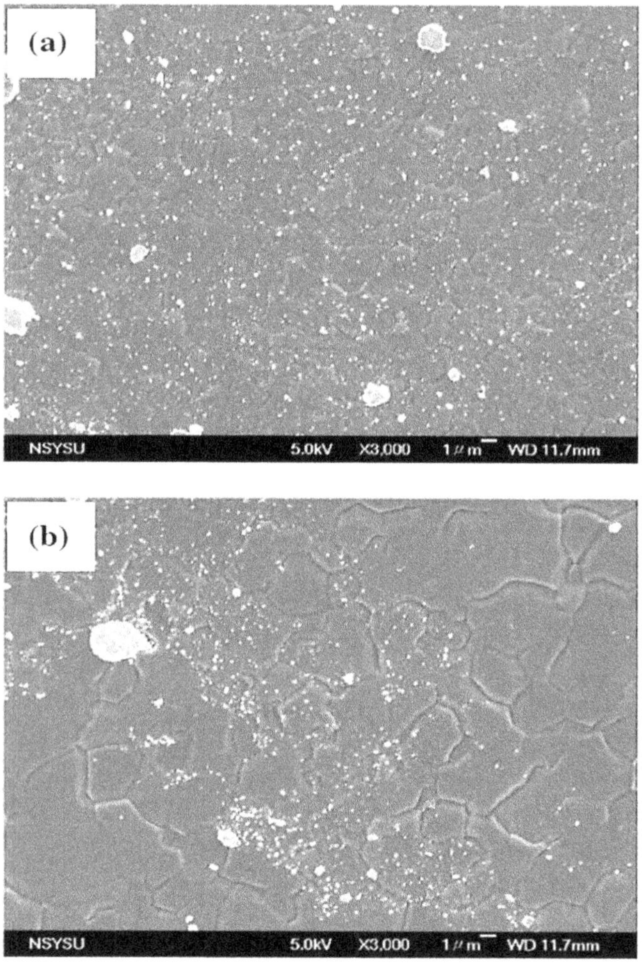

The flow of material below the tool shoulder around the pin is still complex to understand, however studies have elaborated the same regarding a few common materials. There is a scope of study to understand the same for other promising materials.

FSP, being a versatile technique of producing surface composites through refinement of grains at a temperature lower than that of the melting point of the material, has some unique merits. Surface composites with nano crystalline grains can be fabricated along with reinforcements if required. Most of the common metallic, ceramics, CNT and MWCNT can be easily and successfully incorporated without clustering. Also, homogeneous dispersion of reinforcement particles can be ensured with elimination of voids and cracks in the parent metals. FSP has opened a new avenue for Hybrid Composites, which may be of a hard along with a soft reinforcement, can be produced with ease. Concept of micro alloying is also have been addressed by the process and incorporation of Tin and lead are successfully examined. Fabrication of Polymers and composites of polymers were also initialized recently through this technique and successful results are obtained. Further investigations are awaited in the field of polymers and polymer composite with different reinforcements along with tool design and geometry. A.H. Feng and Z.Y. Ma et al. reported the use of Friction Stir processing for property enhancement of castings of Mg-Al-Zn. F.C. Liu et al. 2008 recorded an improvement in superplasticity which was 350–540%, with 200-350 ^{0}C temperature and with a strain rate of 3x10-2 s-1, with a much reduced grain size using FSP on composite of Al-8.9Zn-2.6Mg-0.09Sc. Jana et al. 2009 processed (FSP) a cast alloy of Al-7Si-0.6Mg and recorded a considerable improvement in its cast ability and mechanical properties, however observed a decrease in fatigue properties.

REFERENCES

AfsanehDorri. (2015). Mechanical and tribological properties of self-lubricating metal matrix nano composites reinforced by CNTs. *Composites. Part B, Engineering*, *77*, 402–420. doi:10.1016/j.compositesb.2015.03.014

Akos, M., & Imre, T. (2013). The Importance Of Friction Stir Welding Tool. *Production Processes and Systems.*, *6*(1), 25–34.

Asadi, P., Faraji, G., & Besharati, M. K. (2010). Producing of AZ91/SiC composite by friction stir processing (FSP). *International Journal of Advanced Manufacturing Technology*, *51*(1-4), 247–260. doi:10.1007/s00170-010-2600-z

Asadi, P., Givi, M. K. B., Abrinia, K., Taherishargh, M., & Salekrostam, R. (2011). Effects of SiC particle size and process parameters on the microstructure and hardness of AZ91/SiC composite layer fabricated by FSP. *Journal of Materials Engineering and Performance*, *20*(9), 1554–1562. doi:10.1007/s11665-011-9855-x

Atharifar, H., Lin, D. C., & Kovacevic, R. (2009). Numerical and experimental investigations on the loads carried by the tool during friction stir welding. *Journal of Materials Engineering and Performance*, *18*(4), 339–350. doi:10.1007/s11665-008-9298-1

Attia, A. N. (2001). Surface metal matrix composites. *Materials & Design*, *22*(6), 451–457. doi:10.1016/S0261-3069(00)00081-9

Badarinarayan, H., Yang, Q., & Zhua, S. (2009). Effect of tool geometry on static strength of friction stir spot-welded aluminum alloy. *Int. Mach. Tools Manuf.*, *49*(2), 142–148. doi:10.1016/j.ijmachtools.2008.09.004

Berbon, P. B., Bingel, W. H., Mishra, R. S., Bampton, C. C., & Mahoney, M. W. (2001). Friction stir processing: A tool to homogenize nanocomposite aluminum alloys. *Scripta Materialia*, *44*(1), 61–66. doi:10.1016/S1359-6462(00)00578-9

Besharati, M. K., & Asadi, P. (2014). *Advances in Friction Stir Welding and Processing*. Woodhead Publishing.

Bist, A., Saini, J. S., & Sharma, B. (2016). A review of tool wear prediction during friction stir welding of aluminium matrix composite. *Transactions of Nonferrous Metals Society of China*, *26*(8), 2003–2018. doi:10.1016/S1003-6326(16)64318-2

Chai, F., Zhang, D., & Li, Y. (2014). Effect of Thermal History on Microstructures and Mechanical Properties of AZ31 Magnesium Alloy Prepared by Friction Stir Processing. *Materials (Basel)*, *7*(3), 1573–1589. doi:10.3390/ma7031573 PMID:28788532

Chang, C. I., Lee, C. J., Chuang, C. H., Pei, H. R., & Huang, J. C. (2007). On Mg-Al-Zn Intermetallic Alloys Made by Friction Stir Processing Containing Quasi-Crystals or Amorphous Phases. Trans Tech Publications.

Chang, C. I., Wang, Y. N., Pei, H. R., Lee, C. J., & Huang, J. C. (2006). On the Hardening of Friction Stir Processed Mg-AZ31 Based Composites with 5–20% Nano-ZrO2 and Nano-SiO2 Particles. *Materials Transactions*, *47*(12), 2942–2949. doi:10.2320/matertrans.47.2942

Charit, I., & Mishra, R. S. (2005). Low temperature super plasticity in a friction-stir-processed ultrafine grained Al–Zn–Mg–Sc alloy. *Acta Materialia*, *53*(15), 4211–4223. doi:10.1016/j.actamat.2005.05.021

Choo, S.-H., Lee, S., & Kwon, S.-J. (1999). Effect of flux addition on the microstructure and hardness of TiC-reinforced ferrous surface composite layers fabricated by high-energy electron beam irradiation. *Metallurgical and Materials Transactions. A, Physical Metallurgy and Materials Science*, *30*, 1211. doi:10.1007/s11661-999-0271-x

Chowdhury, S. M., Chen, D. L., Bhole, S. D., & Cao, X. (2010). Effect of spin tool thread orientation on fatigue strength of friction stir welded AZ31B-H34 Mg butt joints. *Proc Eng.*, *2*(1), 825–833. doi:10.1016/j.proeng.2010.03.089

Chowdhury, S. M., Chen, D. L., Bhole, S. D., & Cao, X. (2010, August). Tensile properties of a friction stir welded magnesium alloy: Effect of spin tool thread orientation and weld pitch. *Materials Science and Engineering A*, *527*(21-22), 6064–6075. doi:10.1016/j.msea.2010.06.012

Colegrove, P. A., & Shercliff, H. R. (2004). Two-dimensional CFD modeling of flow round profiled FSW tooling. *Science and Technology of Welding and Joining*, *9*(6), 483–492. doi:10.1179/136217104225021832

Collier, M., Steel, R., Nelson, T., Sorensen, C., & Packer, S. (2003). Grade development of polycrystalline cubic boron nitride for friction stir processing of ferrous alloys. *J. Materials Science Forum, 426–432*, 3011–3016.

Darras, B. M., Khaishesh, M. K., Abu-Faraha, F. K., & Omar, M. A. (2007). Friction stir processing of commercial AZ31 magnesium alloy. *Journal of Materials Processing Technology*, *191*(1-3), 77–81. doi:10.1016/j.jmatprotec.2007.03.045

Dash Wood, R. J., Grimes, R., Harrison, A. W., & Flower, H. M. (2001). High Strain Rate Super plastic Aluminium Alloys: The Way Forward. *Materials Science Forum*, *339*, 357–359.

De Giorgi, M., Scialpi, A., Panella, F. W., & De Filippis, L. A. C. (2009). Effect of shoulder geometry on residual stress and fatigue properties of AA6082 fsw joints. *Journal of Mechanical Science and Technology*, *23*(1), 26–35. doi:10.1007/s12206-008-1006-4

Elangovan, K., & Balasubramanian, V. (2008). Influences of tool pin profile and welding speed on the formation of friction stir processing zone in AA2219 aluminium alloy. *Journal of Materials Processing Technology*, *200*(1-3), 163–175. doi:10.1016/j.jmatprotec.2007.09.019

Farrokhi, H., Heidarzadeh, A., & Saeid, T. (2013). Friction stir welding of copper under different welding parameters and media. *Science and Technology of Welding and Joining*, *18*(8), 697–702. doi:10.1179/1362171813Y.0000000148

Feng, A. H., & Ma, Z. Y. (2007). Formation of Cu2FeAl7 phase in friction stir welded Si Cp/Al-Cu-Mg composite. *Scripta Materialia*, *57*(12), 1113–1116. doi:10.1016/j.scriptamat.2007.08.020

Feng, A. H., & Ma, Z. Y. (2007). Enhanced mechanical properties of Mg–Al–Zn cast alloy via friction stir processing. *Scripta Materialia*, *56*(5), 397–400. doi:10.1016/j.scriptamat.2006.10.035

Fujii, H., Cui, L., Maeda, M., & Nogi, K. (2006). Effect of tool shape on mechanical properties and microstructure of friction stir welded aluminum alloys'. *Materials Science and Engineering A*, *419*(1–2), 25–31. doi:10.1016/j.msea.2005.11.045

Fujii, H., Cui, L., Maeda, M., Sato, Y. S., & Nogi, K. (2006). Effect of threads on tool in friction stir welding of aluminum alloys. *Adv. Struct. Funct.Mater.Des. Proc.*, *512*, 389–394.

Gan, Y. X., Solomon, D., & Reinbolt, M. (2010). Friction Stir Processing of Particle Reinforced Composite Materials. *Materials (Basel)*, *3*(1), 329–350. doi:10.3390/ma3010329

Hattingh, D. G., Blignault, C., van Niekerk, T. I., & James, M. N. (2008). Characterization of the influences of FSW tool geometry on welding forces and weld tensile strength using an instrumented tool. *Journal of Materials Processing Technology*, *203*(1–3), 46–57. doi:10.1016/j.jmatprotec.2007.10.028

Hodowany, J., Ravichandran, G., Rosakis, A. J., & Rosakis, P. (2000). Partition of plastic work into heat and stored energy in metals. *Experimental Mechanics*, *40*(2), 113–123. doi:10.1007/BF02325036

Hu, C., & Baker, T. N. (1997). A new aluminium silicon carbide formed in laser processing. *Journal of Materials Science*, *32*, 504–764.

Jana, S., Mishra, R. S., Baumann, J. B., & Grant, G. (2009). Effect of stress ratio on the fatigue behavior of a friction stir processed cast Al-Si-Mg alloy. *Scripta Materialia*, *61*(10), 992–995. doi:10.1016/j.scriptamat.2009.08.011

Kapoor, R., Kumar, N., Mishra, R. S., Huskamp, C. S., & Sankaran, K. K. (2010). Influence of fraction of high angle boundaries on the mechanical behavior of an ultra fine grained Al-Mg alloy. *Materials Science and Engineering A*, *527*(20), 5246–5254. doi:10.1016/j.msea.2010.04.086

Kapoor, R., & Nemat-Nasser, S. (1998). Determination of temperature rise during high strain-rate deformation. *Mechanics of Materials*, *27*(1), 1–12. doi:10.1016/S0167-6636(97)00036-7

Karthikeyam, L., & Senthilkumar, V. S. (2011). Relationship between process parameters and mechanical properties of friction stir processed AA 6063-T3 aluminium alloy. *Materials & Design*, *32*(5), 3085–3091. doi:10.1016/j.matdes.2010.12.049

Kawanami, S. (2010). Analysis of effect of tool geometry on plastic flow during friction stir spot welding using particle method. *Journal of Materials Processing Technology*, *210*(11), 1455–1463. doi:10.1016/j.jmatprotec.2010.04.003

Kumar, K., Kailas, S. V., & Srivatsan, T. S. (2008). Influence of tool geometry in friction stir welding. *Materials and Manufacturing Processes*, *23*(2), 189–195. doi:10.1080/10426910701774734

Kumar, N., Mishra, R. S., Huskamp, C. S., & Sankaran, K. K. (2011). Microstructure and mechanical properties of friction stir processed ultra fine grained Al-Mg-Sc alloy. *Materials Science and Engineering A*, *528*(18), 5883–5887. doi:10.1016/j.msea.2011.03.109

Lee, C. J., Huang, J. C., & Hsieh, P. J. (2006). Mg based nano-composites fabricated by friction stir processing. *Scripta Materialia*, *54*(7), 1415–1420. doi:10.1016/j.scriptamat.2005.11.056

Lee, C. J., Huang, J. C., & Hsieh, P. L. (2006). Mg based nano-composites fabricated by friction stir processing. *Compos.Mater.*, *313*, 69.

Lienert, T. J., Stellwag, W. L. Jr., Grimmett, B. B., &Warke, R. W. (2003). Friction Stir Welding Studies on Mild Steel. *Welding Journal*, 48-59.

Liu, F. C., & Ma, Z. Y. (2008, October). Achieving exceptionally high superplasticity at high strain rates in a micro grained Al–Mg–Sc alloy produced by friction stir processing. *Scripta Materialia*, *59*(8), 882–885. doi:10.1016/j.scriptamat.2008.06.035

Lorrain, O., Favier, V., Zahrouni, H., & Lawrjaniec, D. (2010). Understanding the material flow path of friction stir welding process using unthreaded tools. *Journal of Materials Processing Technology*, *210*(4), 603–609. doi:10.1016/j.jmatprotec.2009.11.005

Ma, Z. Y., Mishra, R. S., & Mahoney, M. W. (2002). Super Plastic Deformation Behavior of Friction Stir processed 7075 Aluminium alloy. *Acta Materialia*, *50*(17), 4419–4430. doi:10.1016/S1359-6454(02)00278-1

Mahoney, M. W., Binget, W. H., & Mishra, R. S. (2003). Microstructural Modification and Resultant Properties of Friction Stir Processed Cast NiAl Bronze. *Materials Science Forum*, *426*(432), 2843–2853. doi:10.4028/www.scientific.net/MSF.426-432.2843

Miranda, R. M., Gandra, J., & Vilaça, P. (2013). *Surface Modification by Friction Based Processes*. Intech Open.

Mishra, R. S., & Kumar, N. (2014). *Friction Stir Welding and Processing Science and Engineering*. Springer.

Mishra, R. S., & Ma, Z. Y. (2005). Friction stir welding and processing. *Materials Science and Engineering R Reports*, *50*, 1–78.

Mishra, R. S., & Ma, Z. Y. (2005). Friction stir welding and processing. *Materials Science and Engineering A*, *50*(1–2), 1–78.

Mishra, R. S., Ma, Z. Y., & Charit, I. (2003). Friction stir processing: A novel technique for fabrication of surface composite. *Materials Science and Engineering A*, *341*(1), 307–310. doi:10.1016/S0921-5093(02)00199-5

Mishra, R. S., Mahoney, M., McFadden, S., Mara, N., & Mukherjee, A. (1999). High strain rate superplasticity in a friction stir processed 7075 Al alloy. *Scripta Materialia*, *42*(2), 163–168. doi:10.1016/S1359-6462(99)00329-2

Mishra, R. S., & Mahoney, M. W. (2001). Friction stir processing: A new grain refinement technique to achieve high strain rate superplasticity in commercial alloys. In Materials Science Forum. Trans Tech Publ. doi:10.4028/www.scientific.net/MSF.357-359.507

Mishra, R. S., & Mahoney, M. W. (Eds.). (2007). *Friction stir welding and processing*. Materials Park, OH: ASM International.

Mishra, R. S., Mahoney, M. W., McFadden, S. X., Mara, N. A., & Mukherjee, A. K. (2000). High strain rate super plasticity in a friction stir processed 7075 Al alloy. *Scripta Materialia*, *42*(2), 163–171. doi:10.1016/S1359-6462(99)00329-2

Morisada, Y., Fujii, H., Nagaoka, T., & Fukusumi, M. (2006). MWCNTs/AZ31 surface composites fabricated by friction stir processing. *Materials Science and Engineering A*, *419*(1-2), 344–348. doi:10.1016/j.msea.2006.01.016

Najafi, M., Nasiri, A. M., & Kokabi, A. H. (2008). Microstructure and hardness of friction stir processed AZ31 with Si Cp. *International Journal of Modern Physics B*, *22B*(18–19), 2879–2885. doi:10.1142/S0217979208047717

Nandan, R., Deb Roy, T., & Bhadeshia, H. (2008). Recent advances in friction-stir welding – process, weldment structure and properties. *Progress in Materials Science*, *53*(6), 980–1023. doi:10.1016/j.pmatsci.2008.05.001

Newkirk, J.W., Mishra, R.S., & Ply, J. (2005). Advances in Powder Metallurgy and Particulate Materials. *MPIF, 6*, 62.

Park, S., Sato, Y., Kokawa, H., Okamoto, K., Hirano, S., & Inagaki, M. (2009). Boride formation induced by pcBN tool wear in friction-stir-welded stainless steels. *Metallurgical and Materials Transactions. A, Physical Metallurgy and Materials Science*, *40A*(3), 625–636. doi:10.1007/s11661-008-9709-9

Pirondi, A., Collini, L., & Fersini, D. (2008). Fracture and fatigue crack growth behavior of PMMC friction stir welded butt joints. *Engineering Fracture Mechanics*, *75*(15), 4333–4342. doi:10.1016/j.engfracmech.2008.05.001

Prater, T., Strauss Alvin, M., Cook George, E., Gibson Brian, T., & Cox Chase, D. (2013). A;Comparative evaluation of the wear resistance of various tool materials in friction stirs welding of metal matrix composites. *Journal of Materials Engineering and Performance*, *22*(6), 1807–1813. doi:10.1007/s11665-012-0468-9

Rai, R., De, A., Bhadeshia, H., & Deb Roy, T. (2011). Review: Friction stir welding tools. *Science and Technology of Welding and Joining*, *16*(4), 325–342. doi:10.1179/1362171811Y.0000000023

Ramachandran, K. K., Murugan, N., & Shashi Kumar, S. (2014). An assessment on friction stir welding of high melting temperature materials. *AMM*, *592*, 43–47. doi:10.4028/www.scientific.net/AMM.592-594.43

Reshad Seighalani, K., Besharati Givi, M. K., Nasiri, A. M., & Bahemmat, P. (2010). Investigations on the effects of the tool material, geometry and tilt angle on friction stir welding of pure titanium. *Journal of Materials Engineering and Performance*, *19*(7), 955–962. doi:10.1007/s11665-009-9582-8

Rhodes, C. G., Mahoney, M. W., Bingal, W., Spurling, R. A., & Bamption, C. C. (1997). Effects of friction stir welding on microstructure of 7075 aluminum. *Scripta Materialia*, *36*(1), 69–75. doi:10.1016/S1359-6462(96)00344-2

Sabbaghian, M., Shamanian, M., Akramifard, H. R., & Esmailzadeh, M. (2014). Effect of friction stir processing on the microstructure and mechanical properties of Cu–TiC composite. *Ceramics International*, *40*(8), 12969–12976. doi:10.1016/j.ceramint.2014.04.158

Saito, N., Shigematsu, I., Komaya, T., Tamaki, T., Yamauchi, G., & Nakamura, M. (2001). Grain refinement of 1050 aluminum alloy by friction stir processing. *Journal of Materials Science Letters*, *20*(20), 1913–1918. doi:10.1023/A:1012882413401

Santella, M. L., Engstrom, T., Storjohann, D., & Pan, T. Y. (2005). Effect of friction stir processing on mechanical properties of the cast aluminium alloys A319 and A356. *Scripta Materialia*, *53*(2), 201–206. doi:10.1016/j.scriptamat.2005.03.040

Sathiskumar, R., Murugan, N., Dinaharan, I., & Vijay, S. J. (2013). Role of friction stir processing parameters on microstructure and micro hardness of boron carbide particulate reinforced copper surface composites. *Sadhana*, *38*(6), 1433–1450. doi:10.1007/s12046-013-0184-7

Shafiei-Zarghani, A., Kashani-Bozorg, S. F., & Zarei-Hanzaki, A. (2009). Microstructures and mechanical properties of Al/Al2 O 3 surface nano-composite layer produced by friction stir processing. A. *Materials Science and Engineering A*, *500*(1-2), 84–91. doi:10.1016/j.msea.2008.09.064

Sorensen, C. D., & Nelson, T. W. (2007). Friction stir welding of ferrous and nickel alloys in friction stir welding and processing. ASM International, 111–121.

Sorensen, C. D., & Nielsen, B. (2009). Exploring geometry effects for convex scrolled shoulder, step spiral probe FSW tools. *Proc. TMS Annual Meet.*, 85–92.

Sorensen, C. D., & Stahl, A. L. (2007). Experimental measurements of load distributions on friction stir weld pin tools. *Metallurgical and Materials Transactions. B, Process Metallurgy and Materials Processing Science*, *38B*(3), 451–459. doi:10.1007/s11663-007-9041-6

Sorenson, C. D., Nelson, T. W., Packer, S. M., & Steel, R. J. (2004). Innovative technology applications in FSW of high softening temperature materials. *Proc. 5th Int. Conf. on 'Friction stir welding'*.

Spowart, J. E., Ma, Z. Y., & Mishra, R. S. (2003). The effect of friction stir processing (FSP) on the spatial heterogeneity of discontinuously-reinforced aluminum (DRA) microstructures. Friction Stir Welding and Processing II, 243.

Sun, N., Yin, Y. H., Gerlich, A. P., & North, T. H. (2009). Tool design and stir zone grain size in AZ31 friction stir spot welds. *Science and Technology of Welding and Joining*, *14*(8), 747–752. doi:10.1179/136217109X12518083193559

Tewari, A., Spowart, J.E., Gokhale, A.M., Mishra, R.S., & Miracle, D.B. (n.d.). Characterization of the effects of friction stir processing on microstructural changes in DRA composites. *Mater. Sci. Eng. A: Struct. Mater. Prop. Micro struct. Process., 428*, 80.

Thomas, W. M., Nicholas, E. D., Needham, J. C., & Church, M. G. (1991). *International Patent Application No. PCT/GB92/02203 and GB Patent Application No. 9125978.8*. TWI.

Vijay, S. J., & Murugan, N. (2010). Influence of tool pin profile on the metallurgical and mechanical properties of friction stir welded Al– 10 wt.% TiB2 metal matrix composite. *Materials & Design*, *31*(7), 3585–3589. doi:10.1016/j.matdes.2010.01.018

Wang, Y. S., Zhang, X. Y., Zeng, G. T., & Li, F. C. (2001). In situ production of Fe–VC and Fe–TiC surface composites by cast-sintering. *Composites. Part A, Applied Science and Manufacturing*, *32*, 281.

Zhang, Y., Sato, Y. S., Kokawa, H., Park, S. H. C., & Hirano, S. (2008). Stir zone microstructure of commercial purity titanium friction stir welded using pcBN tool. *Materials Science and Engineering A*, *488*(1-2), 25–30. doi:10.1016/j.msea.2007.10.062

Zheng, J., Mishra, R. S., Berbon, P. B., & Mahoney, M. W. (2001). *Microstructure and mechanical behavior of friction stir processed Al-Ti-Cu alloy. Friction Stir Welding and Processing.*

Chapter 9
Development of Natural Bio-Filler-Based Epoxy Composite for Wind Turbine Blade Application:
Wood Epoxy Composite as Wind Turbine Blade Material

Rahul Kumar
National Institute of Technology Silchar, India

Sumit Bhowmik
National Institute of Technology Silchar, India

ABSTRACT

Nowadays, the research and engineering attention stimulated towards development of environmentally gentle materials to satisfy the energy needs of the society through renewable resources. The growing cognizance in the production and consumption of renewable energy for civilization necessities has directed towards the growth in the wind energy utilization. The wind energy is a leading renewable and sustainable energy resource and key answer to the global energy problem. The rotor blades of wind turbines are its integral parts and traditional materials used for blade manufacturing are carbon or glass fibre reinforced polymer composites owing to their low density and high strength to stiffness ratio. But the non renewability and adverse environmental effect during their processing and disposal forced the researchers to look out for some biodegradable and light weight natural plant fibres for reinforcement in polymeric resin to produce required polymer composites. In the present work, application of bio filler based epoxy composite is proposed to be used as wind turbine rotor blades.

DOI: 10.4018/978-1-5225-3401-3.ch009

INTRODUCTION

For the last few decades, the composite materials based on natural cellulosic fibers and polymer resins are increasingly being developed for various applications as replacements for synthetic materials derived from different kinds of petroleum resources. Interest in natural fibers has increased worldwide owing to their excellent price performance ratio, high strength to weight ratio, abundant availability and recyclability, good thermal and mechanical resistivity and to their environmental friendliness [Thakur et al., 2012; and Bhowmick et al., 2012]. The environmental problems associated with the production and use of synthetic fibers has changed the scenario. Natural fibers are now considered as favourable alternative to their manmade counterparts for use in various fields. Natural fibres reinforced polymer composites enjoyed a cult status in this ecological-economical balance owing to the biodegradable and sustainable origin of natural fibres. The natural fibres are obtained either directly from cultivated agricultural plants and crops or in form of secondary by-products after processing of primary crops [Kumar et al., 2014; and Courtenay et al., 2017]. Examples of different natural fibres utilized in polymer composites are wood, sisal, hemp, bamboo, coir, flax, pineapple leaf, kenaf, jute and so forth. The natural polymer based materials are found abundantly in nature and also renewable and biodegradable in nature [Nopparut and Amornsakchai, 2016; and Prabhakaran & Thirumalai, 2014]. The biodegradability of the natural fibers contributes towards formation of a healthy ecosystem, while their lower expense favours the economic aspect of industries. Fibre reinforced polymer composites have been acquainted with the wind industries since a long ago for the manufacturing of wind turbine blades. These fibre reinforced polymer composites are principally utilized for the production of small to large size wind turbine blades. Fibre reinforced plastics are the obvious material selection for wind turbine rotor blades due to their large strength and stiffness to weight ratio compared with conventional structural materials [Shah et al., 2013; and Park, 2015]. During the recent decade, much effort has been put on the improvement of the mechanical properties of these composite materials in order to reduce the blade weight by materials savings.

The some basic properties of materials to be used as wind turbine blade material, is like greater specific stiffness and tensile strength, lower mass per unit volume and ample endurance limit. The rotor blade should have adequate value of strength and resistance to deflection to endure under severe environmental condition without breaking and bending [Nielsen & Landel, 1994]. Also, the time varying cyclic loading has crucial life defining impact on projected life cycle of produced rotor blades, therefore blades must have sufficient fatigue strength to sustain. These basic material criteria in addition to eco-friendly nature for rotor blades can be met by natural fibre reinforced polymer composites. The discussed literature background suggests the potential application of natural fibre/filler reinforced polymer composites as rotor blade material and the determination of optimal value of fibre and matrix content to satisfy such strength requirement is need of the hour. Therefore, the correlation between the required mechanical properties and the characteristic parameters, e.g., the composition of the composite and the working environments is of major importance for designing proper composites in order to satisfy various practical necessities. The optimization of such characteristic parameters is very much important.

Therefore in the present chapter, application of bio filler based epoxy composite was proposed to be used as wind turbine rotor blades. In Section. 2, brief introduction of composite materials and along with natural fibre and their properties is given. In Section. 3, wind turbine, its components followed by exhaustive reviews on parts manufactured with synthetic fibre polymer composites by the previous researchers is discussed. In Section. 4, detailed fundamental scientific aspects of wind turbine and its blade preparation including their working principle, manufacturing methods, parts assembly, advantages, limitation and its applications is presented. At last, before ending up the chapter with conclusions and future scope, an experimental case study on the development of natural filler based epoxy composites and its parametric optimization for optimum mechanical properties is discussed in Section. 5.

COMPOSITE MATERIALS

Composite materials are formed by combining two or more component materials with considerably dissimilar physical or chemical properties, when combined; the characteristics of the material obtained differ from the singular constituents. Composites are the advanced materials that overcome the conflicting properties of their constituent materials to produce more synergetic material [Herrera-Franco and Valadez-Gonza´lez, 2004; and Ashori, 2008]. The two distinct phase namely, discontinuous phase or reinforcement material and a continuous phase or matrix material constitutes the formation of composite material. The position, orientation and alignment of reinforcing phase affects the properties of composite material. Usually, the fibres as reinforcement are blended with the matrix to get the easy load transfer, resulting improved modulus and mechanical strength of resulting composite material. The matrix phase acts a binder for distinctly placed and oriented fibres. In this way, it shields fibrous material from chemical and environmental attacks. The fibre reinforced polymer composites employed in various industrial and general fields like marine, construction, aerospace and automobile sectors due to their light weight and higher strength-modulus combination [Netravali and Chabba, 2003; and Sapuan and Maleque, 2005]. The reinforcement in the composite materials may be of synthetic fibre, natural fibre or some kind of nano or micro sized particle fillers.

The term "natural fiber" conceals an extensive collection of plant, mineral and mammal fibers. Nevertheless, for the reinforcement in polymer matrices, generally plant based agro-waste fibres are used. The plant based bio-fibres can be derived from different parts of a plant like stem, leaf, seed, bast and fruits. These fibres are of significant mechanical strength due to their contribution towards the structural development of trees. Also on reinforcement, they can transfer substantial strength to synergise the final composite properties. Despite of this awareness and ecological charm, natural fibres are restricted to indoor and non-structural application due to their moisture absorption tendency, lower strength and thermal stability in comparison with their synthetic counterparts. This moisture absorption drawback can be substantially overcome by different physical and chemical surface treatment methods. Also, strength deficiencies of fibre reinforced polymer composites can be improved by orienting and placing a number of fibres in the required loading direction [Doan et al., 2012; Kumar et al., 2017; Mittal et al., 2016; Nagarajan et al., 2016; and Joseph et al., 2006].

PROPERTIES OF NATURAL FIBRES

- Natural fibres are lighter in weight which makes them suitable to be used as reinforcement in polymer matrix for the production of aircraft and aerospace materials
- High strength/stiffness to weight ratio;
- Abundant availability;
- Low CO_2 emission;
- Relative ease of waste disposal;

- Biodegradability and recyclability
- Low cost and processing time.
- Compostable and environment friendly
- Great flexibility and can be used to reinforce in any matrix material.

NATURAL FIBRE REINFORCED POLYMER COMPOSITES

Polymeric composites being the one of the most important engineering materials, has witnessed remarkable shift in the reinforcing element from synthetic fibre to natural cellullosic fibres to produce environmentally gentle composite materials. These natural cellulosic fibres traditionally used as construction materials, clothing and as source of energy but nowadays, due to the noticeable increase on awareness for environment and energy consumed, greater devotion is being transferred to natural fibers with a view to conserving energy and protecting the environment [Ishizaki et al., 2008; and Fu et al., 2008]. The main governing factor of mechanical properties of synthetic/natural fibre/filler reinforced polymer composites are weight/volume fraction of fibre/fillers, fibre size and aspect ratio, fibre placement and orientation, interfacial adhesion and load transfer between fibres to matrix. Maximum number of research works on natural fibre/filler composites concentrated on the variation of mechanical properties as a function of fibre/filler loading, fibre surface modifications and usage of several coupling agents. The synergetic combination and individual properties of fibres and matrix both equally contribute in increasing mechanical properties of the polymer composites [Rozman, 2000]. The reinforcing fibre properties mainly determine the stiffness and matrix properties control the ultimate tensile strength of produced composite materials. A strong interface between fibre-matrix and proper fiber orientation are required to improve the tensile strength, whereas weight fraction of fibers, fiber dispersion in the matrix phase, and greater fiber aspect ratio regulate the tensile modulus. The composite fracture properties is highly dependent on fiber aspect ratio.

The critical fibre length in case of short fibre reinforcement in polymer matrix composite is the most critical parameter determining the stress developing ability of the material. The fibre length comparatively smaller than this critical length resulted in failure of material relatively at lower stress value, owing to the debonding of fibre-matrix interface. Quite the reverse, greater fiber length than the critical length, the fiber is strained under applied load and thus lead to higher mechanical strength of the composite [Ku et al., 2011; and Ghenai, 2012]. The recent trends show the lignocellulosic materials have been the subject of exhaustive studies, either in substituting the current wood species in building traditional panel products or fabricating polymeric composite materials for turbine blade application.

Wood has been extensively used in development of wind turbine blades since the early stage. The rotor blades are comprehensively manufactured from plain wood and laminated wood materials. The boat building concept and procedure is the source of inspiration for plain wood rotor blades [Kumar et al., 2014; Courtenay et al., 2017; and Nopparut & Amornsakchai, 2016]. However the hydrophilic character of wood based material created a severe problem for life long service apart from performance point of view. This moisture absorption problem of plain wood rotor blades can be overcome by using laminated wood-epoxy composite, which also has low weight and adequate tensile strength. Another important natural fibres utilized in blade manufacturing is bamboo fibres, also known natural glass fibre.

The decomposable nature and higher specific mechanical strength of bamboo forces researcher to use it in development of bamboo-epoxy laminate for rotor blade materials [Prabhakaran & Thirumalai, 2014]. The bamboo fibres are also used in the aero shells and for sandwich type structure within the rotor blades.

WIND TURBINES AND ITS COMPONENTS

The growing cognizance in the production and consumption of renewable energy for civilization necessities has directed towards the growth in the wind energy utilization. The wind energy is a leading renewable and sustainable energy resource and key answer to the global energy problem. Several wind turbines or mills have been installed for converting the useful kinetic energy associated with the wind into first mechanical and then electrical energy. The rotor blades of wind turbines are its integral parts and the traditional materials used for blade manufacturing are carbon or glass fibre reinforced polymer composites owing to their low density and high strength to stiffness ratio. The basic structure of wind turbines comprises of the three rotor blades revolve around a horizontal hub, fixed on slender structure column. These rotor blades rotates as blowing wind exerts forces and their rotation transforms wind energy into mechanical and subsequently into electrical energy. The following are the different components of wind turbines.

- Tower and foundations
- Shaft
- Gear box
- Generator
- Control and protection system
- Rotor blade

The wind generator in wind turbine setup is supported by a concrete or steel tower in condition of strong wind wave, much desirable for producing significant amount of electricity [Puglia, 2013]. Generally, a tubular construction of concrete or steel is used. The wind-turbine shaft is connected to the centre of the rotor. When the rotor spins, the shaft spins as well. The gearbox is coupled with high speed shaft and in this way, motorised power of rotor is transferred to the generator [Cox and Echtermeyer, 2012]. The forged or cast iron is mainly used for building the main shaft of wind turbine. A wind turbine may need a gear system which consists of several types of gears, depending on its generator type [Abdin and Xu, 1998]. The rotational speed of the input torque is increased by means of a gearbox system. The torque to gearbox is delivered by the low speed shaft, which is connected through hub. At its most basic, a generator is a pretty modest device. It usages the properties of electromagnetic induction to produce electrical voltage - a difference in electrical charge. An intelligent control system is installed to optimize the functionality of the wind turbine. The controller restricts the applied loads on associated structural member and improves the power generation [Wind Energy Technology, 2009]. The rotor blades of a wind turbine are similar to the wings of a glider. The wind turbine utilizes the lift of the rotor blades to make the rotor turn.

Rotor Blades

The blades are basically the navigator of the system; in their humblest form, they perform as obstructions to the wind (more modern blade designs go beyond the barrier method) (as shown in Figure 1). When the wind compels the blades to rotate, it has conveyed some of its energy to the rotor. A blade has an air foil shape which directs the wind forces to the turbine low-speed shaft. The typical structure of wind turbine is Three-bladed horizontal-axis, but two blades downwind configured wind turbines are also used. The lift forces are the driving force for creating torque in wind turbine rotors. These lift forces created over the rotor blades due the pressure difference.

Two different construction methodologies for wind turbine blade production are used such as through single piece or multi-piece adhesive construction [Griffin and Ashwill, 2003; and Watson and Serrano, 2011]. For the single piece construction, inner shell is fixed with a spar and then with the help of adhesive joining, it attached to the outer shell, whereas for multi-piece adhesive construction, the shear webs are adhesively joined with the shells.

The traditional baseline material selection for wind turbine rotor blades is glass, which has been used since the early days of blade manufacturing. In recent years as blades are getting longer, the need for a stiffer and stronger fibre material has emerged. Carbon is the most apparent option, but the additional cost compared with glass needs to be balanced with the increase in mechanical performance [Holmes, 2009]. Hybrid materials (like combinations of glass and carbon) are also coming into play. However, the sustainability and ecological concern has forced the private organizations and governments to concentrate towards the development of more environment friendly materials for rotor blades. In this regard, the most obvious alternative for this potential application is natural fibres like sisal, wood, flax, bamboo, laminated wood [Rocha et al., 2007; and Chikhradze et al., 2015].

Figure 1. Wind turbine and rotor blades

Challenges and Suggestions

The importance of the use of natural fibre reinforced epoxy composite material as wind turbine blade has previously been deliberated in the aforementioned segment. The research and engineering communities are constantly concentrating their research accomplishments on the development of commercially viable high quality natural fibre reinforced plastics for rotor blades. But, there are many challenges associated with the usage of natural fibre reinforced polymer composites as wind turbine blades [Al-Khudairi and Ghasemnejad, 2015; and Debnath et al., 2013]. The major difficulties came across for the utilization of natural fibre reinforced polymer composites are highlighted as follows:

- Availability of varieties of natural fibres and matrix material leading to their blending, processing, fabrication and curing depending on different set input parameter, which in turn increases the complexity for production of natural fibre reinforced composites.
- The change in geographical location and surrounding environment of growth of natural fibre plants modifies the required physical and mechanical properties.
- The weak adhesion and moisture absorption property of natural fibres in compare to synthetic glass/aramid/carbon fibres resulted in the weak bonding between fibres and matrix. This lead to internal defects and lower mechanical properties of the final composite parts produced.
- The lower thermal stability of natural fibres is also a grave concern for its utilization in structural applications.
- The presence of different phases and interfaces in fibre reinforced polymer composites make the machining operation a very challenging task. The machining performance of composite is also significantly affected by weight fraction and orientation of fibres, type of matrix material.

To overcome the several challenges discussed above, some procedures have been established as following.

- Use of diverse range of plasma, alkali, silane treatment and so forth to modify the natural fibre surfaces with the aim of improving the bond between fibre and matrix and increase thermal stability of natural plant based fibres.
- Use of different coupling agent to improve the load transfer capacity between fibre and matrix interfaces. Coating of natural fibre surfaces and addition of fillers and catalysts can also be potential solution for different problems.
- Use of adhesive joining methodology in place of machining operation. Mainly two types of adhesives namely thermosetting polymer based adhesive and thermoplastic based adhesive are used for joining purpose.

EXPERIMENTAL STUDY OF WOOD FILLER REINFORCED EPOXY COMPOSITE AND ITS PARAMETRIC OPTIMIZATION

Background

It is a true fact that the scientific and industrial development be governed by the developments in the area of materials. It is not compulsory be an expert to realise the most advanced rocket design or boiler in power plants are of no usage if appropriate materials to endure the working stresses and surrounding environments are not available. The advancement in any field depends on proper materials selection. Composite materials in this consideration represent nothing but a giant step in the ever constant endeavour of optimization in materials. The concept of use and development composite materials is not a novel or fresh one. Nature has plenty of illustrations wherein the knowledge of composite materials is incorporated. The coconut palm leaf is one of the most popular example of a cantilever using the concept of fiber reinforcement. Wood is a fibrous composite consists of cellulose fibers in a lignin matrix. The cellulose fibers have high tensile strength but are very flexible (i.e. low stiffness), while the lignin matrix combines the fibers and set up the stiffness. Bone is yet another example of a natural composite that holds the weight of various parts of the body. It consists of short and soft collagen fibers embedded in a mineral matrix called apatite. Therefore, composite materials plays an important role in the modern industry. The ever increasing concern for ecological problem and requirement of light weight at the same time environmentally gentle polymer based materials resulted in the renewed attention towards utilization of natural plant based organic fillers as reinforcement in polymer matrices. Also known as green or bio composites, these natural fibre/filler based composites have witnessed an exponential growth owing to their ample availability as agro-waste and biodegradability. The renewability and biodegradability of the natural fibers contributes in the creation of a healthy ecosystem, whereas lower price due to agro-waste favours the monetary concern of manufacturing sector.

Materials

For the present work, the room temperature curing epoxy adhesive resin (Araldite LY 556, density 1.26 gm/cc) of the 'epoxide' family and recommended hardener (HY 951IN) are blended in a ratio of 10:8 by weight. The wood filler is selected as reinforcing medium owing to its greater availability, cellulose and lignin content, heat resistance, whereas epoxy as matrix material is chosen because of its high viscosity, negligible dimensional shrinkage after curing and higher tensile strength and modulus. The wood filler is obtained from local resources during the cutting of tree wood and sieved with multiple stacked test sieves to obtain the required filler particle size. The particle fillers are then washed with double distilled water to remove the dust and other unwanted components mixed during cutting. The matrix material utilized in the present work, has viscosity of 30-50 Pa.s at 25 °C, which is sufficient to prevent fillers from settling down at the bottom after mixing. The wood filler particles with particle size lesser than 2 μm and density of 0.779 gm/cc after washing and oven drying are mixed in epoxy adhesive resin to manufacture the composites specimen. The malleable mixture of wood filler and epoxy resin is then stirred with mechanical stirrer and slowly emptied into the glass mould. The glass mould is placed in the vacuum chamber and left for curing for 24 hours at room temperature. The glass mould was prepared with the required sample dimension of 125 mm × 25 mm × 5 mm and smeared with transparent paper for easy removal of specimen after curing. Afterward the wood filler reinforced epoxy composite samples are taken out from the mould and subjected to different mechanical test.

Mechanical Tests and Parametric Optimization

The tensile and flexural tests for produced wood filler based epoxy composite samples are carried out according to ASTM D 638-03 and ASTM790-03 using Universal testing machine (UTM) with load cell of 50 kN. Three different crosshead movement speed of 1 mm/min, 2 mm/min and 3 mm/min are taken to study the strain rate effect on mechanical properties. The filler wt. % used for reinforcement in epoxy matrix are 0%, 2.5%, 5%, 7.5%, 10% and 12.5%. For tensile test, gauge length and width of narrow section was considered as 50 mm and 13 mm respectively. There are six number of specimen subjected to test for each filler wt. % to attain the repeatability in results. Standard laboratory condition is 24 °C temperature and 53% relative humidity. The span length used for flexural test is 50 mm.

The genetic algorithm is a class of evolutionary algorithms (EA) which is based on the natural selection, a process that defines biological evolution. It is used to find the optimal solution by following the bio-inspired operations such as mutation, crossover and selection. The genetic algorithm repeatedly modifies a population of individual solutions. At each step, the genetic algorithm selects individuals at random from the current population to be parents and uses them to produce the children for the next generation [Mishra et al., 2002]. Each individual (or candidate solution) is assigned a fitness value (based on its objective function value) and the fitter individuals are given a higher chance to mate and yield more "fitter" individuals. This is in line with the Darwinian Theory of "Survival of the Fittest". Over successive generations, the population "evolves" toward an optimal solution [Mylsamy and Rajendran, 2011].

Genetic algorithm (GA) was utilized to find the optimum values of process parameters [Cheng, 1997; and Michalewicz, 1999] i.e., filler content (f1) and speed (f2) using nonlinear regression models for mechanical properties to get the maximum load, tensile stress and flexural stress values for the developed composite material. Hence the optimization problem can be framed as shown in equation 1.

$$Max\, f(X),\ \text{Subjected to}\ x_i^{(L)} \le x_i \le x_i^{(U)},\ x_i = f_1, f_2 \tag{1}$$

Initial randomly produced chromosome population at the start of optimization process is 20. The following linear mapping rule is used to decode the genes of each chromosome (Xi) (equation 2),

$$X_i = x_i + \frac{x_i^{(U)} - x_i^{(L)}}{2^{l_i - 1}} decoded\,value\left(s_i\right) \tag{2}$$

where, x_i = filler content (f_1) and speed (f_2), L = minimum limit, U = maximum limit, l_i= number of binary bits, and s_i = decimal equivalent of binary values.

The chromosome fitness is obtained by linear mapping with the following general equation (equation 3),

$$F(X) = f(X)\ \text{for maximization problem} \tag{3}$$

The objective functions are subjected to the following constraints,

Lower bound = [0 1], Upper bound = [12.5 3].

RESULTS AND DISCUSSIONS

The tensile properties variation with respect to change in crosshead speed and filler loading are presented in Figure 2 and Figure 3. The load and stress variation of composite samples along with neat epoxy specimen in tensile loading are depicted in Figure 2 and Figure 3 respectively. It can be observed from the presented results that as the filler loading increases, tensile strength of composite samples as well as neat epoxy specimen increase and becomes highest at 10 wt. % of filler content. After that aforementioned properties started to drop and attained lowest value at 12.5% of filler content except for elastic modulus values which is minimum for neat polymer sample.

The tensile loads achieved its maximum and minimum values for crosshead speed of 1 mm/min at 10% and 12.5% of filler loading respectively. Their individual values are 3.55 kN and 1.48 kN. The highest and lowest values of ultimate tensile strengths at the speed of 1 mm/min are 28.29 MPa and 10.83 MPa at 10% and 12.5% filler wt. respectively. The greatest and lowest values for elastic modulus are 19.10 Mpa and 2.69 MPa respectively for 10 wt. % and 0 wt. % of filler loading. As the speed increased from 1 mm/min to 3 mm/min, strength of prepared material increases with higher filler loading. The reason may be associated with the time the fillers get to orient themselves in loading direction. Similar values

Figure 2. Variation of load at different speeds with % filler wt

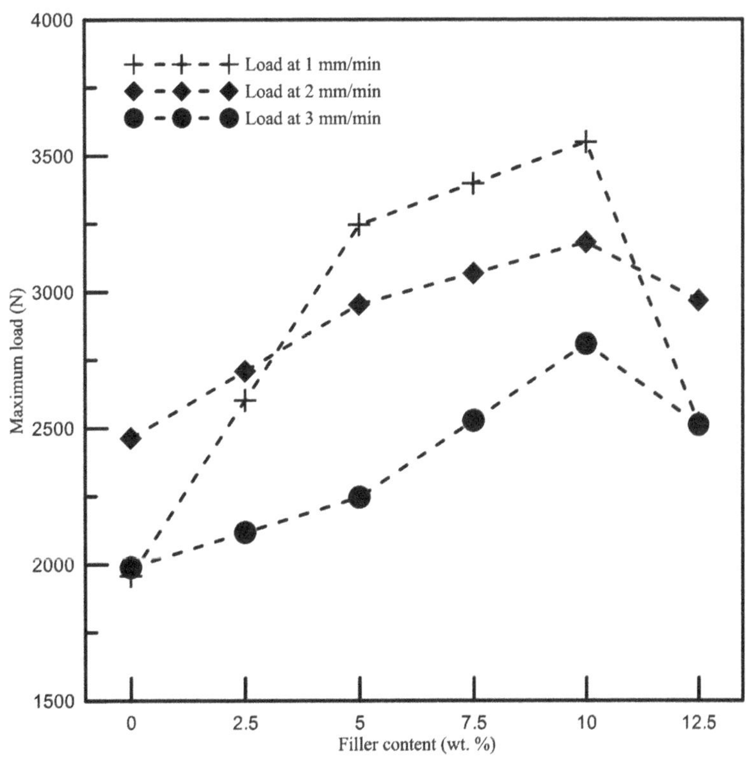

Figure 3. Tensile stress vs. % filler wt. at different speed

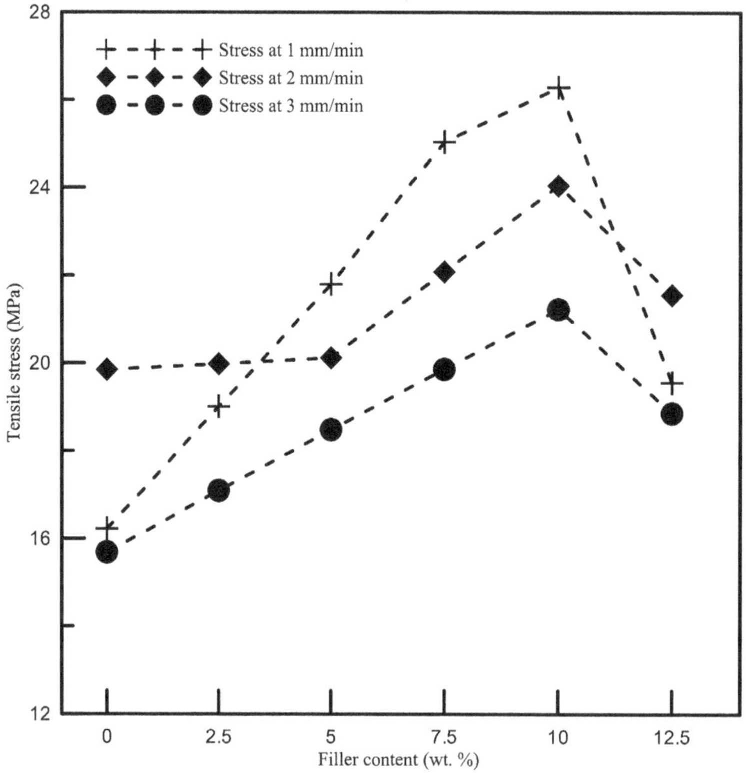

of tensile strengths in the range of 13 – 28 MPa were reported for different natural fibre based polymer composites such as polypropylene and high density polyethylene (HDPE).

The variation of flexural properties of wood epoxy composite in three point bending mode is demonstrated in Figure 4 as a function of flexural stress vs. filler loading. It can be perceived from the diagram that with the increase in filler wt. %, flexural stress values increase and attained highest value for 10% of filler content. After that the properties started to fall for further filler addition. The highest and lowest values of flexural stress are 18.24 MPa and 47.65 MPa respectively at filler content of 12.5% and 10% with crosshead movement speed of 1 mm/min. The highest and lowest values of bending modulus are 1.33 GPa and 435.70 MPa separately obtained for filler loading of 10% and 0%. As deliberated for tensile properties variation, the filler orientation mechanism and easy load transfer mechanism are responsible for properties alternation in three point bending mode.

Optimization Results

The mechanical properties are modeled using Minitab 14 statistical software on the basis of F-test value. Three nonlinear regression equations (4), (5) and (6) are the established to model the load, tensile stress and flexural stress values, respectively. The nonlinear relationship between an output or response variable and one or more independent input variables is generated by the nonlinear regression analysis. The successive approximation technique is used to fit the obtained data. Also, the Pearson's correlation coefficient

Figure 4. Variation of flexural stress at different speed with % filler wt

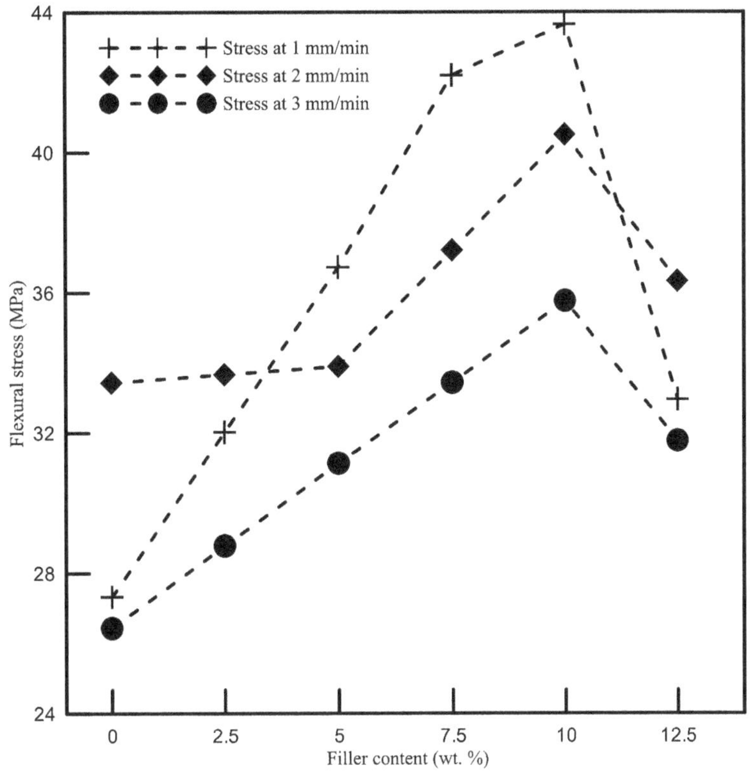

is measured to find the extent of linear relationship between the paired data. The Pearson's correlation coefficient for load and stresses values are found close to zero indicating strong nonlinear correlation.

$$\text{Load} = 1 \Big/ \begin{pmatrix} -0.024942 + 0.00282366 \times f_1 + 0.00374498 \times f_2 + 0.000324303 \\ \times f_1^2 - 3.08302e^{-007} \times f_1 \times f_2 - 0.00115131 \times f_2^2 \end{pmatrix}^2 \tag{4}$$

$$\text{Tensile stress} = 1 \Big/ \begin{pmatrix} -0.07982 + 0.0103532 \times f_1 + 0.0013794 \times f_2 + 0.00114515 \\ \times f_1^2 - 6.37541e^{-005} \times f_1 \times f_2 - 0.00408146 \times f_2^2 \end{pmatrix} \tag{5}$$

$$\text{Flexural stress} = 1 \Big/ \begin{pmatrix} -0.00192825 + 0.000393474 \times f_1 + 0.000536508 \times f_2 \\ +4.40931e^{-005} \times f_1^2 - 7.4176e^{-008} \times f_1 \times f_2 - 0.000160588 \times f_2^2 \end{pmatrix}^{0.5} \tag{6}$$

The decoded values of genes are engaged with regression models of equations (4)-(6) to maximize load, tensile stress and flexural stress. The GA randomly generated a real valued population of possible process parameter sets (chromosomes). A new population for the next generation is created based on old population using different operator functions viz. number of generations: 200, population size: 20, crossover rate: 0.4, crossover mechanism: two point and mutation rate: 0.005. MATLAB R2013a is used for the analysis. A number of GA runs has been carried out to identify the range of process parameters for maximum output. The results of optimum process parameters for load, tensile stress and flexural stress values are presented in Table 1. The comparison between the maximum values of function obtained experimentally and optimized function value using GA for load and stresses values are presented in Table 2. The percentage error associated with them is also given in Table 6 and it is observed that the error lies in the range of 0-8% which establishes the validity of the result. The graph obtained at the end of the iteration process for load, tensile stress and flexural stress values are shown in Figure 5, 6 and 7 respectively.

CONCLUSION AND FUTURE SCOPE

This chapter provides an insight into the wind turbine blades made of natural fibre polymer composites and associated challenges and suggestions. An experimental investigation on the development of wood fillers reinforced polymer composite composites is discussed along with the evaluation and optimization of its mechanical properties through genetic algorithm techniques. It is summarized that wood filler reinforced epoxy composite sample is a viable and effective alternate to the conventional synthetic fibre reinforced polymer composite with good quality and high productivity. The parametric analysis of the experimental data reveals that filler content and crosshead movement speed are found to be the major influential parameter for mechanical properties. The results of properties optimization through genetic algorithm found that the optimum value for load, tensile stress and flexural stress is 3339.40 N, 23.477 MPa and 40.725 MPa respectively. These optimum values of load and stresses were found at 5% of filler

Table 1. Optimum process parameters for load and stress values

Process Parameters		Output	Response
Filler Content (wt. %)	Speed (mm/min)		
4.3612	1.6246	Load	3339.40
4.687	1.639	Tensile stress	23.477
4.458	1.688	Flexural stress	40.725

Table 2. Comparison between experimental value and optimized value

Output	Experimental Value	Optimum Value	Error
Load	3550.36	3339.40	-5.94
Tensile stress	25.290	23.477	-7.16
Flexural stress	43.65	40.725	-6.70

Figure 5. Fitness value vs. number of generations for load

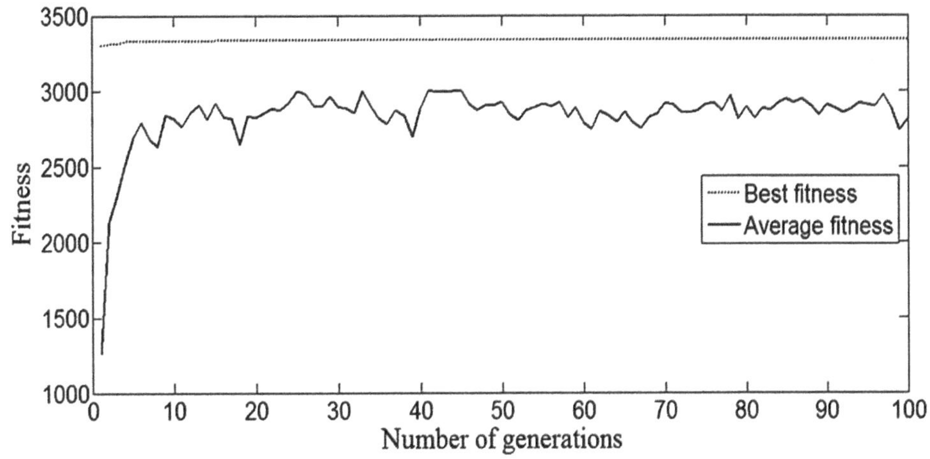

Figure 6. Fitness value vs. number of generations for tensile stress

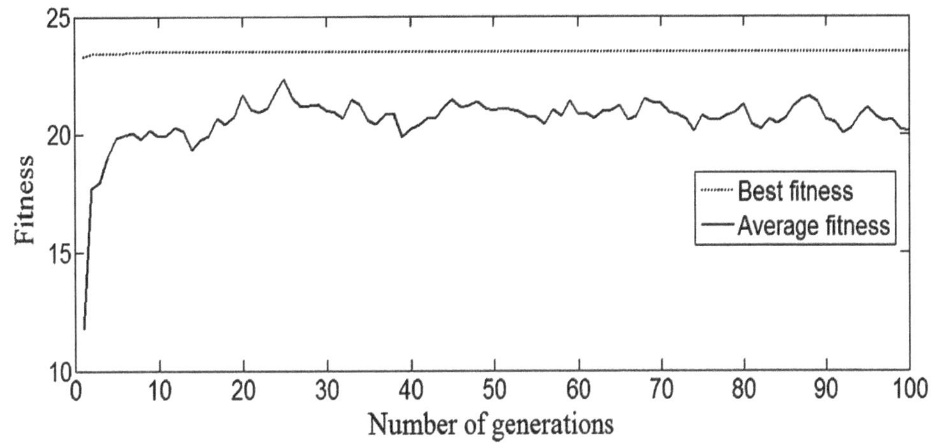

Figure 7. Fitness value vs. number of generations for flexural stress

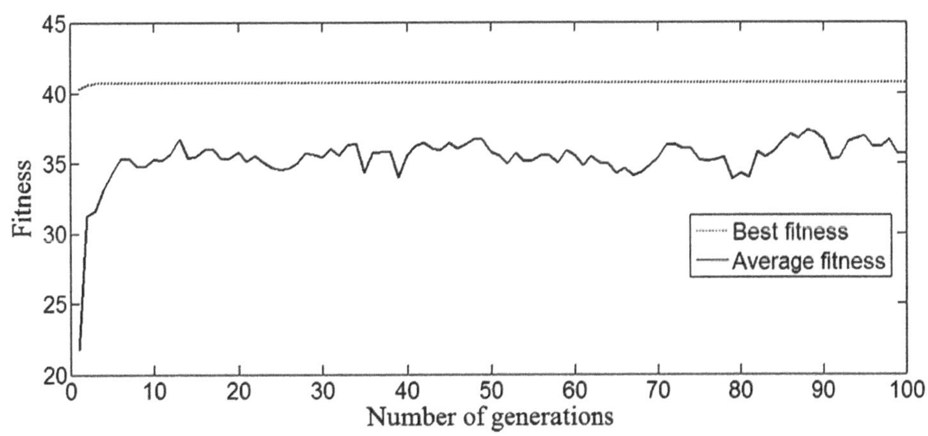

content and 2 mm/min of speed. The mechanical properties of wood fillers reinforced polymer composites for wind turbine blade application is proven satisfactory regarding the tensile strength and the fabrication process is worth further development for industrial applications. Although, this chapter provides on experimental study on mechanical behaviour of wood filler reinforced polymer (WFRP) composites and its application as wind turbine blade. However, the development of other composites materials such as fiber reinforced polymer (FRP) composites, natural filler reinforced polymer (NFRP) composites etc. and its machining behaviour are yet to investigate using abrasive eater jet machining process.

REFERENCES

Abdin, E. S., & Xu, W. (1998). Control Design and Dynamic Performance Analysis of a Wind Turbine-Induction Generator Unit. IEEE, 1198 – 1202.

Al-Khudairi, O., & Ghasemnejad, H. (2015). To improve failure resistance in joint design of composite wind turbine blade materials. *Renewable Energy*, *81*, 936–951. doi:10.1016/j.renene.2015.04.015

Ashori, A. (2008). Wood–plastic composites as promising green-composites for automotive industries. *Bioresource Technology*, *99*(11), 4661–4667. doi:10.1016/j.biortech.2007.09.043 PMID:18068352

Bhowmick, M., Mukhopadhyay, S., & Alagirusamy, R. (2012). Mechanical properties of natural fibre reinforced composites. *Textile Progress*, *44*(2), 85–140. doi:10.1080/00405167.2012.676800

Cheng, G. R. (1997). *Genetic Algorithm and Engineering Design*. New York: John Wiley.

Chikhradze, N. M., Marquis, F. D., & Abashidze, G. S. (2015). Hybrid fiber and nanopowder reinforced composites for wind turbine blades. *Journal of Materials Research and Technology*, *4*(1), 60–67. doi:10.1016/j.jmrt.2015.01.002

Courtenay, J. C., Johns, M. A., Galembeck, F., Deneke, C., Lanzoni, E. M., Costa, C. A., & Sharma, R. I. (2017). Surface modified cellulose scaffolds for tissue engineering. *Cellulose (London, England)*, *24*(1), 253–267. doi:10.1007/s10570-016-1111-y

Cox, K., & Echtermeyer, A. (2012). Structural design and analysis of a 10MW wind turbine blade. *Energy Procedia*, *24*, 194–201. doi:10.1016/j.egypro.2012.06.101

Debnath, K., Singh, I., Dvivedi, A., & Kumar, P. (2013). Natural Fibre-Reinforced Polymer Composites for Wind Turbine Blades: Challenges and Opportunities. In Recent Advances in Composite Materials for Wind Turbine Blades. The World Academic Publishing Co. Ltd.

Doan, T. L., Brodowsky, H., & Mäder, E. (2012). Jute fibre/epoxy composites: Surface properties and interfacial adhesion. *Composites Science and Technology*, *72*(10), 1160–1166. doi:10.1016/j.compscitech.2012.03.025

Fu, S. U., Feng, X. Q., Lauke, B., & Mai, Y. W. (2008). Effects of particle size, particle/matrix interface adhesion and particle loading on mechanical properties of particulate-polymer composites. *Composites. Part B, Engineering*, *39*(6), 933–961. doi:10.1016/j.compositesb.2008.01.002

Griffin, D. A., & Ashwill, T. D. (2003). Alternative Composite Materials for Megawatt-Scale Wind Turbine Blades: Design Considerations and Recommended Testing. *Journal of Solar Energy Engineering*, *125*(4), 515. doi:10.1115/1.1629750

Herrera-Franco, P. J., & Valadez-Gonza'lez, A. (2004). Mechanical properties of continuous natural fibre-reinforced polymer composites. *Composites. Part A, Applied Science and Manufacturing*, *35*(3), 339–345. doi:10.1016/j.compositesa.2003.09.012

Holmes, J. W., Brøndsted, P., Sørensen, B. F., Jiang, Z. H., Sun, Z. H., & Chen, X. H. (2009). Development of a Bamboo-Based Composite as a Sustainable Green Material for Wind Turbine Blades. *Journal of Wind Engineering*, *33*(2), 197–210. doi:10.1260/030952409789141053

Ishizaki, M. H., Visconte, L. L., Furtado, C. R., Oliveira, M. G., & Leblanc, J. L. (2008). Characterization of Green Coconut Fiber Composites. *International Journal of Polymer Analysis and Characterization*, *13*(5), 319–330. doi:10.1080/10236660802351409

Joseph, S., Joseph, K., & Thomas, S. (2006). Green Composites from Natural Rubber and Oil Palm Fiber: Physical and Mechanical Properties. *International Journal of Polymeric Materials*, *55*(11), 925–945. doi:10.1080/00914030600550505

Ku, H., Wang, H., Pattarachaiyakoop, N., & Trada, M. (2011). A review on the tensile properties of natural fiber reinforced polymers composites. *Composite Part B*, *42*(4), 856–873. doi:10.1016/j.compositesb.2011.01.010

Kumar, R., Bhowmik, S., & Kumar, K. (2017). Establishment and Effect of Constraint on Different Mechanical Properties of Bamboo Filler Reinforced Epoxy Composite. *International Polymer Processing*, *32*(3), 308–315. doi:10.3139/217.3311

Kumar, R., Kumar, K., Sahoo, P., & Bhowmik, S. (2014). Study of Mechanical Properties of Wood Dust Reinforced Epoxy Composite, *Procedia. Materials Science*, *6*, 551–556.

Michalewicz, Z. (1999). *Genetic algorithms + data structures = evolution programs*. Berlin: Springer-Verlag.

Mishra, S., Misra, M., Tripathy, S. S., Nayak, S. K., & Mohanty, A. K. (2002). The influence of chemical surface modification on the performance of sisal-polyester biocomposites. *Polymer Composites*, *23*(2), 164–170. doi:10.1002/pc.10422

Mittal, V., Saini, R., & Sinha, S. (2016). Natural fiber-mediated epoxy composites: A review. *Compos Part B*, *99*, 425–435. doi:10.1016/j.compositesb.2016.06.051

Mylsamy, K., & Rajendran, I. (2011). The mechanical properties, deformation and thermo mechanical properties of alkali treated and untreated Agave continuous fibre reinforced epoxy composites. *Materials & Design*, *32*(5), 3076–3084. doi:10.1016/j.matdes.2010.12.051

Nagarajan, V., Mohanty, A. K., & Misra, M. (2016). Perspective on Polylactic Acid (PLA) based Sustainable Materials for Durable Applications: Focus on Toughness and Heat Resistance. *ACS Sustainable Chemistry & Engineering*, *4*(6), 2899–2916. doi:10.1021/acssuschemeng.6b00321

Netravali, A. N., & Chabba, S. (2003). Composites get greener. *Materials Today: Proceedings*, *6*(4), 22–29. doi:10.1016/S1369-7021(03)00427-9

Nielsen, L. E., & Landel, R. F. (1994). *Mechanical Properties of Polymers and Composites* (2nd ed.). New York: Marcel Deckker.

Nopparut, A., & Amornsakchai, T. (2016). Influence of pineapple leaf fiber and it's surface treatment on molecular orientation in, and mechanical properties of, injection molded nylon composites. *Polymer Testing*, *52*, 141–149. doi:10.1016/j.polymertesting.2016.04.012

Park, H. (2015). A study on structural design and analysis of small wind turbine blade with natural fibre (flax) composite. *Advanced Composite Materials*, *25*(2), 125–142. doi:10.1080/09243046.2015.1052186

Prabhakaran, D., & Thirumalai, R. (2014). A critical review of future materials for wind turbine blades, *Int. J. Materials Engineering Innovation*, *5*(2), 2014.

Puglia, G. (2013). *Life cycle cost analysis on wind turbines* (Master's thesis). Department of Energy and Environment, Division of Electric Power Engineering, Chalmers University of Technology, Gothenburg, Sweden.

Rocha, I. B. C. M., Raijmaekers, S., Nijssen, R. P. L., van der Meer, F. P., & Sluys, L. J. (2017). Hygrothermal ageing behaviour of a glass/epoxy composite used in wind turbine blades. *Composite Structures*, *174*, 110–122. doi:10.1016/j.compstruct.2017.04.028

Rozman, H. D., Lai, C. Y., Ismail, H., & Mohd-Ishak, Z. A. (2000). The effect of coupling agents on the mechanical and physical properties of oil palm empty fruit bunch polypropylen composites. *Polymer International*, *49*(11), 1273–1278. doi:10.1002/1097-0126(200011)49:11<1273::AID-PI469>3.0.CO;2-U

Sapuan, S. M., & Maleque, M. A. (2005). Design and fabrication of natural woven fabric reinforced epoxy composite for household telephone stand. *Materials & Design*, *26*(1), 65–71. doi:10.1016/j.matdes.2004.03.015

Shah, D. U., Schubel, P. J., & Clifford, M. J. (2013). Can flax replace E-glass in structural composites? A small wind turbine blade case study. *Composites. Part B, Engineering*, *52*, 172–181. doi:10.1016/j.compositesb.2013.04.027

Thakur, V. K., Singha, A. S., & Thakur, M. K. (2012). Green Composites from Natural Fibers: Mechanical and chemical aging properties, *International Journal of Polymer Analysis and. Characterization*, *17*, 401–407.

Watson, J. C. & Serrano, J. C. (2011). Composite Materials for Wind Blades. *Wind Systems Magazine*, 46-51.

Wind Energy Technology. (2009). *50th AIAA Structures*. Structural Dynamics, & Materials Conference Palm Springs.

Chapter 10
Functionally Graded Varying Geometry Annular Fins

Vivek Kumar Gaba
National Institute of Technology Raipur, India

Anil Kumar Tiwari
National Institute of Technology Raipur, India

Tarakant Bhanu
National Institute of Technology Raipur, India

Shubhankar Bhowmick
National Institute of Technology Raipur, India

ABSTRACT

The research work aims at investigating the performance of functionally graded annular fin. The work involves computation of efficiency and effectiveness of such fins and compares the fin performances for different geometry and grading parameters. It is also assumed that the temperature gradient only exists in one direction i.e. along the radius. A general second order governing differential equation has been derived for all the profiles and grading. The thickness of fin is considered to be a power function of radial co-ordinate. The functional grading of thermal conductivity is assumed to be a power function of radial co-ordinate and also consists of parameters, namely grading parameters, varying which different grading combinations can be carried out. The efficiency and effectiveness of the annular fin of different geometry and grading combinations have been calculated and plotted. The performance analysis reveals the dependence of thermal behaviour of annular fins on geometry and grading parameter.

INTRODUCTION

The thermal analysis of annular fins has been the area of interest for many researchers across the world due to their widespread applications in industry and provides many fronts of investigation to be carried out. A substantial amount of research endeavors have been made to find the best dimension of the annular fins so that the fin performance can be maximized for a given fin volume or the fin volume can

DOI: 10.4018/978-1-5225-3401-3.ch010

be maximized for given heat duty. The chapter presents one dimensional analysis of annular fins under steady state condition subject to geometry variation and material variation along spatial co-ordinate under insulated boundary condition.

BACKGROUND

Although there are abundant reports available on thermal performance of annular fin made of isotropic material, the studies of fin made of materials having varying thermal conductivity is scare. Jensen et. al. (1999) presented the work for finding the efficiency of annular fin made of two materials using Gardner's expression. Bahadur and Bar-cohen (2004) gave an optimization methodology for a thermally conductive polyphenylene sulphide (PPS) polymer staggered pin fin heat sink, for an advanced natural convection cooled microprocessor application. The coefficient of thermal performance, COP, that relates cooling capability to the energy invested in the formation of the heat sink, has been determined and compared with conventional aluminium heat sinks.

An analytical equation to optimize thermally conductive PPS polymer fins used in electronic heat sink of microprocessor application has been described by Bahadur and Bar-cohen (2005) the geometric dependence of heat dissipation and relationships between the pin fin height, pin diameter, horizontal spacing, and pin fin density for a fixed base area and excess temperature has been discussed. The thermal calculation of composite metallic fins of variable thickness has been obtained by Campo et. al. (2008) wherein analytical solution for the two-dimensional, two-material conduction problem, under the form of an infinite series of orthogonal eigen functions has been obtained and Bi number limitation has been checked to express the fin efficiency in closed form.

The effects of radiation and convection heat transfer in porous media have been discussed by Gorla and Bakier (2011). The effects of the porosity parameter (Sh), radiation parameter (G) and temperature ratio (CT) on the dimensionless temperature distribution and heat transfer rate discussed in detail. Results show that the radiation transfers more heat compared to a similar model without radiation. Heat transfer and temperature distribution equations for longitudinal convective–radiative porous fins have been presented by Hatami and Ganji (2014). Al,SiC and Si_3N_4 has been considered ceramic porous materials, and it has been observed that exponential section fin with Si_3N_4 material has the maximum amount of transferred heat.

Turkyilmazoglu (2014) reported the combined heat and mass transfer mechanisms owing to the temperature and humidity ratio differences considering their influence on the efficiency of different exponential type porous fin configurations. A comparison with the traditional straight porous fins proves that the fin tip temperature and efficiency of the exponential porous wet fins are much better than those corresponding to the straight porous wet fins.

Lee et. al. (2004) reported the solution for estimating the unknown heat flux at the pin fin base by the conjugate gradient method. The accuracy of the inverse analysis has been examined by simulated exact and inexact measurements of temperature at interior locations of the pin fin.

Recursive numerical formulation technique has been applied by Kou et. al. (2005) to analyze the longitudinal fin of variable thermal properties with discrete model dividing the singular into many sections. Finally combining each section together, the complete solution of the fin has been obtained.

Naphon (2006) investigated annular fin under dry-surface conditions, partially wet-surface conditions, and fully wet-surface conditions and a mathematical models has been developed based on the conservation equations of energy and mass. Central finite difference numerical technique has been used to obtain temperature distribution along the fin.

Correlation equations are obtained by Arslanturk (2008) to optimize the design of annular fins with temperature-dependent thermal conductivity. The nonlinear fin equation which is associated with variable thermal conductivity condition is solved by Adomian decomposition method that provides an analytical solution in the form of an infinite power series.

Bayat et. al. (2009) minimized the weight of functionally graded (FG) disk with variable thickness under a steady temperature field. Results reported that for a given value of grading index, the FG disk having concave thickness profile is the lightest in weight whereas the FG disk with uniform thickness profile is the heaviest. Lai et. al. (2009) introduced a discrete model for calculating the thermal performance of annular fins with variable thermal properties. The fin was divided into various circular sections such that each section can take its variable thermal properties like heat transfer coefficient and thermal conductivity into account. The results from each section was combined and calculated together by recursive formula.

The homotopy analysis method is used by Arslanturk (2010) to analyze the thermal performance of annular fins with temperature dependent thermal conductivity. The fin efficiency is obtained as a function of thermo geometric fin parameter, thermal conductivity and radius ratio. Exact solutions for thermal diffusion in a straight fin with varying exponential shape with temperature depended thermal conductivity and heat transfer coefficients are presented by Turkyilmazoglu (2012). Analysis shows that the performance and of the exponential profiles are higher than the rectangular fin. The performance of functionally graded parabolic annular fin having constant weight is reported in Gaba et. al. (2014), wherein the effects of grading and geometry parameter have been investigated and dependence of fin performance on profile and grading is reported.

In a recent work Gaba et. al. (2015), the performance of the rectangular annular fins made of functionally graded material subjected to rotation is reported. It is observed that for increase in fin parameter and rotational speed parameter the performance of fin decreases for same aspect ratio of the fin. The study of grading parameter on fin performance indicates that both temperature distribution and heat transfer are enhanced with decrease in grading parameter. Aziz (2016) investigated hyperbolic annular fin with temperature dependent thermal conductivity by pseudo spectral method. The effect of fin dimension, surface convection characteristics, thermal conductivity parameter on the thermal performance of the fin is reported. Temperature distribution and the fin efficiency both increases when thermal conductivity increases with temperature.

Zubair (2016) studied the thermal performance of hyperbolic annular fins subjected to dehumidifying operating condition. Analytical solution for completely wet fin is derived by using an approximate linear temperature humidity relation and numerical solution using actual psychometric relation for completely and partially wet operating conditions. A numerical solution is more convenient for partially wet hyperbolic fin and for completely wet fin regression equation is developed to calculate the optimum fin geometry. Huang (2016) considered the nonlinear optimum shape design problem of fully wet annular fins. The study shows that the optimum annular fin has the highest fin efficiency among six annular fins with the same operating fin conditions. The optimum fin efficiency and optimum fin shape changed significantly when Biot number for ambient air is varied.

MATHEMATICAL FORMULATION

An annular fin of variable geometry is shown in Figure 1. It is assumed that the effect of external environment on the surface convection is negligible and hence a constant convective coefficient, ***h***, is considered for the study. It is also assumed that temperature gradient along the thickness of the fins does not exist.

The governing equation is derived from the heat balance applied over an element of length 'dr' of the fin shown in Figure 1(b) is considered. The second order differential equation for the heat transfer through the fins is developed to find the temperature profile.

Applying thermal energy balance to the element, the differential equation of heat flow is derived as follows:

$$\frac{d}{dr}\left\{2\pi r k_r \delta \frac{d\theta}{dr}\right\} dr = 4\pi r h \theta dr \sqrt{1+\left(\frac{1}{2}\frac{d\delta}{dr}\right)^2} \tag{1}$$

Equation (1) is normalized using the following expressions;

Normalized radius, $\xi = \dfrac{R-R_0}{R_1-R_0} = \dfrac{R-R_0}{L}$;

Normalized thickness, $\bar{\delta} = \dfrac{\delta}{\delta_0}$

Aspect ratio, $R_f = \dfrac{R_1}{R_0}$;

Normalized temperature field, $\phi = \dfrac{\theta}{\theta_0} = \dfrac{T-T_f}{T_{base}-T_f}$,

Figure 1. Sectional isometric and front view of annular fin geometry

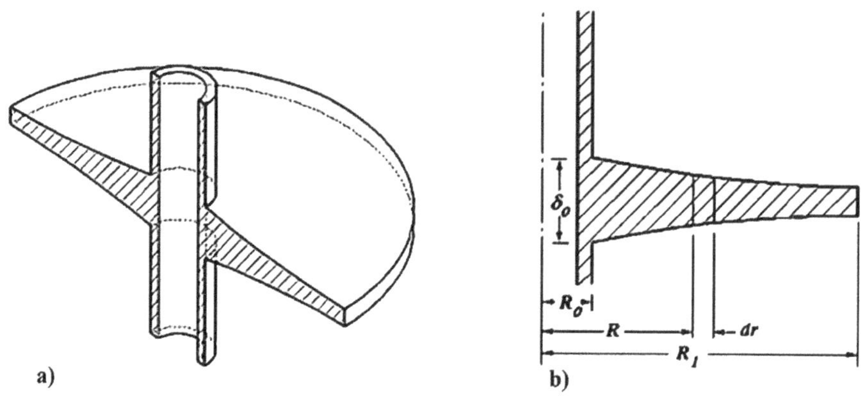

Fin parameter $m_f^2 = \frac{2hL^2}{k_0\delta_0}$

In the above natural expressions, δ is thickness at any radius R and δ_0 is the thickness at base radius R_0, R_1 is the tip radius, θ is the excess temperature of the element over ambient fluid and θ_0 is the excess temperature at the base over ambient fluid.

Upon introducing the normalized variables, the governing differential equation becomes,

$$\frac{d^2\phi}{d\xi^2} + A_1\frac{d\phi}{d\xi} + A_2\phi = 0 \tag{2}$$

Equation (2) is a generalized governing differential equation in which the coefficient $\boldsymbol{A_1}$ and $\boldsymbol{A_2}$ are as follows:

$$A_1 = \left(\frac{1}{\bar{\delta}}\frac{d\bar{\delta}}{d\xi} + \frac{1}{k_r}\frac{dk_r}{d\xi} + \frac{L}{L\xi + R_0}\right) \tag{3}$$

$$A_2 = -\frac{2hL^2}{k_r\delta_0\bar{\delta}}\sqrt{1 + \frac{\delta_0^2}{4L^2}\left(\frac{d\bar{\delta}}{d\xi}\right)^2} \tag{4}$$

In Equation (2), first and the second terms represent the variation of thickness and of thermal conductivity respectively along the radius of the fin. For annular fins of rectangular profile with constant thermal conductivity, these two terms will not exist. In Equation (4) the expression inside the root also represents thickness variation. Equation (4), hence as mentioned above is a general expression of the governing differential equation of constant weight annular fins of varying thickness having functionally graded thermal conductivity. At this point, functional grading of thermal conductivity of the fin material is introduced into the mathematical model using the following relations respectively;

$$k = ar^b = a\left(L\xi + R_0\right)^b \tag{5}$$

So, $k_0 = a\left(R_0\right)^b$

Now in Equation (3), the first term belongs to variation of thickness, so in this work thickness has been varied using power function is normalized co-ordinate (ξ).

For power profile thickness variation is taken as

$$\bar{\delta} = \left(1 - n\xi\right)^{m} \tag{6}$$

Upon substituting, Equation (5) and (6) in Equation (2), the governing equation takes the following form;

$$\frac{d^2\phi}{d\xi^2} + \left(-\frac{m.n.\xi^{m-1}}{1-n\xi} + \frac{L}{(L\xi + R_0)} + \frac{bL}{(L\xi + R_0)}\right)\frac{d\phi}{d\xi} - \frac{m_f^2}{\bar{\delta}.}\left(\sqrt{\frac{\delta_0^2}{4L^2}\cdot\left(m.n.\xi^{m-1}\right)^2 + 1}\right)\left(\frac{R_0}{L\xi + R_0}\right)^{b}\phi = 0 \tag{7}$$

Similarly, efficiency and effectiveness of the fin is obtained from the general equations given below

$$\eta = \frac{-\left(\phi'\right)_{\xi=0}}{m_f^2 \int_0^1 \left[\left(\left(R_f - 1\right)\xi + 1\right)\sqrt{\left(\frac{\delta_0 mn}{2L}\right)^2 e^{-2n\xi^m}\xi^{2m-2} + 1}\,d\xi\right]} \tag{8}$$

$$\varepsilon = \frac{-4L\left(\phi'\right)_{\xi=0}}{m_f^2\delta_0} \tag{9}$$

RESULT AND DISCUSSION

The main objective of the present research is to investigate the performance of functionally graded annular fin of varying geometry and thermal properties. The mathematical technique used to carry out the above mentioned investigation is of approximate nature with methods based on polynomial collocation techniques. The governing equation (7) is obtained by applying the energy conservation on the elemental strip. The effect of geometry variation, thermal conductivity variation along the radial co-ordinate on the performance of the annular fin is reported. The results are reported in dimensionless form and presented graphically.

The general second order differential equation (7) is derived and solved using ***bvp4c*** subroutine in MATLAB. Both thickness variation and thermal conductivity variation using the power law is reported. The effect of geometry parameters ***n*** and ***m*** on the performance of annular fin is investigated for the aspect ratio ($\boldsymbol{R_f}$) at 3. The efficiency and effectiveness of annular fin is reported for different parameters. The observations are found to provide some interesting insight into the design of annular fin with specific industrial applications such as heat exchangers.

The dimensionless temperature $\boldsymbol{\phi}$ depends upon normalized variables, ***n*** and ***m*** due to chosen fin shape, grading parameter, ***a*** and ***b*** for thermal conductivity variation, $\boldsymbol{m_f}$ and $\boldsymbol{\xi}$ due to the fin geometry. Considering $\boldsymbol{\xi}$ as the only independent variable and keeping the other variables constant, the Eq. (7-9), have been solved for the various values of $\boldsymbol{n}$ ranging between 0 to 0.9, $\boldsymbol{m}$ ranging between 0.5 to 1.5, grading parameter $\boldsymbol{b}$ ranging between -2 to 2, thermal conductivity at the base $\boldsymbol{k_0}$ ranging between 100 to 300 W/m-K and for all the cases the values of $\boldsymbol{R_f}$ is kept constant i.e. 3. The numerical values of system properties are $\boldsymbol{h}$=25 W/m^{2}K, $\boldsymbol{\delta_0}$=0.01 m, and $\boldsymbol{R_0}$ = 0.1 m.

The first derivative of temperature at the fin base i.e. $\boldsymbol{\phi'}_{\xi=0}$ helps to calculate the fin efficiency and effectiveness and is calculated by solving second order differential equation in MATLAB. The values of $\boldsymbol{\phi'}_{\xi=0}$ can be obtained for different values of $\boldsymbol{R_f}$, $\boldsymbol{n}$, $\boldsymbol{m}$ and $\boldsymbol{b}$. Temperature variation along the radius has been depicted in Figure 2 for the rectangular annular fin i.e. n=0 for different values of $\boldsymbol{m_f}$ at grading parameter $\boldsymbol{b}$ =-1, it has been observed that results obtained from the numerical coding is having good agreement to the Aziz and Rahman (2009). The validation is presented for fin variables defined $\boldsymbol{m_{ff}}$ where,

$$m_{ff}^2 = m_f^2\left(\frac{R_0^b}{L^2}\right)$$

Figure 3 clearly shows the variation of efficiency and effectiveness with the parameter $\boldsymbol{m}$ for different values of geometry parameter $\boldsymbol{n}$ keeping the other variable constant (i.e. $\boldsymbol{b}$= -1 $\boldsymbol{k_0}$=200W/m-K and $\boldsymbol{m_f}$=1). It is evident from Figure 3, both efficiency and effectiveness decreases with increase in parameter $\boldsymbol{m}$. Effectiveness decreases sharply with the parameter $\boldsymbol{m}$, fin performance is higher for rectangular fins i.e. $\boldsymbol{n}$=0, this is due to the constraint of constant weight. Both efficiency and effectiveness decreases sharply with geometry parameter $\boldsymbol{n}$ as shown in Figure 4 and similar trends have also been obtained for different values of geometry parameter $\boldsymbol{m}$ as shown in Figure 4.

Figure 2. Excess temperature over radial co-ordinate for Rectangular annular fin

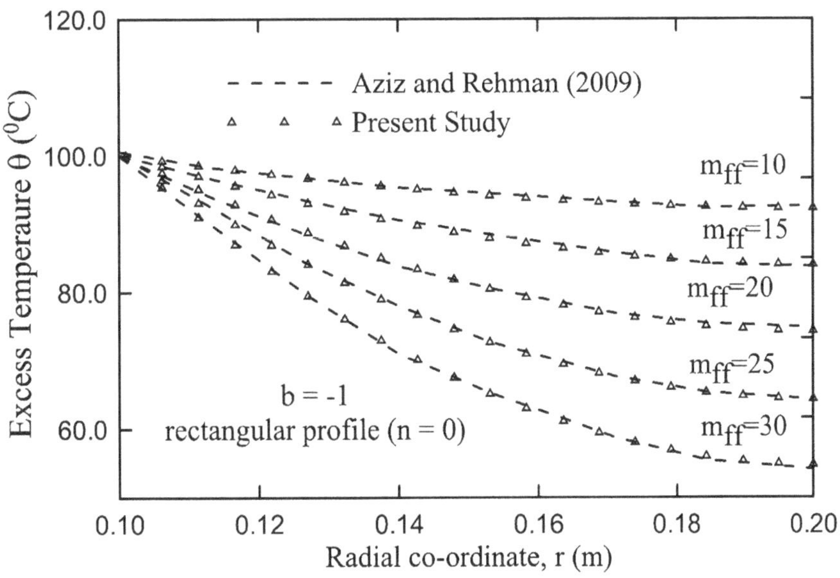

Excess temperature over radial co-ordinate for rectangular annular fin.

Figure 3. Variation of efficiency and effectiveness with parameter **m** *for different values of* **n**

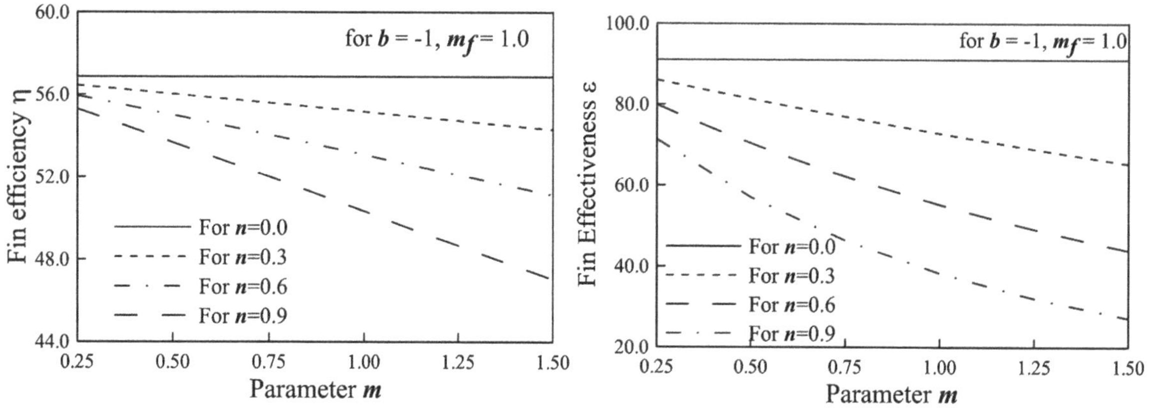

Figure 4. Variation of efficiency and effectiveness with **n** *for different values of* **m**

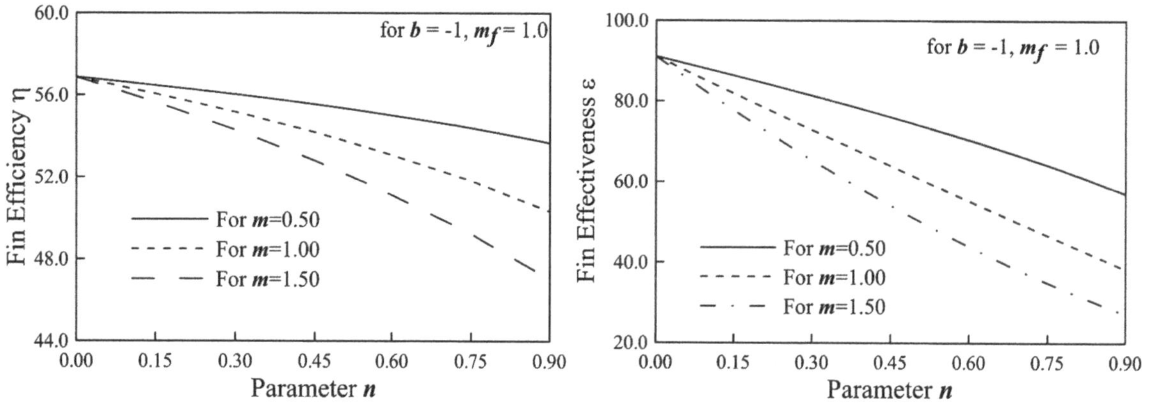

Very interesting results have been obtained by varying the base thermal conductivity k_0 for the fixed geometry and grading in Figure 5 and Figure 6. As the base thermal conductivity increases, the performance of the fin increases due to more conduction of the heat from base to tip. Also fin efficiency of varying geometry is higher compared to the rectangular annular fin, this is due to the fact that when the geometry is varying for any fixed value of k_0 then base thickness increases (for keeping weight of the fin constant), this leads to smaller parameter m_f. This effect of increasing thickness at the base is compensating with increase in convective coefficient of the fin for the constant value of the m_f as shown in Figure 3.

The effect of grading parameter b on fin performance is also investigated in Figure 7 and it is observed temperature distribution over the surface decreases as b value shifted from negative to positive and it is highest for b=-2. As the performance of the fin depends that how close the temperature over the surface is with the base temperature. So it can be concluded that highest thermal conductivity is required at the base of the fin to dissipate the more amount of heat (due to highest temperature difference available).

Figure 5. Variation of efficiency and effectiveness with $\mathbf{k_0}$ *for different values of* **n**

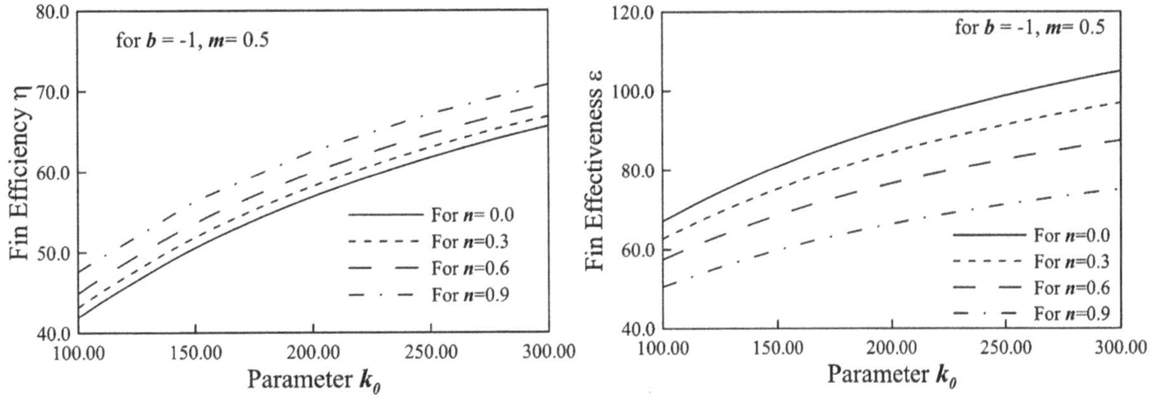

Figure 6. Variation of efficiency and effectiveness with $\mathbf{k_0}$ *for different values of* **m**

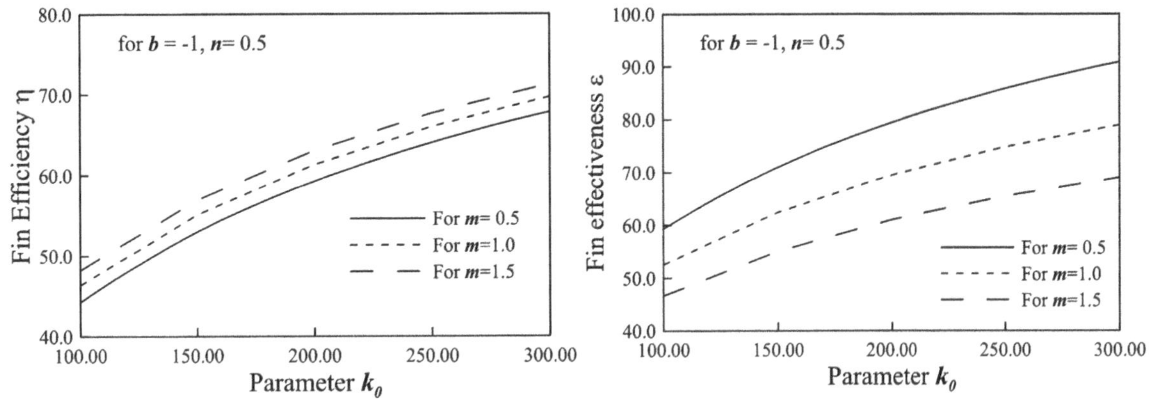

Figure 7. Excess temperature distribution over the fin length

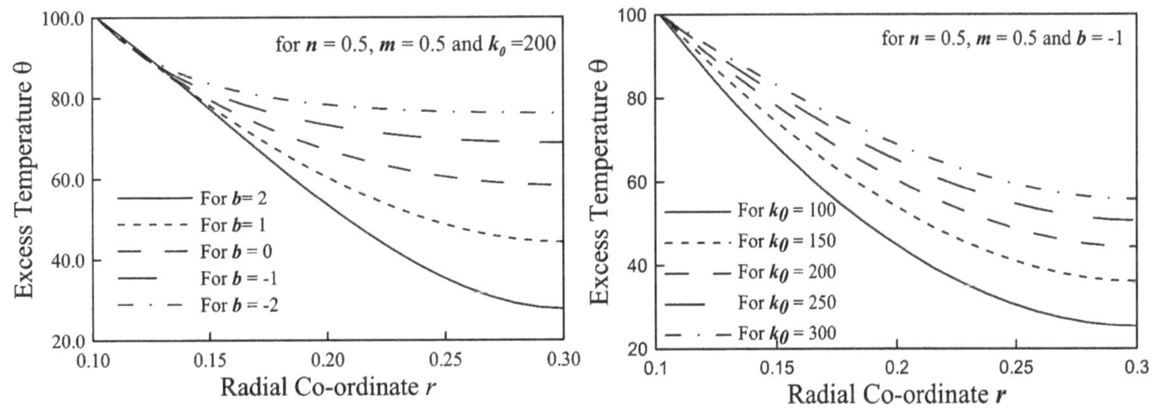

CONCLUSION

The efficiency and effectiveness of the annular fins of different profile and grading combinations keeping the weight of the fin constant have been calculated and plotted and results reveal the dependence on profile and grading parameter. The important conclusions are:

1. It is observed that effectiveness and efficiency both decreases as m_f increases for any geometry and grading.
2. Performance of the fin increases with increase in thermal conductivity at the base k_0 due to higher penetration of the heat form base to tip.
3. Functionally grading enhances the performance if the thermal conductivity along the radial direction decreases.

REFERENCES

Arslantürk, C. (2008). Correlation equations for optimum design of annular fins with temperature dependent thermal conductivity. *Heat and Mass Transfer*, *45*(4), 519–525. doi:10.1007/s00231-008-0446-9

Arslantürk, C. (2010). Analysis of thermal performance of annular fins with variable thermal conductivity by homotopy method. *Journal Of Thermal Science And Technology*, *30*(2), 1–7.

Aziz, A., & Rahman, M. M. (2009). Thermal performance of a functionally graded radial fin. *International Journal of Thermophysics*, *30*(5), 1637–1648. doi:10.1007/s10765-009-0627-x

Bahadur, R., & Bar-cohen, A. (2005). Thermal Design and Optimization of Natural Convection Polymer Pin Fin Heat Sinks. *IEEE Transactions on Components and Packaging Technologies*, *28*(2), 238–246. doi:10.1109/TCAPT.2005.848498

Bahadur, R., & Bar-cohen, A. (2007). Orthotropic thermal conductivity effect on cylindrical pin fin heat transfer. *International Journal of Heat and Mass Transfer*, *50*(5-6), 1155–1162. doi:10.1016/j.ijheatmasstransfer.2006.04.025

Campo, A., & Cui, J. (2008). Temperature/Heat Analysis of Annular Fins of Hyperbolic Profile Relying on the Simple Theory for Straight Fins of Uniform Profile. *Journal of Heat Transfer*, *130*(5), 0545011–0545014. doi:10.1115/1.2885162

Darvishi, M. T., Khani, F., & Aziz, A. (2016). Numerical Investigation For a hyperbolic Annular Fin with Temperature dependent Thermal Conductivity. *Propulsion and Power Research*, *5*(1), 55–62. doi:10.1016/j.jppr.2016.01.005

Gaba, V. K., Tiwari, A. K., & Bhowmick, S. (2014). Thermal Performance of Functionally Graded Parabolic Annular Fins Having Constant Weight. *Journal of Mechanical Science and Technology*, *28*(10), 4309–4318. doi:10.1007/s12206-014-0945-1

Gaba, V. K., Tiwari, A. K., & Bhowmick, S. (2015). A Parametric Study of Functionally Graded Rotating Annular Fin. *Procedia Engineering*, *127*, 126–132. doi:10.1016/j.proeng.2015.11.436

Gorla, R. S. R., & Bakier, A. Y. (2011). Thermal analysis of natural convection and radiation in porous fins. *International Communications in Heat and Mass Transfer*, *38*(5), 638–645. doi:10.1016/j.icheatmasstransfer.2010.12.024

Hatami, M., & Ganji, D. D. (2014). Thermal behavior of longitudinal convective – radiative porous fins with different section shapes and ceramic materials (SiC and Si3N4). *Ceramics International*, *40*(5), 6765–6775. doi:10.1016/j.ceramint.2013.11.140

Huang, C. H., & Chung, Y. L. (2016). A nonlinear fin design problem in determining the optimum shapes of fully wet annular fins. *Applied Thermal Engineering*, *103*, 195–204. doi:10.1016/j.applthermaleng.2016.03.066

Kou, H., Lee, J., & Lai, C. (2005). Thermal analysis of a longitudinal fin with variable thermal properties by recursive formulation. *International Journal of Heat and Mass Transfer*, *48*(11), 2266–2277. doi:10.1016/j.ijheatmasstransfer.2004.12.038

Lai, C.-Y., Kou, H.-S., & Lee, J.-J. (2009). Recursive formulation on thermal analysis of an annular fin with variable thermal properties. *Applied Thermal Engineering*, *29*(4), 779–786. doi:10.1016/j.applthermaleng.2008.04.007

Lee, H., Chou, H., & Yang, Y.-C. (2004). The function estimation in predicting heat flux of pin fins with variable heat transfer coefficients. *Energy Conversion and Management*, *45*(11-12), 1749–1758. doi:10.1016/j.enconman.2003.09.017

Naphon, P. (2006). Study on the heat transfer characteristics of the annular fin under dry-surface, partially wet-surface, and fully wet-surface conditions. *International Communications in Heat and Mass Transfer*, *33*(1), 112–121. doi:10.1016/j.icheatmasstransfer.2005.08.009

Pashah, S., Moinuddin, A., & Zubair, S. M. (2016). Thermal Performance and Optimization of hyperbolic annular fins under dehumidifying operating conditions-analytical and numerical solutions. *International Journal of Refrigeration*, *65*, 42–54. doi:10.1016/j.ijrefrig.2016.01.006

Turkyilmazoglu, M. (2012). International Journal of Thermal Sciences Exact solutions to heat transfer in straight fins of varying exponential shape having temperature dependent properties. *International Journal of Thermal Sciences*, *55*, 69–75. doi:10.1016/j.ijthermalsci.2011.12.019

Turkyilmazoglu, M. (2014). Efficiency of heat and mass transfer in fully wet porous fins: Exponential fins versus straight fins. *International Journal of Refrigeration*, *46*(1970), 158–164.

Section 3

Product and Process Optimization

Chapter 11
Optimizing Material Removal Rate Using Artificial Neural Network for Micro-EDM

Ananya Upadhyay
Birla Institute of Technology Mesra, India

Ved Prakash
Central Mechanical Engineering Research Institute (CSIR), India

Vinay Sharma
Birla Institute of Technology Mesra, India

ABSTRACT

Machining can be classified into conventional and unconventional processes. Unconventional Machining Process attracts researchers as it has many processes whose physics is still not that clear and they are highly in market-demand. To predict and understand the physics behind these processes soft computing is being used. Soft computing is an approach of computing which is based on the way a human brain learns and get trained to deal with different situations. Scope of this chapter is limited to one of the soft computing optimizing techniques that is artificial neural network (ANN) and to one of the unconventional machining processes, electrical discharge machining process. This chapter discusses about micromachining on Electric Discharge Machining, its working principle and problems associated with it. Solution to those problems is suggested with the addition of powder in dielectric fluid. The optimization of Material Removal Rate (MRR) is done with the help of ANN toolbox in MATLAB.

MANUFACTURING AND MACHINING

Manufacturing is the process of value addition to raw material. It is broadly classified into four processes which are casting, forming, machining and joining. Machining is one of the four basic manufacturing processes in which material removal takes place in the form of chips. Machining is used almost in manufacturing of all the products. (Rao, 2009) It is one of the most expensive and energy consuming

DOI: 10.4018/978-1-5225-3401-3.ch011

manufacturing process as material is removed from workpiece to complete the process. Machining process has a machine tool and a workpiece. Machining processes can be classified as traditional and non-traditional processes. The traditional or conventional machining processes include a machine tool and workpiece which are in direct contact during the machining. Some of the traditional processes are drilling, milling, grinding, polishing etc. The nontraditional methods came into picture to fix some of the drawbacks that conventional methods had which were the high cutting force due to which the material of tool should always be harder than that of the workpiece. With this constraint, it is difficult to machine materials like titanium alloy, tungsten carbide, and many others. Nonconventional processes can machine hard materials too. Most of them are contactless processes which remove the constraint of high cutting force impact on workpiece due to tool. Some of nontraditional processes are electro discharge machining, electrochemical machining, abrasive water jet machining, ultrasonic machining etc.

Micromachining

The rapidly developing technology aims at development of miniaturized products with multiple functions embedded in them. In the last few years, CIRP (College International Pour La Researche En Productique) has contributed significantly towards spreading awareness of micromachining and its different methods (Dornfeld et.al.,2006). The continuous demand to follow the trend of miniaturization requires advancement in micro manufacturing techniques.It has wide applications in the field of aerospace, automobile, biomedical, electronics and many others where miniaturization of parts is on high demand. Micro-holes on turbine blades, MEMS, Micro- gears, are few examples of applications of micro-manufacturing processes. Micro-EDM is a nontraditional process which can be an alternative to fulfill these demands of miniaturization of parts to an extent (Hong et. al., 2016).

The basic principle of EDM (Electro Discharge Machining) and Micro-EDM for metal removal is melting and vaporization. The basic difference between EDM and Micro- EDM is the size of tool, the limits of movements of the X, Y and Z axes and the amount of discharge energy supplied.

Electro- Discharge Machining (EDM)

In 1770, English chemist Joseph Prisetly found the effect of erosion from electrical spark (S.Webzell, 2001). Later in 1943 at Moscow University, Dr. B.R. Lazarenko and Dr. N.I. Lazarenko invented the resistance-capacitance (RC) circuit which is also known as Lazarenko circuit for EDM. This was for the first time when the erosion effect was used for machining difficult to cut material. At around same time, three American employees came up with this application of electrical spark for removal of broken tapes and drill in hydraulic valve. This became the basis for electrically controlled servo system for automatically maintaining the gap between tool and electrode (Jameson, 2001). In 1980, with the development of Computer Numerical Control (CNC) in EDM that most of the things became automatic and the process got many more modifications and still the researchers are contributing towards development of this process through modifications (Houman, 1983).

One of the latest and most popular developments of EDM is micro-EDM. It is gaining popularity and attention of researchers because of its diverse application in machining. Some of its application is manufacturing of micro-holes and shafts of 5μm diameter, intricate shapes and complex features of 3D cavities (Ho, 2003). Micro-EDM process has a wide range of applications in aerospace, electronics and automotive industries as it is capable of producing intricate shapes.

It is one of the unconventional processes which have a wide range of application due to its ease and accurate machining performance. The contactless nature of this process avoids vibrations, chatter and mechanical stresses. This process can be used to machine materials with high hardness, high strength and brittleness. The principle of working that EDM follows is the conversion of electrical energy to thermal energy and material removal takes place due to melting and vaporization. The basic requirement of the process is an electrode and workpiece with opposite polarities, a dielectric fluid and power source. The power source is a DC generator that supplies electrical energy to the system. There is a gap between workpiece and electrode which creates a potential difference due to which an electric field is developed. During discharge the electrical energy is converted into thermal energy and plasma is generated between the electrode and workpiece. The temperature of plasma can shoot up to 20,000°C (McGeough, 1988), normally temperature is in the range of 8000°C to 12000°C (Boothroyd and Winston, 1989). Due to the high temperature of plasma the portion where spark strikes the workpiece and electrode gets melted and when the dielectric passes through gap between them evaporates the removed material.

Micro-EDM

(Jahan, 2013) Micro-EDM came into picture in 1968 when Kurafuji and Masuzawa came up with its first application by drilling a minute hole in a 50 μm thick carbide plate (Masuzawa, 2000). Micro Electro Discharge Machining is an advanced manufacturing process which gives freedom to machine any conducting and semi-conducting material irrespective of its hardness with respect to tool. Micro-EDM process has a wide range of applications in industries as it is capable of producing intricate shapes. It is a non-contact process and so the problems related to impact forces of the tool on workpiece are absent. (Baraskar et.al., 2013) Also this process has a very slow MRR and the tool wear is high.

It follows the same working principle as that of the EDM. The basic difference between EDM and micro-EDM is of the amount of energy, temperature of plasma and dimension of electrode and workpiece. Macro EDM requires a DC source which can provide a pulse energy of 1μJ whereas Micro EDM requires few nJ energy and most of micro EDM work on RC circuit so the energy is stored in the capacitors. The temperature of plasma can shoot up to a maximum of 50,000°C while the minimum temperature being 4000°C. In the case of micro-EDM, polarity of the electrode is kept as negative to reduce tool wear whereas in the case of Macro-EDM the polarity of the electrode is kept as positive.

There are different approaches to perform micro-EDM processes. Only two of them are discussed in details, as these two approaches are used in the experiment that is discussed later in this chapter.

Die-sinking micro-EDM is one of the most frequently used approaches as it reduces the machining time when compared to other EDM approaches. The die sinking micro-EDM has a tank in which workpiece and electrode set up is there. This tank is completely filled with the dielectric fluid during the process. This approach is preferred for complex and intricate shapes.

Micro-Drilling on Micro-EDM is preferred over conventional drilling when the precise hole is required on a hard workpiece. Drilling on Micro-EDM is also preferred when high aspect ratio is requirement of the job. A cylindrical electrode is used for machining in micro-EDM. One of the important differences between micro drilling through EDM and conventional drilling is the tapering effect. Although this defect cannot be totally eliminated but Micro-drilling on EDM has less tapering effect compared to conventional drilling. (Modica et.al., 2017)

Many other approaches are micro-EDM milling; micro-turning, wire electro discharge grinding and wire cut EDM.

Powder Mixed Micro-EDM

To increase and enhance the machining efficiency of the Micro-EDM process, many new modifications were tried by researchers. Powder mixed micro-EDM is one of them. So far it has been concluded that if at the right concentration, particles of a particular size are mixed with the dielectric, it can be helpful in increasing MRR, decreasing TWR and increasing the surface quality.

LITERATURE SURVEY

Powder mixed micro-EDM is used for good surface finish and high MRR. Due to presence of particles in the dielectric fluid bridging effect occurs which increases the gap between workpiece and electrode. This helps the dielectric to flush away debris. The effect on surface quality due to deposition of debris, short circuiting and arcing gets lowered down due to its proper flushing.

Chaudhary et.al., 2017 has investigated the powder mixed EDM process on EN-19 steel with copper tungsten electrode. Tungsten powder was added in the dielectric fluid. The process parameters decided by authors were powder concentration, peak current and pulse on time. The performance parameters were MRR, Surface quality and TWR. They concluded from the paper that by regression analysis that the most significant process parameter is powder concentration.

Tripathy and Tripathy, 2017 have investigated the powder mixed EDM process on H-11 die steel with SiC powder mixed in dielectric fluid and copper as electrode. A separate tank of 20 litres capacity was installed with a pump and stirrer to maintain the circulation of the dielectric fluid and not letting the powder to settle down. It was reported by the authors that MRR and surface finish gets improved with addition of powder in dielectric fluid.

Toshimitsu et. al., 2016 have studied the effect of chromium powder on surface quality enhancement of EDM process. Alloy Tool steel SKD 11 was machined with a cylindrical copper electrode having diameter of 25 mm. Chromium layer is deposited on the workpiece surface and it enhances the surface finish, water repellency and corrosion resistance of the workpiece.

Talla et. al., 2016, investigated the powder mixed EDM process with an Inconel 625 machined with a copper electrode. Graphite powder was mixed in the kerosene dielectric fluid and an arrangement for continuous stirring is done to not let the powder particles settle down. The effect of the powder concentration is that the surface roughness decreases, crater size increases and the thickness of recast layer get reduced.

Kuriachen and Mathew, 2015 have studied the powder mixed micro-EDM process to enhance the surface quality and machining parameters. Silicon Crabide (SiC) micro particles was mixed in EDM-3 oil dielectric to enhance the performance efficiency when working on a titanium alloy Ti-6Al-4V workpiece with a tungsten carbide electrode. It is observed by the authors that powder concentration is one of the significant factors for influencing the performance parameters.

Zain et.al., 2014 have discussed the effect of tantalum carbide powder in kerosene dielectric fluid on EDM of stainless steel with copper electrode. It has been observed by the authors that Tantalum Carbide(TaC) powder gets deposited on the surface of workpiece and increases the micro hardness and corrosion resistance of the machined surface. The corrosion resistance enhancement was achieved due to addition of powder in the dielectric fluid because in conventional EDM process the corrosion resistance decreases.

Adepu and Kolli, 2014, have studied the PMMEDM titanium alloy. For aerospace industry Titanium alloy is important as it has high strength, excellent thermal properties, is light in weight, has low modulus of elasticity, corrosion resistance and fracture resistance. The powder that is added in this case was sorbitan (4g/lt.) and graphite (14g/lt.). The addition of powder was done with continuous stirring to check the uniform mixing. Surfactant was added in order to reduce the surface tension of sediments particles, to distribute uniformly graphite particles and during discharge they get oxidized and will increase the explosion. According to author, powder suspension is found to increase MRR and decrease TWR.

Prihandana et. al., 2012, suggested that if the flushing of debris is not proper it will lead to short circuiting which will result in increase of machining time. It is reported that the introduction of nano-graphite powder in the dielectric fluid was found to be helpful to fix this issue. It is also found that this has resulted in speeding up the process to five times faster. The workpiece is vibrated with the help of PZT actuator in Y- axis direction. Stirrer is used to mix the powder in the dielectric fluid, formation of chain by these nano-particles leads to bridging effect between the electrode and workpiece, which results in increasing the gap between them and removing debris through effective flushing.

Kung et.al., 2009 investigated the process of PMEDM with a conductive aluminum powder on a cobalt-bonded tungsten carbide workpiece with a copper electrode. The aluminum powder was added in different grain size and concentration in the commercially available dielectric fluid mineral oil EDM44. A circulation system is provided with dielectric system to not let the powder settle down. With the increase in grain size MRR and TWR increases. With increase in powder concentration till a certain value MRR increases and TWR decreases after that if further concentration is increased MRR decreases and TWR increases.

Lee and Tzeng, 2001, studied effect of different additives which includes copper, aluminum, silicon carbide and chromium on EDM process. The author has done selection of additives keeping in mind the thermo-physical properties. They concluded that chromium is most effective as the MRR is highest and TWR is lowest and copper should not be used as additive because it had no effect on the process.

The conclusion that can be drawn from the literature review of 10 papers is addition of powder to dielectric is effective for EDM as well as micro-EDM processes. If powder is added in right concentration, the performance parameters will be enhanced and the process will become more efficient. Table 1 shows the different types of powder used in PMEDM process and the research findings associated with covered papers.

The conclusion that can be drawn from the literature review is addition of powder to dielectric is effective for EDM as well as micro-EDM processes. If powder is added in right concentration, the performance parameters will be enhanced and the process will become more efficient.

INTRODUCTION TO ARTIFICIAL NEURAL NETWORK

Neural network is one of the soft computing tools that are basically used for stochastic processes. Artificial Neural Network is a network architecture that can learn from a set of experimental data to describe a non-linear and interaction effect with great success (Chandrashekar et. al., 2010).The advantage of using ANN for making a model is that the model can be easily constructed by giving a set of input and output data and then it can be trained with a set of data to accurately predict process dynamics (Somashekhar et.al., 2010). ANN is useful whenever complete understanding of a physical process is not there. Whenever it is difficult to obtain an analytical model based on the physics of the process, ANN can be used to build a model.

Table 1. Different types of powder used in PMEDM process and the research findings associated with covered papers

S. No.	Authors	Powder Mixed in Dielectric Fluid	Research Findings
1.	Chaudhary et. al., 2017	Tungsten powder	According to Regression analysis powder concentration is most influencing factor for performance parameters.
2.	Tripathy and Tripathy, 2017	SiC Powder	Surface roughness decreases and MRR increases
3.	Toshimitsu et. al., 2016	Chromium powder	Surface Roughness decreases, water repellency and corrosion resistance increases.
4.	Talla et. al., 2016	Graphite powder	Surface roughness decreases, crater size gets enlarged with increase in the concentration of powder, surface cracks are less and a thin recast layer is formed
5.	Kuriachen and Mathew, 2015	SiC micro Particles	High MRR and low TWR are achieved at a fixed powder concentration increasing this can contribute in deceasing MRR and increasing TWR
6.	Zain et.al., 2014	Tantalum Carbide	Increases micro-hardness and corrosion resistance of the machined surface.
7.	Adepu and Kolli, 2014	Graphite and sorbitan	Powder suspension increases MRR and decreases TWR.
8.	Prihandana et.al., 2012	Nano- graphite powder	Machining time can be reduced.
9.	Kung et. al., 2009	Aluminum Powder	MRR and TWR increases with increase in grain size. MRR increases and TWR decreases to o a certain value of powder concentration. If the powder concentration is increased further MRR decreases and TWR increases.
10.	Lee and Tzeng, 2001	Copper, aluminum, silicon carbide and chromium.	Chromium is the most effective powder. Copper should not be used as additive in PMEDM. The concentration of powder is important factor. Too high or low concentration can lead to decrease in the efficiency of the process.

ANN structure is based on the human nervous system. The architecture of ANN is inspired from the neurons of human brain. The network of any ANN model has at least three layers; first is the input layer, second layer is the hidden layer and third is the output layer. The main elements of the network are activation function and the weight factors. The weight factors are adjusted to minimize the error. Error is the difference between the output of network and targeted output of the process. In this work, the error is reduced by using the back-propagation method and the network is constructed by using feed forward network. (Porwal and Yadav, 2012).

ARTIFICIAL NEURAL NETWORK BACKGROUND

ANN acquires knowledge through a learning process. (McCulloch and Pitts, 1943) In this research the first mathematical model of the neuron was outlined. (Hebb, 1949) In this work for the first time the concept of learning of networks was introduced through synaptic weights. (Rumelhart et.al., 1986) researched and found the back propagation algorithm which resulted into a powerful tool to minimize error. Since then, researchers have been found exploring the wide range of applications of ANN in different fields. These fields include pattern recognition, industrial processes, manufacturing processes like welding, machining, molding and many others. Many authors (Coit et.al., 1998, Haykin, 2008, Al-Ahmari, 2007) have chosen ANN to solve problems due to two reasons; the first is that it learns through examples and the second, it can generalize the network for unseen data also. Learning happens with a set of data which is collected through experimental runs. The more the number of runs the better will be the accuracy of network. The generalizing feature of ANN is unique as it can produce results with a data set which was not used for training the network.

Tsai and Wang, 2001, have done estimation of surface finish for EDM process when electrode polarity is varying. The author has done this through six different neural network models and one neural fuzzy model. The network architecture is based on previous semi empirical models developed. The number of input parameters is taken as 5 and output parameter as 1, the number of hidden layer is taken as 1 by the author to start the analysis. The six different neural network models were logistic sigmoid multilayered perceptron (LOGMLP), hyperbolic tangent sigmoid multi-layered perceptron (TANMLP), Radial Basis function network (RBFN), fast error back propagation multi layered network with hyperbolic tangent function (error TANMLP), TANMLP with adaptive learning rate (adaptive TANMLP), Radial Basis function network with adaptive learning rate(adaptive RBFN) and Neuro-fuzzy model was Adaptive network based fuzzy inference system. It has been concluded on the basis of result showed that TAN-MLP, RBFN, adaptive RBFN and ANFIS models have smaller error compared to other models and experimental verifications agree with the obtained results. Tsai and Wang, 2001, followed the research work by estimation of Material Removal Rate for the EDM process keeping all the process parameters same as in the previous work. The estimation is done following the same six neural network models and one neuro-fuzzy model. In the work it is mentioned that for the adaptive network based fuzzy inference system (ANFIS) the input parameters should be less than 7 with one output parameter. It is found that the previous semi empirical models satisfy the conditions for constructing an architecture of ANFIS model. The results show that ANFIS is most accurate amongst all the 7 models considered for estimation of MRR. These models were found to be applicable for industrial applications. Panda and Bhoi, 2005, have used the Multilayer perceptron feed forward network with Back propagation algorithm to predict the MRR for EDM process. The number of neurons and hidden layers is obtained with the help of trial and error method. Three network architectures were selected based on the R- square coefficients which are 3-4-3-1, 3-7-1 and 3-10-1.It is found that 3-7-1 network results has closest R-efficiency and so it is selected for further investigation of this work. The author has mentioned that increase in the number of epochs increases the efficiency of the model but after a certain limit when the epochs are increased it results in decrease of the efficiency of model which indicates the saturation of training program. The model developed has not much sensitivity to noise. Cus and Zuperl, 2006, developed a neural optimization algorithm to develop a model which is helpful to determine optimum cutting parameters for turning operation. The architecture of neural network model is 3-3-6-1. Two neural networks models are used, feed forward and radial basis neural network. It is found that feed forward network results are more accurate but time taken by algorithm to do computation is more when compared to RBNN. Mandal et.al., 2007, has discussed about a feed forward neural network with back propagation learning algorithm which uses gradient search technique to minimize the mean square error of the output of the network. The optimization of the process is done through non-dominating sorting genetic algorithm-II. The authors have also mentioned that ANN is selected for this process as the physics of the process is not clear. The architecture of the network was 3-10-10-2 with learning coefficient of 0.5. It is found that the model is suitable to predict the output responses. Rao et. al., 2008, discusses a multi layered perceptron neural network which is used to build a relationship between process parameters and performance parameters in die-sinking EDM process. The input parameters that were selected are peak current, machining time, type of material and voltage and the output parameter is MRR. Sensitivity analysis is used to estimate the relative influence of process parameters on performance measures. Genetic algorithm is used to adjust the weights as well as for the optimization of process parameters. It is found that type of material has the highest influence on the performance measures.

Joshi and Pande, 2009, In their research work discuss about the thermal modeling of the EDM using Finite Element Method (FEM) and the relationship is being established between process parameters and performance parameters using artificial neural network (ANN). The network architecture used for the analysis is 4-8-12-4. Two ANN models are tried, Radial basis neural network (RBNN) and Back propagation neural network (BPNN). It is found that BPNN produces more accurate results when compared to RBNN. Pradhan and Chandan, 2010, discussed three models namely; artificial neural network and two neuro-fuzzy models to estimate the MRR, TWR and radial overcut in a die sinking EDM process. The artificial neural network used for this work is feed forward neural network with back propagation learning algorithm with architecture of 4-1-1. The learning rate was 0.05 and momentum parameter is 1.05 for back propagation leaning algorithm. The analysis of variance suggests that discharge current is the most influencing factor for MRR than any other parameter. It is found that ANN and two neuro-fuzzy models predict results under acceptable limits. Gaitonde and Karnik, 2012, have optimized the process parameter for burr size during drilling operation with the help of Artificial Neural Network (ANN) and particle swarm optimization (PSO). A multi- layered feed forward network with error back propagation learning algorithm with architecture of 3-13-2is constructed to estimate the burr size of the drilling operation. The input parameters decided by the authors are feed, drill diameter and point angle and the response parameters are burr height and burr thickness. The neural network and PSO combination produce results in an acceptable range. Hence this combination can be used for building relationship and optimizing the process parameters to produce response.

Suganthi et. al., 2013, in this paper, two approaches of soft computing are used for prediction of quality responses of the micro-EDM process. The two approaches, ANFIS- adaptive neuro- fuzzy inference system and Artificial Neural network with back propagation training algorithm. The architecture for ANN model is 4-6-6-6-3, which is decided using trial and error method. Both the models produce results within acceptable limit and are used to build relationship between input and output parameters. According to the authors, ANFIS model is faster and more accurate when compared to BP- based ANN.

Majumder, 2014, In this paper, ANN is used to establish relationship between input parameters and output parameters and three different algorithms, simulated annealing, genetic algorithm and particle swarm optimization are used to optimize the process parameters so as to result in maximum MRR and minimum wear ratio. For training the neural network back propagation algorithm is used. BPNN model acts as the fitness function to optimize the process parameters. It is concluded that Particle Swarm Optimization is better than simulated annealing algorithm and Genetic Algorithm optimization for modeling and optimization of EDM process.

The training of network is done by two ways: supervised and unsupervised training. The supervised training of network is done with a set of input data and a set of targeted output data. Now, the network tries to train itself and produce output close to or equal to targeted output data. Whereas, the unsupervised training technique does not require targeted output data, which means only a set of input data is provided to the network and it learns by its own discovering and adapting to structural features in the input pattern. (Pattanaik, 2017)

Table 2. Different combination of neural networks used for micro-EDM

S. No.	Authors	Neural Network Used	Research Findings
1.	Prabhu and Vinayagam, 2015	Adaptive Neuro Fuzzy Inference System (ANFIS)	Used for predicting Surface Roughness. This approach can be used for on line control and optimization.
2.	Majumder, 2014	ANN with BP training algorithm	Relationship between input and output parameters can be built through ANN.
3.	Suganthi et. al., 2013	ANFIS and BP based ANN	Relation between process and performance parameters is built. ANFIS is faster than BP based ANN.
4.	Gaitonde and Karnik, 2012	BP based ANN with PSO	Optimization of process parameter for burr size of drilled micro holes.
5.	Pradhan and Chandan, 2010	BP based ANN and Neuro Fuzzy Network	It is able to predict optimal values of performance parameters for die-sinking EDM.
6.	Joshi and Pande, 2009	BPNN and RBNN	BPNN is faster than RBNN. Relationship between input and output parameters can be built through neural networks for calculations of optimal performance parameters.
7.	Rao et. al., 2008	MLP-ANN with GA	Optimization of process parameters for die-sinking EDM. Most influencing factor can be predicted.
8.	Mandal et.al., 2007	ANN with BP algorithm	ANN should be used were nature of process is stochastic.
9.	Cus and Zuperl, 2006	Feed Forward Neural Network and RBNN	Optimum cutting parameters for turning operation is predicted. Feed forward neural network is accurate but slow than RBNN.
10.	Panda and Bhoi, 2005	ANN with BP learning algorithm	MRR can be predicted with ANN models for EDM process. Epochs can result in increasing the efficiency of network to a certain limit after that if epochs are increased it can lead to lowering down the efficiency.

ARCHITECTURE OF ARTIFICIAL NEURAL NETWORK

(Fabricio et.al., 2010) In this paper, the network architecture and its working has been explained. The working is each neuron in the input layer is assigned to a parameter from the input data set. In the input layer, the neuron is responsible to produce an output which is a scaled value of each input parameter data. The hidden layer, in case of multilayer feed forward network is present between the input and the output layer. The job allotted to the neurons of the hidden layer is the computation of an output which is a scalar product of neuron's input vector and the vector of weights associated to its input. The result produced is compared with a threshold limit. If the value exceeds the threshold limit then the scalar product is used as an independent variable of an activation function whose output will be neuron's output.

Any neural network can be represented with the help of directed graph (see Figures 1-3).

The ANN architecture can be basically two types:

- Single layer feed forward ANN
- Multilayer feed forward ANN

Figure 1. Representation using directed graph

Figure 2. This symbol indicates the neuron

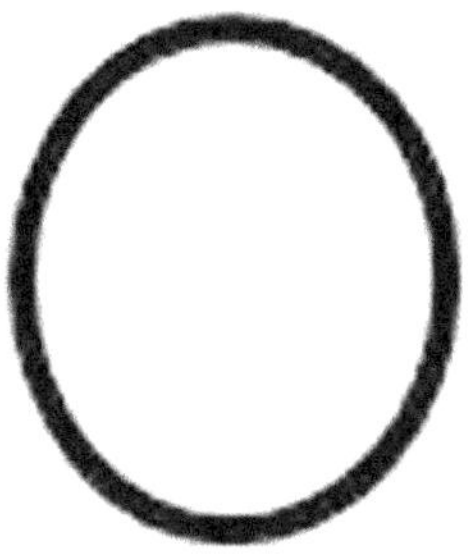

Figure 3. This symbol indicates the direction of flow of network computation

Single Layer Feed Forward ANN

As the name suggests single layer feed forward will have single layer for input to be fed in the network. The structure of the network is shown in Figure 4.

It has two layers, input layer and output layer. All signals move in one direction. It is not reverted back. Although it has two layers the input layer and output layer but since all computation is alone done with output layer so its name is single feed forward network. The input layer has m number of neurons and output layer has n number of neurons. The $x_1, x_2, \ldots, x_m$ are the input parameters and $y_1, y_2, \ldots, y_n$ are the output parameters. The $w_{11}, w_{12}, \ldots, w_{mn}$ are the weights of synaptic link in the format w_{ij} which is the weights of synaptic link between i^{th} and j^{th} neuron where i and j are the index for input and output neurons.

Multilayer Feed Forward ANN

It has a minimum of three layers. The two layers, input and output layers are common to a single feed forward network and one additional layer named as hidden layer is added to increase the calculation

Figure 4. The network architecture of single layer feed forward network

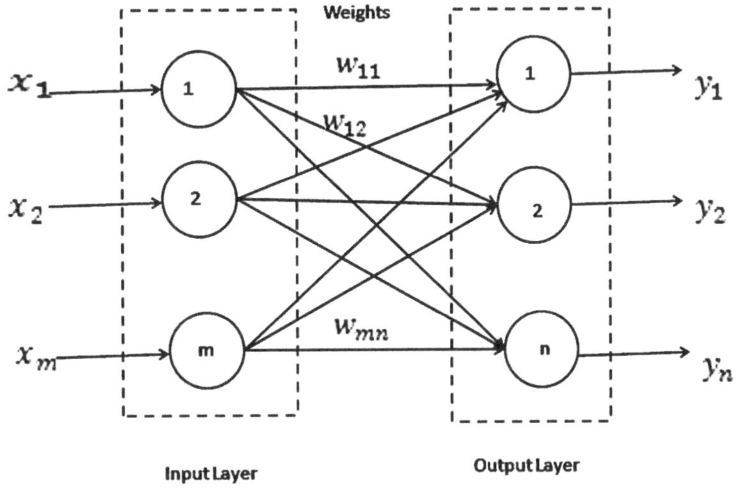

accuracy. Now the hidden layer and output layer neurons will be involved in computational concerns. Again, the input layer will not be involved in any computation. The network structure of a Multilayer feed forward network is shown in Figure 5.

The nomenclature of this network is done (l-m-n) where l, m, n indicate the number of neurons in the input layer, hidden layer and output layer respectively.

Figure 5. The architecture of multilayer feed forward network

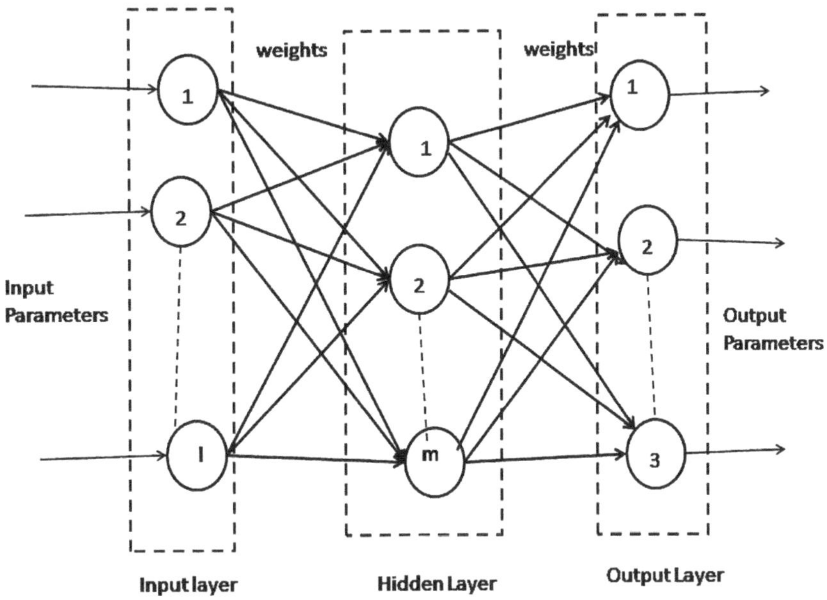

Neural Network on MATLAB

MATLAB – MATrixLABoratory is a software for numerical computation and visualization. It is a versatile and a multi-dimensional software. One of its tools is Neural Network.

Neural Network design steps:

Step 1: Choose the area in which the problem falls. It has four application areas:
Function fitting
Pattern recognition
Clustering
Time series analysis

Step 2: Collect data. For ANN it is suggested to collect more data. The accuracy of network increases with the increase in the amount of data.

Step 3: Creating a network. Choose the number of input parameters and output parameters. In this step the target output is also decided for which the network will be trained.

Step 4: Configuring the network which includes trying and finding the number of hidden layers and number of neurons that will provide an optimized result for the network.

Step 5: Initialize the weights and biases. In MATLAB this can be done automatically by use of the software i.e. selecting the weights randomly. The weights that each input parameter should be given can be decided manually also and then it can be fed to the network.

Step 6: Training the network. A part of the total data is used to train the network. Training the network makes the network adapt to the working condition with given input parameters and target output.

Step 7: Validating the network. This step is done to check the status of training of the network. If training of network is done properly the network can produce output for any randomly selected parameter. Validation of network is done by randomly selecting data from the set of data provided for training. If the result is close to targeted output then the network is trained properly.

Step 8: When the validation step is over and the regression value (a graph is obtained with a black continuous line which indicates the targeted output and a scattered output produced through network. regression value is the value of deviation from targeted slope of line.) is equal to or above 0.93, the network is ready to be used for a generalized data set of that particular working condition.

In this chapter the problem falls under fitting. To solve a problem there are two ways:

- Graphical User Interface (GUI)
- Command line function

In this chapter command line technique is used for optimizing the Material Removal Rate. The GUI can also be used and then automatically command line script can be generated.

Objective

Model Development with the help of Neural Network:

A model is being developed using Artificial Neural Network (ANN) to determine the Material Removal Rate (MRR) for Micro-EDM.

Effect of different types of additives on Micro-EDM process:

Determine the effect of conductive (Aluminum) as well as semi conductive (Silicon Carbide) additives on the Micro-EDM process.

Case Study

In the EDM process the gap between tool & workpiece is more than 10 μm, so during the process of material removal, the gap is enough to let the dielectric flush away debris easily. In contrast to this, in micro-EDM the gap between tool and workpiece is very small, in the range of 1 to 5μm which makes the flushing away of debris difficult.

If the debris is not flushed away properly, it will have the following effects on the process:

- The debris will get accumulated in this gap and deteriorate the surface finish of workpiece.
- The MRR will also get reduced as debris will obstruct the flow of dielectric.
- The tool wear will increase due to the clogging of debris instead of sparking, arcing results.
- The surface finish is also affected as the clogging results in short circuit which eventually leads to arcing.
- All these problems will further increase as the depth of cut will increase.

The solution to these problems is to increase the gap between tool and workpiece which can provide this process with proper flushing away of debris with the help of dielectric.

In this research, additives are added in the dielectric fluid and then it is passed from the gap. It is found that this slight modification in the process solves these problems to some extent.

Process Parameters

Selection of process parameter depends on the following factors, the material of electrode and workpiece, the feature and final dimension required on the product, the surface quality of the finished product, the amount of energy required for the erosion process to occur.

There are many other factors which can be considered other than these for selection of process parameters.

The main process parameters that are used as input parameters for micro-EDM process can be:

- Voltage
- Capacitance
- Current
- Electrode polarity
- Delay time
- Frequency
- Dielectric fluid
- Tool rotation speed
- Tool feed rate

Mostly, current is calculated after measuring other parameters. Frequency is the reciprocal of total time interval within which spark takes place. Electrode polarity can be negative or positive depending upon the work requirement. Choice of dielectric fluid is important as it responsible for the flushing the debris. If the debris are not flushed away properly the efficiency of the process drops as short circuit occurs frequently which increases machining time. The dielectric fluid should be eco-friendly and safe. The tool rotation speed is responsible for uniform tool wear and it can also assist removal of debris. If the removed material gets deposited it lowers down the efficiency of process as well as the dimensions of final product is also disturbed.

In order to understand the effect of different process parameters on MRR following parameters are being investigated:

- Voltage
- Capacitance
- Additives

Performance Parameters

There are some parameters to check the efficiency of the electro discharge machining process. These parameters which help us define the quality of work done are known as performance parameters. Some of the performance parameters for micro-EDM process are Material Removal Rate (MRR), Tool Wear Rate (TWR), surface roughness and dimensional accuracy. (Rahman,2012)

These parameters are checked by different equipment and instruments. The MRR can be checked with the help of a weighing balance. Atomic Force Microscope (AFM), scanning tunnel microscope(STM) can also be helpful to define the profile. Tool Wear Rate can be checked with Laser micrometer or image processing. The surface quality and dimensional accuracy can be checked with the help of profilometer, Scanning Electron Microscope (SEM) and Energy Dispersive Spectroscopy (EDS).

The input parameters were voltage, capacitance, electrode cross section area and additives. The output parameter is Material Removal Rate (MRR).

SELECTION OF RANGE OF PROCESS PARAMETERS

Literature Survey

Mahendran et.al., 2010, has done the review of micro-EDM process. The author has found that the working principle of EDM and Micro-EDM is similar only with difference in the amount of discharge energy. The process and performance parameters has been found to be similar. Some of the process parameters that were discussed are discharge voltage, peak current, pulse duration, pulse interval, pulse waveform, polarity and electrode gap. The performance parameters were MRR, surface finish and tool wear ratio. Banu et.al., 2014, discussed the investigation of micro-EDM process for a ceramic workpiece using assisting electrodes. The electrode used to machine it is tungsten with an assisting electrode. The die-sinking approach of micro EDM is been investigated with focusing the analysis on MRR and hardness of recast layer. It is found that copper adhesive as assisting electrode gives better results as compared to other considered assisting electrodes. The process parameters considered were gap voltage 80-110V, feed rate 1-6 μm/s, and rotational speed of 100-700 rpm. Jahan et. al., 2015, has discussed about micro-EDM drilling process on aluminum alloy (AA 2024) with tungsten electrode of 0.3 μm.. The optimum process parameters found to maximize MRR and minimize the tool wear ratio were voltage, capacitance, resistance and spindle speed. Saxsena et. al., 2016, has investigated machining of ceramic on Micro-EDM. The optimization of the process parameters is done using ANOVA. It has been concluded that the process performance depends on all the three process parameters gap voltage, capacitance and threshold value. The performance parameters considered were MRR, tool wear ratio, surface finish and radial over cut. Voltage was taken in the range of 110-150 V, capacitance 10nF- 0.4μF and threshold as 40 -80%. It is found that the discharge energy depends on the product of capacitance and square of voltage. Any variation in discharge energy is proportional to capacitance and voltage.

From the literature review it is concluded that voltage and capacitance are two process parameters that have major influence on the micro-EDM process performance. So for this study voltage and capacitance are taken as two process parameters and to check the effect of additives on the performance one more factor is considered as additives.

- **Voltage:** The values of voltage can range between 80 – 130 V for the μ- EDM machine considered but for milling and drilling operations, the voltage is taken between 105- 115 V because below 105 V the MRR is very less and above 115 V the deposition of debris on workpiece and electrode is predominant.
- **Capacitance:** The range of capacitance available on this machine is 0- 6 pF. It is observed that below 4 pF the MRR is very low so the optimal range for capacitance is 4-6 pF.
- **Additives:** To understand the effect of additives on the process the following three cases are considered for investigation; without additive, with conductive additive and with semi-conductive additive. The conductive additive used is aluminum powder and the semi-conductive additive used is silicon carbide powder. The semi-conductive SiC powder is chosen as it is used as abrasive in machining processes so in this process also with the addition of this powder the material removal can be enhanced and also the surface finish is expected to improve.

Preparation of Electrode

The electrode is prepared on the hybrid Micro-EDM machine (DT-110 Model) MIKROTOOL using turning process. A 4*mm* brass rod is mounted on the spindle and the turning process occurs. The turning process occurs between the carbide tool present on the machine and the rotating brass rod. The tool preparation is done on the same machine because if not, while mounting on the spindle problems like deflection and centering are likely to occur.

Experimental Set-Up

In this experiment the circulation of powder and flushing is a problem. In this machine, the dielectric flushing has to pass through a filter channel. This filter channel will filter additives also. So to maintain the additives in dielectric an auxiliary arrangement was done. In the whole set-up one extra jar is attached. This jar will collect used dielectric mixed with additive from the dielectric tank. For the circulation two submersible pumps are fixed. One of the pumps is in the dielectric tank and the other in the jar. These pumps regulate the flow of dielectric mixed with additives from tank to jar and jar to tank. (Upadhyay,et. al., 2016)

The input parameters for the process are capacitance and voltage and the output parameter is Material Removal Rate (MRR). Theoretical MRR is calculated with the formula:

$$MRR = \left(Volume\ of\ Machining\right) / \left(Total\ time\ of\ machining\right) \quad (1)$$

Time is recorded with the help of a stopwatch. The operation that is considered for this study is drilling. The dimensions of the machined part are diameter of 0.5mm and depth of 0.1mmand the diameter of the tool is 0.5 mm.The volume of machining is calculated with the formula:

$$Volume\ of\ Machining = depth \times \frac{\pi}{4} \times diameter\ of\ electrode^2 \quad (2)$$

The result shows that the experimental and theoretical values of MRR are not the same. The experimental MRR is obtained with an error of 2.405% when compared to the theoretical MRR.

Table 3. Process parameters and their range considered for this experiment

Parameters	Units	Values
Voltage	Volts (V)	105, 110, 115
Capacitance	pF	4,5,6
Additives	-	No additive, conductive, semi-conductive

Table 4. Different additives used and their properties

Name of Additive	Type of Additive	Concentration (gram/Litre)	Conductivity (S/m)
Al Powder	Conductive	4	3.50×10^7
SiC powder	Semi Conductive	2	310

Table 5. The conditions under which experiment was performed and the equipment used

Machine	Micro-EDM Machine, Model- DT110, MIKROTOOL
Workpiece (Anode)	20×20*mm*, brass sheet
Electrode (Cathode)	0.5*mm* diameter, brass rod
Dielectric Fluid	EDM oil
Thickness of Workpiece	0.1*mm*
Cross Sectional Area of Electrode	$0.19625mm^2$

Table 6. The input parameters, experimental and theoretical MRR

RUN	Voltage	Capacitance	MRR (Experimental)	MRR (Theoretical)
1	110	6	0.00677	0.0062
2	110	6	0.00643	0.0062
3	110	6	0.00735	0.0062
4	110	5	0.00878	0.00889
5	110	5	0.00934	0.00889
6	110	5	0.00835	0.00889
7	105	4	0.0022	0.002
8	105	4	0.00242	0.002
9	105	4	0.00238	0.002
10	110	4	0.00291	0.003
11	110	4	0.003911	0.003
12	110	4	0.00256	0.003
13	115	5	0.00919	0.0093
14	115	5	0.00892	0.0093
15	115	5	0.00878	0.0093

Implementation of ANN for Calculation of MRR

The approach for solving the problem statement isfirst with the help of MATLAB codes number of neurons is selected which can give minimum root mean square error (RMSE) and then a network is created with the optimum number of neurons and number of hidden layers.

The network which has two input parameters (Voltage and Capacitance), four neurons in the hidden layer and one output parameter (MRR). The result is shown in Table 9.

The result shows that the network has 3.52% error. Therefore, the output produced is 96.48% accurate.

RESULTS AND DISCUSSION

In this work, an attempt has been made to find the significant factors that influence the MRR of Micro-EDM process with additives in the dielectric fluid. The ANN result predicts that the network constructed has 96.48% accuracy. The error that is obtained in the network is within acceptable limits. The input parameters for the processes were capacitance and voltage. The output parameter for this network was MRR. Optimum result is obtained with a network structure of 2-4-1 which means 2 neurons in the input layer, 4 neurons in the hidden layer and 1 neuron in the output layer. The network is limited to 4 to 6 pF capacitances and 105 to 115 V voltages. Additionally, the effect of a conductive additive is more than a semi-conductive one. When the material removal rate was compared it was found for conductive additive it is more whereas for semi-conducting it is less.

Table 7. The Root Mean Square Error corresponding to number of neurons

S. No.	Number of Neurons	Root Mean Square Error
1	1	0.011
2	2	3.77E-04
3	3	6.28E-04
4	4	3.42E-04
5	5	3.59E-04
6	6	3.99E-04
7	7	6.08E-04
8	8	0.0022
9	9	3.65E-04
10	10	5.83E-04
11	11	3.50E-04
12	12	3.75E-04
13	13	3.56E-04
14	14	3.64E-04
15	15	3.43E-04

Table 8. Number of hidden layers with 4 neurons and corresponding RSM error

S. No.	No. of Layers	Root Mean Square Error
1	1	3.42E-04
2	2	0.0017
3	3	3.45E-04
4	4	3.44E-04
5	5	3.50E-04
6	6	0.0069
7	7	3.79E-04
8	8	3.64E-04
9	9	0.0032
10	10	3.10E-03

The graph shows that the targeted output and the actual output obtained from ANN has close proximity. On the y-axis is MRR and on the x-axis, the number of experimental runs. Since the results are in close proximity it can be used to calculate the MRR for Micro-EDM process when a brass sheet is machined with an electrode of brass with input parameters as voltage and capacitance.

Error histogram shows the spread of error. The yellow vertical line indicates zero error and it is called zero error line. The bars indicate the data having that error. The bar has three colors, blue, red and green. Blue color indicates training data, red indicates test data and green indicates validation data.

Table 9. MRR targeted and MRR through ANN

S.No.	MRR (Tagret)	MRR (ANN)
1	0.0062	0.0062
2	0.0062	0.0062
3	0.0062	0.0062
4	0.00889	0.0091
5	0.00889	0.0091
6	0.00889	0.0091
7	0.002	0.0025
8	0.002	0.0025
9	0.002	0.0025
10	0.003	0.0025
11	0.003	0.0025
12	0.003	0.0025
13	0.0093	0.0091
14	0.0093	0.0091
15	0.0093	0.0091

Figure 6. The plot between MRR and no. of experimental runs; the target output and ANN output comparison

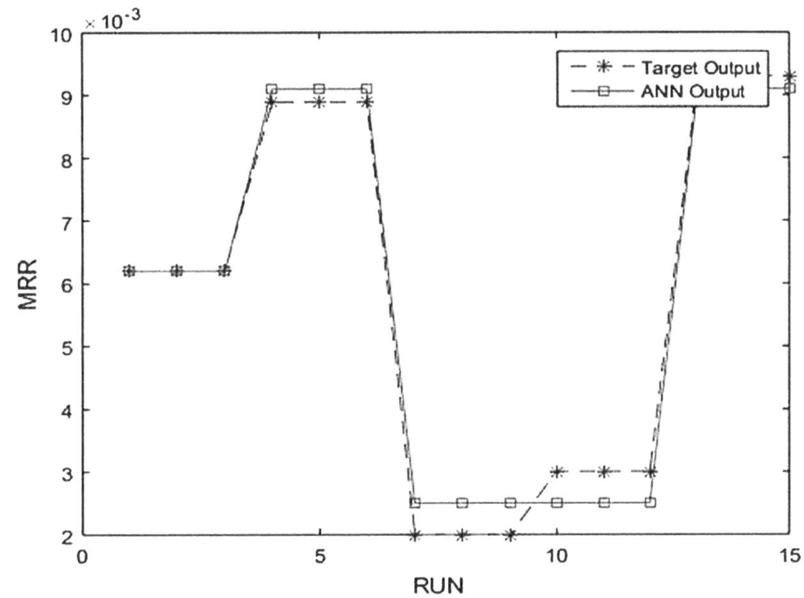

Figure 7. Error histogram for network 2-4-1

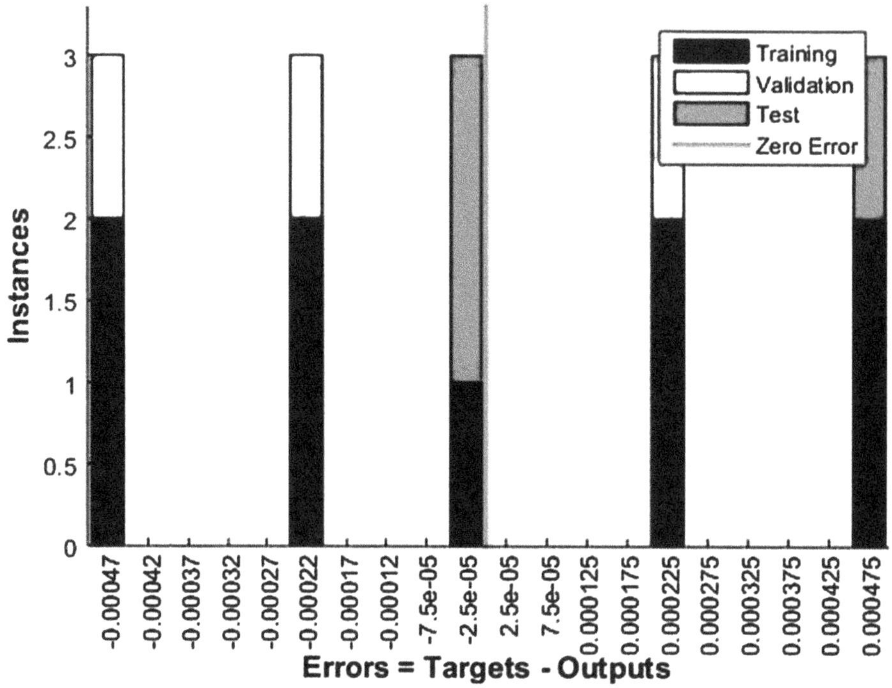

Figure 8. Plot between the mean square error and epochs for network 2-4-1

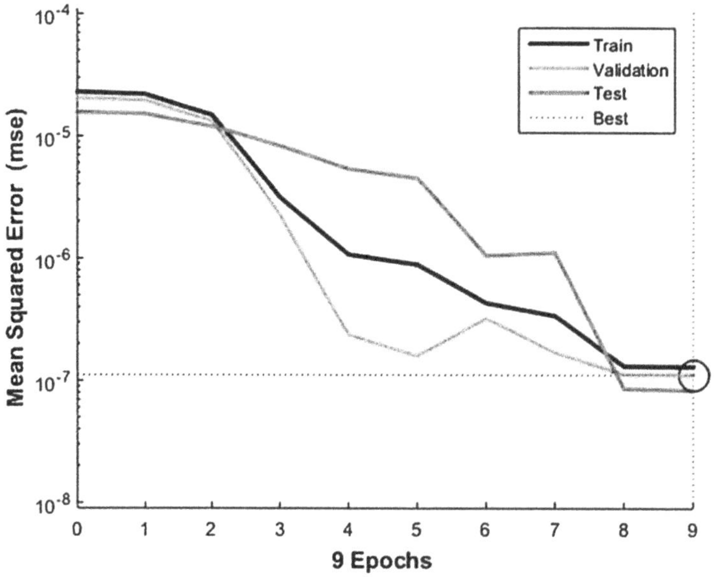

The bar which is closest to the histogram indicates that has the least error. The network 2-4-1 has least error of -2.5e-05 with training data for 1 instance and testing data till 3 instances starting from 1.

The lines indicate the variation of mean square data with epochs. The colors blue, red and green indicate the training, testing and validation data respectively. The plot above shows that validation and training data are close at epoch 9. Close to epoch 8 all the three data lines come close to each other and hence show that the network is trained and is giving output close to the target output.

FUTURE SCOPE

- Use of Back Propagation Neural Network learning algorithm can improve the result.
- The network accuracy can be increased if the number of experimental runs is increased. So,the same network architecture can be checked for calculation of MRR with more data.
- Energy distribution among tool, work-piece and dielectric can be done which will be helpful in thermal analysis of this process.
- The thermal analysis can be used to predict tool wear.
- This work is limited to one-dimensional dielectric flow, it can be further analyzed with multi-dimensional flow.
- In this work the work piece is a conductive material, this work can be extended to a work piece which is semi-conducting or insulating.
- The machining of ceramics and glass on micro-EDM can be thought of, as machining of hard and brittle materials with conventional machining techniques is extremely difficult.

REFERENCES

Adepu, K., & Kolli, M. (2014). Effect of Additives Added in Dielectric Fluid on Electrical Discharge Machining of Titanium Alloy. *International Conference onAdvanced Research & innovations in Mechanical, Material science, Industrial Engineering & Management*, 247-250.

Al-Ahmari, A. M. A. (2007). Predictive machinability models for a selected hard material in turning operations. *Journal of Materials procesing. Technology (Elmsford, N.Y.)*, *190*, 305–311.

Banu, A., Ali, M. Y., & Rahman, M. A. (2014). Micro-Electro Discharge Machining of non-conductive zirconia ceramic: Investigation of MRR & recast layer hardness. *International Journal of Advanced Manufacturing Technology*, *75*(1-4), 257–267. doi:10.1007/s00170-014-6124-9

Baraskar, S. S., Banwait, S. S., & Laroiya, S. C. (2013). Multiobjective Optimization of Electrical Discharge Machining Process Using a Hybrid Method. *Materials and Manufacturing Processes*, *28*(4), 348–354. doi:10.1080/10426914.2012.700152

Bembde, M. S., & Sawale, J. K. (2014). Review of Powder Mixed Dielectric on Material Removal Rate in EDM. *International Journal of Science and Research*, *12*(4), 821–824.

Boothroyd, G., & Winston, A. K. (1989). *Non-Conventional Machining Processes. In Fundamentals of Machining & Machine Tools* (p. 491). New York: Marcel Dekker Inc.

Chandrashekar, M., Muralidhar, M., Murali Krishna, C., & Dixit, U. S. (2010). Application of soft computing techniques in machining performance prediction & optimization: A literature review. *International Journal of Advanced Manufacturing Technology*, *46*(5-8), 445–464. doi:10.1007/s00170-009-2104-x

Chaudhury, P., Samantaray, S., & Sahu, S. (2017). Multi Response Optimization of Powder Additive Mixed Electrical Discharge Machining by Taguchi Analysis. *Materials Today: Proceedings*, *4*(2), 2231–2241. doi:10.1016/j.matpr.2017.02.070

Coit, D. W., Jackson, B. T., & Smith, A. E. (1998). Static Neural Network process models: Considerations and case studies. *International Journal of Production Research*, *36*(11), 2953–2967. doi:10.1080/002075498192229

Cus, F., & Zuperl, U. (2006). Approach to optimization of cutting conditions by using Artificial Neural network. *Journal of Materials Processing Technology*, *173*(3), 281–290. doi:10.1016/j.jmatprotec.2005.04.123

Dornfeld, D., Min, S., & Takeuchi, Y. (2006). Recent Advances in Mechanical Micromachining. *Annals of the CIRP*, *55*(2), 1–25. doi:10.1016/j.cirp.2006.10.006

Fabricio, J. P., Joao, R. F., Messias, B. S., Anderson, P. P., & Pedro, P. B. (2010). Artificial neural networks for machining processes surface roughness modeling. *International Journal of Advanced Manufacturing Technology*, *49*(9-12), 879–902. doi:10.1007/s00170-009-2456-2

Gaitonde, V. N., & Karnik, S. R. (2012). Minimizing burr size in drilling using Artificial Neural Network-Particle Swarm optimization approach (PSO). *Journal of Intelligent Manufacturing*, *23*(5), 1783–1793. doi:10.1007/s10845-010-0481-5

Gunawan, S. P., Tutik, S., & Muslim, M. (2012). Improvement of Machining time through workpiece vibration and graphite powder mixed in dielectric fluid. *Indian Journal of Engineering and Materials Sciences*, *19*, 375–378.

Haykin, S. (1999). *Introduction. In Neural networks and learning machines* (3rd ed.). Pearson Prentice Hall, New Jersy.

Hebb, D. O. (1949). *The organization of behavior A Neuropsychological theory*. New York: John Wiley& Sons,Inc.

Ho, K. H., & Newman, S. T. (2003). State of the art electrical discharge machining (EDM). *International Journal of Machine Tools & Manufacture*, *43*(13), 1287–1300. doi:10.1016/S0890-6955(03)00162-7

Hong, Y., Chunmei, W., Wujun, F., & Xuyang, C. (2017). Analysis of mechanism based on two types of pulse generators in micro-EDM using single pulse discharge. *International Journal of Advanced Manufacturing Technology*, *89*(9), 3217–3230.

Houman, L. (1983). Total EDM. In E. C. Jameson (Ed.), *Electrical Discharge Machining: Tooling, Methods and Applications* (pp. 5–19). Dearborn, MI: Society of Manufacturing Engineers.

Jahan, M. P. (2013). Micro-Electrical Discharge Machining. In J.P. Davim (Ed.), Non traditional Machining Processes. Springer-Verlag.

Jahan, M. P., Kakavand, P., Kwang, E. L. M., Rahman, M., & Wong, Y. S. (2015). An experimental investigation into the micro-electro-discharge machining behavior of aluminium alloy (AA 2024). *International Journal of Advanced Manufacturing Technology*, *78*(5-8), 1127–1139. doi:10.1007/s00170-014-6712-8

Jameson, E. C. (2001). Description and development of electrical discharge machining (EDM). In *Electrical Discharge Machining* (Vol. 12). Dearborn, MI: Society of Manufacturing Engineers.

Joshi, S. N., & Pande, S. S. (2009). Development of an intelligent process model for Electro-Discharge Machining. *International Journal of Advanced Manufacturing Technology*, *45*, 300–317. doi:10.1007/s00170-009-1972-4

Kung, K. Y., Horng, J. T., & Chiang, K. T. (2009). Material removal rate and electrode wear ratio study on the powder mixed electrical discharge machining of cobalt-bonded tungsten carbide. *International Journal of Advanced Manufacturing Technology*, *40*(1-2), 95–104. doi:10.1007/s00170-007-1307-2

Kuriachen, B., & Mathew, J. (2015). Effect of Powder Mixed Dielectric on Material Removal and Surface Modification in Micro Electric Discharge Machining of Ti-6Al-4V. *Materials and Manufacturing Processes*, 1–30.

Mahendra, S., Devarajan, R., Nagarajan, T., & Majdi, A. (2010). A review of Micro-EDM. *Proceedings of International Multiconference of Engineers & computer scientists,* 2.

Majumder, A. (2014). Comparative study of three evolutionary algorithms coupled with neural network model for optimization of electric discharge machining process parameters. *Proceedings of the Institution of Mechanical Engineers, Part B: Journal of Engineering Manufacture*, 1-13.

Mandal, D., Pal, S. K., & Saha, M. (2007). Modelling of EDM process using back propagation neural network and multi-objective optimization using non-dominating sorting genetic algorithm-II. *Journal of Materials Processing Technology*, *186*, 154–162. doi:10.1016/j.jmatprotec.2006.12.030

Masuzawa, T. (2000). State of the art micromachining. *Annual CIRP*, *49*(2), 473–488. doi:10.1016/S0007-8506(07)63451-9

McCulloch, W. S., & Pitts, W. (1943). A logical calculus of the ideas immanent in nervous activity. *Bulletin of Mathematical Biology*, *52*(1/2), 99–115. PMID:2185863

McGeough, J. A. (1988). *Electro Discharge Machining. In Advanced Methods Machine* (p. 130). London: Chapman & Hall.

Modica, F., Morrocco, V., & Fassi, I. (2017). *Micro-Electrodischarge Machining. In Micro-Manufacturing Technologies & their Applications*. Springer International Publishing.

Panda, D. K., & Bhoi, R. K. (2005). Artificial Neural Network Prediction of Material Removal Rate in Electro Discharge Machining. *Materials and Manufacturing Processes*, *20*(4), 645–672. doi:10.1081/AMP-200055033

Pattanaik, L. N. (2017). Artificial Neural Network(MATLAB). In Analytical Tools in Research (pp. 250-261). Dwarka, India: Educreation Publishing.

Porwal, R.K., & Yadava, V. (2012). ANN Modelling For The Prediction Of Material Removal Rate And Machined Hole Overcut In Hole Drilling Electro Discharge Micro Machining. *International Journal of Mechanical Engineering and Robotics Research, 1*(2), 1-16.

Prabhu, S., & Vinayagam, B. K. (2015). Adaptive neuro fuzzy inference system modelling of multi-objective optimisation of electrical discharge machining process using single-wall carbon nanotubes. *Australian Journal of Mechanical Engineering*, *13*(2), 97–117. doi:10.7158/M13-074.2015.13.2

Pradhan, M. K., & Chandan, B. K. (2010). Neuro- fuzzy and Neural Network Based prediction of various responses in EDM of AISI D2 steel. *International Journal of Advanced Manufacturing Technology*, *50*, 591–610. doi:10.1007/s00170-010-2531-8

Prihandana, G. S., Sriani, T., & Mahardika, M. (2012). Improvement of machining time in micro-EDM with workpiece vibration and graphite powder mixed in dielectric fluid. *Indian Journal of Engineering and Materials Sciences*, *19*, 375–378.

Rahman, M. M. (2012). Modeling of Machining Parameters of Ti-6Al-4V for EDM: A neural network approach. *Scientific Research and Essays*, *7*(8), 881–890. doi:10.5897/SRE10.1116

Rao, G. K. M., Janardhana, G. R., Rao, D. H., & Rao, M. S. (2008). Development of hybrid model & optimization of Material Removal Rate in EDM using Artificial neural Network and Genetic Algorithm. *Journal of Engineering and Applied Sciences (Asian Research Publishing Network)*, *3*(1), 19–30.

Rao, P. N. (2009). Introduction. In Manufacturing Technology (vol. 2, pp. 1-5). New Delhi: Tata McGraw-Hill.

Rumelhart, D. E., Hinton, G. E., & Williams, R. J. (1986). Learning representations by back-propagating errors. *Nature*, *323*(6088), 533–536. doi:10.1038/323533a0

Saxsena, K. K., Srivastava, A. S., & Agarwal, S. (2016). Experimental Investigation into micro- EDM characteristics of conductive SiC. *Ceramics International*, *42*(1), 1597–1610. doi:10.1016/j.ceramint.2015.09.111

Somashekhar, K. P., Ramchandran, N., & Mathew, J. (2010). Optimization of Material Removal Rate in Micro-EDM Using Artificial Neural Network and Genetic Algorithms. *Materials and Manufacturing Processes*, *25*(6), 467–475. doi:10.1080/10426910903365760

Suganthi, X. H., Natarajan, U., Sathiyamurthy, S., & Chidambaram, K. (2013). Prediction of quality responses in micro-EDM process using adaptive neuro- fuzzy inference system (ANFIS) model. *International Journal of Advanced Manufacturing Technology*, *68*(1-4), 339–347. doi:10.1007/s00170-013-4731-5

Talla, G., Gangopadhyay, S., & Biswas, C. K. (2016). Influence of Graphite Powder Mixed EDM on the Surface Integrity Characteristics of Inconel 625. *Particulate Science and Technology*, 1–26.

Toshimitsu, R., Okada, A., Kitada, R., & Okamoto, Y. (2016). Improvement in Surface Characteristics by EDM with Chromium Powder Mixed Fluid. *Procedia CIRP*, *42*, 231–235. doi:10.1016/j.procir.2016.02.277

Tripathy, S., & Tripathy, D. K. (2017). Surface Characterization and Multi-response optimization of EDM process parameters using powder mixed dielectric. *Materials Today: Proceedings*, *4*(2), 2058–2067. doi:10.1016/j.matpr.2017.02.051

Tsai, K. M., & Wang, P. J. (2001). Predictions on surface finish in EDM based upon Neural Network Models. *International Journal of Machine Tools & Manufacture*, *41*, 1385–1403. doi:10.1016/S0890-6955(01)00028-1

Tsai, K. M., & Wang, P. J. (2001). Comparisons of neural network models on Material Removal Rate in EDM. *Journal of Materials Processing Technology*, *117*, 111–124. doi:10.1016/S0924-0136(01)01146-3

Tzeng, Y. F., & Lee, C. Y. (2001). Effects of Powder Characteristics on EDMing Efficiency. *International Journal of Advanced Manufacturing Technology*, *17*, 586–592. doi:10.1007/s001700170142

Upadhyay, A., Prakash, V., Pandey, V., & Sharma, V. (2016, November). *Effect of different types of Additives Added in Dielectric Fluid on Micro-EDM of Brass sheet.* Paper presented at International Conference on Evolutions in Manufacturing: Technologies and Business strategies for Global Competitiveness, Mesra, India.

Webzell, S. (2002). That first step into EDM. Machinery, (4040), 41.

Zain, Z. M., Ndaliman, M. B., Khan, A. A., & Ali, M. Y. (2014). Improving micro-hardness of stainless steel through powder-mixed electrical discharge machining. *Proceedings of the Institution of Mechanical Engineers, Part C: Journal of Mechanical Engineering Science,* 1-8. doi:10.1177/0954406214530872

Chapter 12
Multi-Criteria Decision Making for Optimization of Product Development Under Green Manufacturing Environment

Sumit Bhowmik
National Institute of Technology, India

Jagadish
National Institute of Technology, India

ABSTRACT

At present, increasing environmental awareness and stringent environmental regulations of the competent authority, the use of green manufacturing (GM) technique is an important topic for the present industries. To overcome the environmental issues, the manufacturing industries use various advanced manufacturing processes for optimal product development. The advance manufacturing process generates large amounts of toxic substances results in various environmental issues during the optimal product manufacturing. Minimization of environmental issues and the amount of waste generated are strongly depends on its process and response parameters. Thus, optimization of process parameters for GM is essential and is a multi-criteria decision making (MCDM) optimization problems. This chapter provides an overview of applications of some MCDM methods for optimization followed by detailed fundamental aspects of optimization issues in green manufacturing. The work proposed an integrated method consisting of AHP coupled with MOORA and validated through an experimental case study.

INTRODUCTION

In the process of product development, the manufacturing system plays an important role. In the development of the product from the raw material, the various manufacturing processes such as electrical discharge machining (EDM), laser machining (LM), ultrasonic machining (USM), abrasive water jet machining (AWJM) etc. have been employed. During product manufacturing, the various harmful

DOI: 10.4018/978-1-5225-3401-3.ch012

components such as solid, liquid, and gaseous wastes have been produced which affects the manufacturing environment (Sivapirakasam et al. 2011, Shokrani et al. 2012). Besides, the waste generation, manufacturing processes also use various forms of power sources that directly or indirectly affects the machining atmosphere. Due to government guidelines and public cognizance, ecological aspects in the manufacturing processes become utmost important consideration (Sheng and Srinivasan 1995, Yeo et al. 1998, Abbas et al. 2007). Furthermore, the performance of the product manufacturing processes strongly depends on its process and response parameters (Melngk and Smith 1996, Liu et al. 1999). The influence of process parameters and their corresponding performance characteristics results in the optimal result and affects the performance, quality, as well as manufacturing cost. Hence, process parameter optimization of the manufacturing process is essential during product development from the environment and cost effective point of view.

The optimization of the process parameters for the manufacturing process is further considered to be a complex MCDM optimization problem. It involves several criteria's, i.e. process parameters and environmental response parameters, and several alternatives, i.e. various experimental settings. In actual practice, the response/process parameters depend on the priority weights and the degree of importance of the various parameters. The assignment of priority weights also depends on the preference of the decision maker. An improper selection of priority weights yields the error in the selection of optimal process parameters and corresponding response parameters, thus it affects the performance, quality of the product, manufacturing costs and efficiency.

In the traditional approach, choices of optimal process parameters for the manufacturing processes are carried out through data handbook values or operator knowledge. However, this mode of selection does not ensure the optimal performance of manufacturing process (Machining Data Center, 1980). On the other hand, selection of process parameter based on operator or expert judgment is also not an appropriate approach because it does not provide the guarantee that process setting obtained is optimal. In spite of these reasons, the appropriate selection of process parameters for any advanced manufacturing process is further considered to be a still ill-defined optimization problem and involves imprecise data and subjective plus objective data and vagueness information (Yadav and Patel 2013), which results in complexity in the optimization of manufacturing processes. Thus, development of an exact mathematical model to control the machining process is a difficult and essential task. Therefore, a systematic, efficient decision-making approach based on MCDM methods is a necessary requirement for effective modeling and process parameter optimization for product manufacturing processes.

In the past, many investigators have proposed various MCDM based decision-making approaches for effective modeling and optimization of some of the advanced manufacturing processes such as an integrated method, i.e. utility function approach with the analytical hierarchical process (AHP) (Yeo and New, 2000), grey relational analysis (GRA) with Taguchi method (Lin et al. 2002, Lin and Lee 2005, Doniavi et al. 2008, Patel et al. 2010, Kao et al. 2010, Singh et al. 2012, and Tang and Du 2014), Fuzzy with TOPSIS (Sivapirakasam et al. 2011) and visekriterijumska optimizacija kompromisno resenje (VIKOR) (Yadav and Patel, 2011) for modeling and optimization of EDM process. Similarly, same methods like GRA with Taguchi (Mehul et al. 2013, Gupta et al. 2014, Laxman and Raj, 2014 and Mhamunkar and Rau, 2016), AHP (Gaidhani and Kalamani, 2013), and VIKOR (Vedansh Chaturvedi, 2015) have been employed in AWJM process considering the different machining and corresponding response parameters. However, the above literature review reveals that considerations of environmental factors during the optimization of the product manufacturing process have ignored. Similarly, consideration of precise priority weights and the degree of importance of the environmental parameters as well as subjective plus objective data and vagueness information during the optimization have also ignored by the existing strategies. In addition, existing strategies have failed to consider the effect of both beneficial and non-beneficial parameters of the environmental parameters separately while ranking the experimental runs during the process parameter selection. Hence, the existing decision-making strategies could not produce the optimal results for a given product material and process. Thus, a development of decision-making methods using MCDM techniques for effective modeling and process parameter optimization of the manufacturing process is essential (Rao and Kalyankar 2014).

In this context, the present paper proposed an integrated approach, i.e. AHP coupled with MOORA for optimization of product development under green manufacturing environment. In this chapter, first, a brief introduction to the MCDM methods followed by an integrated decision-making method, i.e. AHP with MOORA method is presented. Second, application of integrated approach is carried out by taking an experimental case study. The selected case study is on modeling and optimization of AWJM considering wood dust based filler reinforced polymer (WDFRP) composites under green manufacturing environment. Finally, the detailed conclusions and scope of future work is highlighted.

MULTI-CRITERIA DECISION-MAKING METHODS

Multiple-criteria decision-making (MCDM) is a sub-discipline of operations research and gained importance as a potential decision-making tool for complex optimization problems. MCDM is a concern with the development of computational and mathematical decision tool for evaluation of complex real life problems considering the both subjective and objective criteria (Zavadskas, Turskis and Kildiene, 2014). On the other hand, MCDM is considered to be a generic tool which helps the decision makers to take the optimal decision as per the preferences among the one or more conflicting criterion (Ho, 2008). The schematic procedure of MCDM methods is shown in Figure.1.

In the late 1960s, the MCDM techniques become the most popular and performed vigorous research in all the areas of applications as well as in the development of an adequate number of hypothetical models, books, and articles, etc. (Zavadskas and Turskis, 2010; Zavadskas and Zakarevicius, 2012, Roy, 2005).

Figure 1. General MCDM decision process

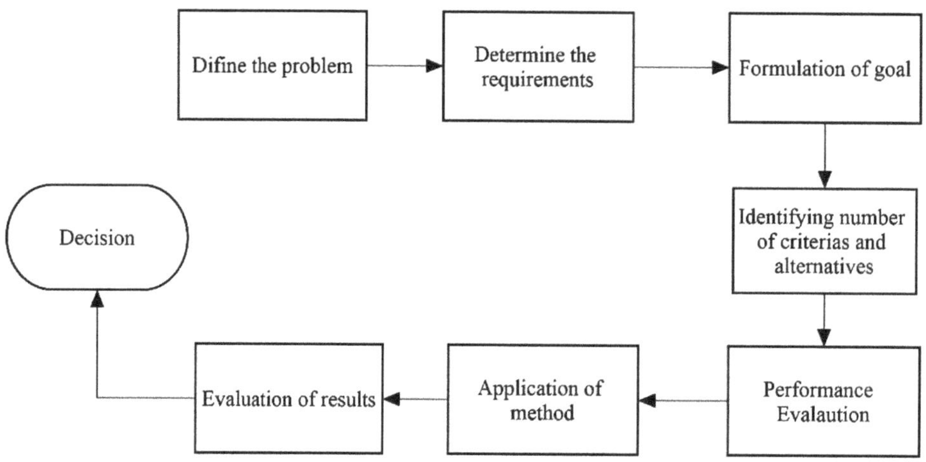

The most commonly developed MCDM methods were, analytical hierarchical process (AHP), analytical network process (ANP), data envelopment analysis (DEA), goal programming, simple additive weighting (SAW), simple multi-objective rating technique (SMART), ELECTRE, PROMETHEE, and technique for order of preference by similarity to ideal ratio analysis (TOPSIS) etc. These MCDM methods have been established to select the optimal alternatives, classify, or rank the alternatives with consideration of subjective and objective criteria. Among these MCDM methods, the present chapter presented AHP with MOORA methods for optimization of product development under green manufacturing scenario.

INIGRATED MCDM METHOD

An integrated MCDM method consists of an analytical hierarchical process (AHP) (Saaty, 1989) combined with multi-objective optimization on the basis of ratio analysis (MOORA) (Brauers and Zavadskas 2006, Chakraborty 2011, Karande and Chakraborty 2012, Sarkar et al. 2015, Singaravel et al. 2016) is presented for optimization of product development under green manufacturing environment. The detailed steps of AHP-MOORA based decision-making method are as follows:

Step 1: Development of decision matrix.

Optimization process starts with the decision matrix, which includes a number of response parameters and corresponding process parameters. First, formulation of decision matrix is done based on a number of process/response parameters of the system. Various designs of experimental techniques are used for the formulation of the decision matrix. In the decision matrix, performance values for each of the response parameters with to the process parameters are listed. The decision matrix is formulated using the following expression.

$$D_{ij} = \begin{bmatrix} D_{11} & D_{12} & \ldots\ldots\ldots\ldots & D_{1q} \\ D_{21} & D_{22} & \ldots\ldots\ldots\ldots & D_{2q} \\ \ldots\ldots & \ldots\ldots & \ldots\ldots\ldots\ldots & \ldots\ldots \\ D_{p1} & D_{p2} & \ldots\ldots\ldots\ldots & D_{pq} \end{bmatrix} \text{For } \left(i = 1,2,\ldots.p, j = 1,2,\ldots.q\right) \quad (1)$$

where D_{ij} represents the response value of ith responses on jth process parameter, p represents the number of experiments and q represents the number of response parameters.

Step 2: Formulation of pairwise comparison decision matrix and computation of precise priority weights of the response parameters using AHP

To find out the priority weights of the response parameters, the pairwise comparison matrix is required. Here, precise priority weights for each of the response parameters are determined using AHP method (Saaty, 1989). Initially, pairwise comparison decision matrix between each of the response parameters is developed. Simultaneously, the precise priority weights for each of the response parameters are determined using the fundamental scale of the AHP process with the ith response and jth criteria of the response parameters using the following expression (Tzeng and Huang, 2011).

$$C_{ij} = \begin{bmatrix} C_{11} & C_{12} & \dots\dots\dots & C_{1j} \\ C_{21} & C_{22} & \dots\dots\dots & C_{2j} \\ \dots\dots & \dots\dots & \dots\dots\dots & \dots\dots \\ C_{i1} & C_{i2} & \dots\dots\dots & C_{ij} \end{bmatrix} C_{ii} = 1, C_{ij} = \frac{1}{C_{ji}}, C_{ji} \neq 0 \tag{2}$$

where, C_1, C_2...C_j denotes the response parameters, C_{ij} is the relative importance of *i*th response parameter with *j*th response parameter. The relative importance of two response parameters is rated using the fundamental scale of AHP with the digits 1, 3, 5, 7 and 9. In which digit, 1 denotes "equally important", 3 for "slightly more important", 5 for "strongly more important", 7 for "demonstrably more important" and 9 for "absolutely more important" and 2, 4, 6, 8 for "slightly differing judgment" (Tzeng and Huang, 2011). Thereafter, priority weights (ϕj) for each response parameter are determined as a row average of the resulting pairwise comparison matrix using Eqs. (3) and (4) respectively (Saaty, 1989).

$$_{Mj} = \left[\prod_{j=1}^{n} C_{ij}\right]^{1/n} \text{ For } i, j = 1, 2, \dots n \tag{3}$$

$$\varphi_j = \frac{_{Mj}}{\sum_{j=1}^{n} {_{Mj}}} \tag{4}$$

where M_j represent the geometric mean of the pairwise decision matrix (C_{ij}) for *j*th response parameter, ϕ_j is the priority weights of *j*th response parameter and *n* denotes the number of response parameters.

After that, consistency of the pairwise comparison decision matrix and precision of the priority weights (Prasad and Chakraborty, 2012) obtained from AHP method are verified by consistency index (I) and consistency ratio (R) (Saaty, 1989):

$$I = \frac{\left(\lambda_{Max} - n\right)}{n - 1} \tag{5}$$

$$R = \frac{I}{R_I} \tag{6}$$

where λ_{Max} represented the largest eigenvalues of the pairwise comparison decision matrix (C_{ij}), *n* represents the number of experimental runs considered in this study, RI denotes the random index and is depends on several response parameters.

Step 3: Establishment of weighted decision matrix and its normalization

In this step, weighted decision matrix using Eq. (7) is developed. It involves the list of priority weights for each of the response parameters obtained by the AHP method with respect to the different experimental runs. This measures the performance values for each of *i*th experimental runs with respect to the *j*th response parameters. After that, conversion of different units of the performance values in the decision matrix into a comparable unit of performance values in the normalization process. The calculation of normalized decision matrix using the following expression:

$$WD_{ij} = \begin{bmatrix} \varphi_1 \times D_{11} & \varphi_1 \times D_{12} & \ldots\ldots\ldots\ldots & \varphi_j \times D_{1j} \\ \varphi_2 \times D_{21} & \varphi_2 \times D_{22} & \ldots\ldots\ldots\ldots & \varphi_j \times D_{2j} \\ \ldots\ldots & \ldots\ldots & \ldots\ldots\ldots\ldots & \ldots\ldots \\ \varphi_j \times D_{i1} & \varphi_j \times D_{i2} & \ldots\ldots\ldots\ldots & \varphi_j \times D_{ij} \end{bmatrix} \tag{7}$$

$$WD_{ij}^{*} = \frac{WD_{ij}}{\sum_{i=1}^{n} WD_{ij}} \tag{8}$$

where, WD_{ij}^{*} represent the performance weights of *i*th experimental runs on *j*th response parameters and *n* represent the number of response parameters. WD_{ij} represents a dimensionless number in the range [0, 1] interval representing the normalized performance weights of *i*th experimental runs on *j*th response parameters.

Step 4: Evaluation of assessment values for optimal setting

Next, assessment values for each of the response parameters are evaluated in order to assess the optimal selection of process parameters. The evaluation of assessment values for *i*th experimental runs to convert the multi-attribute problem into a single attribute subjected to certain constraints with consideration of both beneficial and non-beneficial parameters into account using the following expression (Brauers and Zavadskas 2006, Prasad and Chakraborty, 2012).

$$AV_i = \sum_{j=1}^{g} WD_{ij}^{*} - \sum_{j=g+1}^{n} WD_{ij}^{*} \tag{9}$$

where *g* represents the number of response parameters to be maximized, *(n-g)* represents the number of response parameters to be minimized, AV_i represents the assessment value of i*th* experimental setting with respect to all j*th* response parameters, and ϕ_j represent the precise priority weights of the *j*th response parameters.

Step 5: Rank and establish the optimal setting for the green manufacturing process

At last, calculation of optimal selection of process parameters by ranking the experimental runs based on assessment values using Eq.(9). The experimental run with maximum assessed value is optimal among other experimental runs. The optimal setting or optimal combination of process parameters provides the optimal responses, which have less environmental impact, improves the product quality, reduce the manufacturing cost, and increase the productivity of the green manufacturing process.

CASE STUDY: OPTIMIZATION OF AWJM PROCESS UNDER GREEN MANUFACTURING ENVIRONMENT

Background

In this section, a case study on modeling and optimization of process parameters of the AWJM process of machining of nature filler reinforced polymer (NFRP) composites under green manufacturing environment is given. NFRP composite used in this study is wood dust filler based reinforced polymer (WDFRP) composites (Markarian 2002, Pritchard 2004, Jagadish et al. 2015a, Jagadish et al. 2015b). Due to the anisotropic and inhomogeneous properties, machining of these composites by conventional processes is extremely difficult and expensive (Abrate 1997, Patel and Shaikh 2013, Hutyrová, et al. 2015). Therefore, present work utilized AWJM for machining of these composites.

Furthermore, the performance of the AWJM process on the machining of WDFRP composites and amount of waste generated from it depends on various process parameters such as a stand of distance, work pressure, nozzle speed, abrasive grain size etc. The influence of these parameters results in improper responses like higher material removal rate, higher surface roughness with larger taper angle and plays an important role in the generation of environmental issues such as solid, liquid, and gaseous wastes and pollutes the machining environment results in effecting the performance of AWJM process. On the other hand, machining process i.e. AWJM known as environmentally friendly process and material used for machining from natural fillers. Hence, machining process including (AWJM process + material used) considered as a green manufacturing process. Therefore, present work studied the effect of MRR, SR, and TA on the machining environment as well as the performance of AWJM process by optimizing the process parameters.

In order to optimize the AWJM process parameters, the present work proposed an integrated approach, i.e. AHP with MOORA methods. In which, AHP is employed to determine the response parameter weights and optimal selection of AWJM parameters by MOORA. At last, an optimum combination of the process parameters for practical applications is recommended. The recommended optimal parameters reduce the environmental impact, increase the product quality, and improve the performance of the AWJM process during the machining of WDFRP composites.

Experimental Procedure

The abrasive water jet (AWJ) machine (Figure. 2) used for actual experimentation of the WDFRP composite specimen. The WDFRP composite specimen was prepared by hand lay-up process taking the wood dust fillers (2 μm size and ρ (density) 0.799 g/cc) as a reinforcement material and epoxy resin (Araldite LY 556) and a corresponding hardening agent (HY 951) as a matrix material. The specimen of size 150 mm x 100 mm x 6 mm is prepared for the experimental work. Although, AWJM includes many numbers

Figure 2. (a) AWJM experimental setup; (b) AWJM nozzle head setup

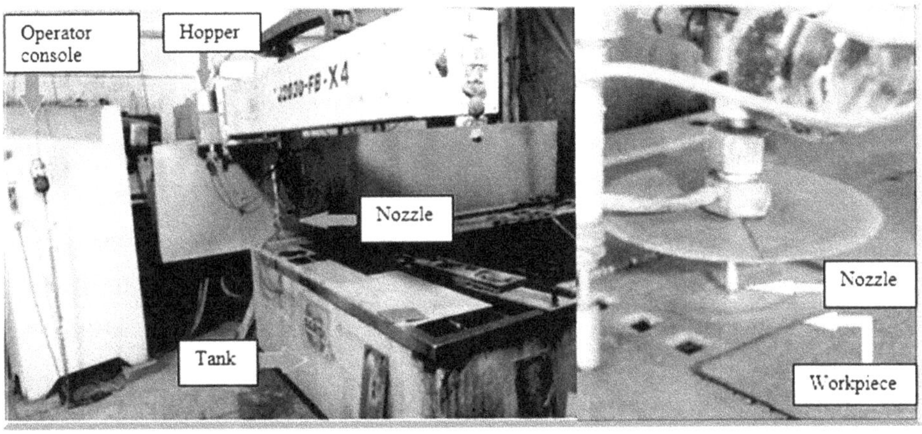

of process parameters as suggested by Hashish (1984) and has an impact on the response parameters during machining. However, the present work utilized only major and easy-to-adjustable process variables as tabulated in Table 1. Thereafter, deciding the parameter levels based on their effect on the response parameters. As the selected parameters (Table 1) shown a linear and a quadratic effect on the response parameters so therefore level 3 for each of the process parameters have been considered instead of level 2. The process parameters with their levels of AWJM process is shown in Table 1.

During the experiment, drilled a square hole of 15 mm x 15 mm based on the experimental design with varying the four independent parameters as tabulated in Table 1. The corresponding performance characteristics or response parameters such as the material removal rate (MRR), the surface roughness (Ra) and a taper angle (TA) were determined. Each experiment was performed three intervals and corresponding response parameters were measured. In order to get the accuracy and consistency in the response data, the average values of the response parameters were taken for the analysis. For this purpose, three experiments were performed at same interval setting subsequently measured the response parameters and averaged them. The result of averaged experimental results of AWJM process is depicted in Table 2.

Modeling of AWJM Process Using Integrated Method

In this section, the modeling of AWJM process is presented using a proposed decision-making method, i.e. AHP with MOORA. First, the development of decision matrix based on the Taguchi (L_9-orthogonal array) design and tabulated in Table 2. Second, the precise priority weights of the response parameters

Table 1. Machining parameters and their levels

Process Parameters	Symbol	Units	Level 1	Level 2	Level 3
Stand of Distance	SoD	mm	1	2	3
Work Pressure	WP	MPa	100	125	150
Nozzle Speed	NS	mm/min	100	200	300
Abrasive Grain Size	AGS	Mesh	70	80	90

Table 2. Experimental results of AWJM process

Exp. No.	Process Parameters				Response Parameters		
	SoD (mm)	WP (MPa)	NS (mm/min)	AGS (mesh)	MRR (g/mm)	SR (μm)	TA (Degree)
1	1	100	100	70	5.412	0.112	0.184
2	1	125	200	80	10.723	0.194	0.125
3	1	150	300	90	21.251	0.208	0.148
4	2	100	200	90	20.327	0.304	0.202
5	2	125	300	70	42.781	0.249	0.236
6	2	150	100	80	15.654	0.129	0.167
7	3	100	300	80	50.687	0.258	0.721
8	3	125	100	90	18.327	0.552	0.129
9	3	150	200	70	47.871	0.178	0.177

such as MRR, SR, and TA are determined using the AHP method (Saaty, 1989). In the AHP method, first, pairwise comparison matrix between the each of the response parameters is formulated using Eq. (2) and subsequently, precise priority weights for each of the response parameters of AWJM process are determined using the Eqs. (3) and (4). The pairwise comparison matrix between response parameters is depicted in Table 3.

After that, consistency of the decision matrix and precision of the priority weights (Das et al. 2012) are verified through consistency index (I) and consistency ratio (R) (Eqs. 5-6). The results of the precise priority weights of the AWJM response parameters are tabulated in Table 4.

The precise priority weights (ϕ) for the AWJM response parameters such as MRR, SR, and TA are calculated to be 0.0953, 0.2499, and 0.6548 respectively. The result indicates that response parameter TA is most influencing parameter followed by SR and MRR. Also, consistency index (I) and consistency ratio (R) for AWJM response parameters are calculated to be 0.00017 and 0.00029 which less than 10%. This indicates that the priority weights of the response parameters for AWJM process and the corresponding pairwise decision matrix are shown higher accuracy and better consistency.

Table 3. Pairwise comparison matrix using AHP

Response Parameters	MRR	SR	TA
MRR	1	0.333	0.166
SR	3	1	0.333
TA	6	3	1

Table 4. Priority weights of the AWJM response parameters

Priority Weights (ϕ)	Response Parameters		
	MRR	SR	TA
	0.0953	0.2499	0.6548

Thereafter, formulation of weighted decision matrix using Eq. (7), in which priority weights for each of the AWJM response parameters are obtained by an AHP method with respect to the different experimental runs are listed. This measure the performance values for each of the response parameters with respect to the each of the experimental runs and the results of the weighted decision matrix are tabulated in Table 5.

After that, normalization of performance values of the weighted decision matrix for each of the AWJM response parameters is performed to convert the original sequence data to comparable sequence values using Eq. (8). The normalization result using MOORA method is depicted in Table. 6.

At last, assessment values (*AV*) for each of the response parameters are computed using Eq. (9). This converts the multi-criteria problem into a single criteria problem subjected to the certain constraints with consideration of both beneficial and non-beneficial parameters (Brauers and Zavadskas 2006, Prasad and Chakraborty, 2012). The assessment values for each of the experimental runs of AWJM process are tabulated in Table 7.

Table 5. Weighted decision matrix of AWJM process

Exp. No.	Response Parameters		
	MRR	SR	TA
1	0.516	0.028	0.120
2	1.022	0.048	0.081
3	2.026	0.052	0.096
4	1.937	0.076	0.132
5	4.078	0.062	0.154
6	1.492	0.032	0.109
7	4.832	0.064	0.472
8	1.747	0.137	0.084
9	4.563	0.044	0.115

Table 6. Normalized decision matrix of AWJM process

Exp. No.	Response Parameters		
	MRR	SR	TA
1	0.023	0.051	0.088
2	0.046	0.088	0.059
3	0.091	0.095	0.070
4	0.087	0.139	0.096
5	0.183	0.114	0.113
6	0.067	0.059	0.079
7	0.217	0.118	0.345
8	0.078	0.252	0.061
9	0.205	0.081	0.084

Table 7. Assessment values of AWJM process

Exp. No.	Assessment Values (*AV*)
1	0.162
2	0.194
3	0.257
4	0.323
5	0.412
6	0.206
7	0.682
8	0.393
9	0.377

RESULTS AND DISCUSSIONS

Optimization of AWJM Process Parameters

The assessment values for each of the experimental runs are used to optimize the process parameters of the AWJM process. The optimization is performed using an integrated decision-making approach, i.e. AHP-MOORA method (Saaty, 1989), Brauers and Zavadskas 2006, Chakraborty 2011). The optimization results of AWJM process are tabulated in Table 8. Thereafter, a ranking of the experimental runs is carried out based on the assessment value obtained from the proposed AHP-MOORA method.

The optimization results as depicted in Table 7 shows that, the exp. no. 7 is optimal experimental setup for optimal response among the nine experiments. The assessment value obtained for the optimal setting is 0.682 and shows the highest assessment value among the nine experiments.

Table 8. Rank the experimental runs of AWJM process

Exp. No.	Process Parameters				Assessment Valus	Rank
	SoD (mm)	WP (MPa)	NS (mm/min)	AGS (mesh)		
1	1	100	100	70	0.162	9
2	1	125	200	80	0.194	8
3	1	150	300	90	0.257	6
4	2	100	200	90	0.323	5
5	2	125	300	70	0.412	2
6	2	150	100	80	0.206	7
7	3	100	300	80	0.682	1
8	3	125	100	90	0.393	3
9	3	150	200	70	0.377	4

The optimal setting obtained are as the stand-off distance (SoD) as 3 mm with level 3, working pressure (WP) as 100 MPa with level 1, nozzle speed (NS) as 300 mm/min with level 3 and abrasive grain size (AGS) as 80 mesh with level 2. The optimal setting obtained using proposed AHP-MOORA method provides the most optimal response parameter for AWJM process. The optimal setting produces less influence on the performance of AWJM process during the processing of WDFRP composites under green manufacturing environment. In addition, the optimal response parameters reduce the generation of wastes and pollution due to the different chemical composition, different materials behavior and diverse properties of WDFRP composites machining. Furthermore, the recommended optimal experimental setting minimizes ecological disputes, improves product quality as well as minimizes the manufacturing cost and progresses the machining efficiency of the AWJM process. The research also suggested that the integrated method could be employed as effective decision-making framework model for modeling and multi-response optimization of the AWJM process on WDFRP composites machining under green manufacturing scenarios.

CONCLUSION AND FUTURE SCOPE

This chapter provides for the application of MCDM methods for optimization of product development under green manufacturing environment. An integrated approach, i.e. AHP coupled with MOORA based MCDM method is proposed. To show the effectiveness and working significance of the proposed methodology, an experimental case study on machining of WDFRP polymer composite in AWJM process under green manufacturing environment is discussed. The work summarized that AWJM process is a viable and effective alternative to the conventional cutting processes for machining of WDFRP composite with good quality and high productivity. The optimal result shows that exp. No. 7 is the optimal experimental setup for optimal response among the nine experiments and response TA is the most influencing response parameter among the response parameters MRR and SR.

The optimal combination of process parameters obtained is stand-off distance (SoD) as 3 mm with level 3, working pressure (WP) as 100 MPa with level 1, nozzle speed (NS) as 300 mm/min with level 3 and abrasive grain size (AGS) as 80 mesh with level 2. The optimal setting provides a higher removal of MRR, good surface finish, and lesser taper angle during machining of WDFRP composite by AWJM. Furthermore, the recommended optimal experimental setting minimizes environmental issues produced due machining responses, i.e. MRR, and SR expands product quality and efficiency of the AWJM process.

Although, the present chapter provided the application of AHP-MOORA based MCDM method for modeling and optimization of product development under green manufacturing environment. However, development of other MCDM methods such as TOPSIS, PROMETHEE, VIKOR etc., and their applications in green manufacturing optimization have been investigated yet. Apart from the MCDM methods, the research on the machinability of other than wood filler based polymer composite materials by AWJM process and their process optimization are yet to be explored.

REFERENCES

Abbas, N. M., Solomon, D. G., & Bahari, M. F. (2007). A review of current research trends in electrical discharge machining (EDM). *International Journal of Machine Tools & Manufacture*, *47*(7-8), 1214–1228. doi:10.1016/j.ijmachtools.2006.08.026

Abrate, S. (1997). Machining of composite materials. In P. K. Mallick (Ed.), *Composites engineering handbook*. New York: Marcel Dekker.

Benedict, G. F. (1987). *Non-traditional manufacturing processes*. New York: Marcel Decker Inc.

Brauers, W. K. M., & Zavadskas, E. K. (2006). The MOORA method and its application to privatization in a transition economy. *Control and Cybernetics*, *35*(2), 445–469.

Chakraborty, S. (2011). Applications of the MOORA method for decision making in a manufacturing environment. *International Journal of Advanced Manufacturing Technology*, *54*(9-12), 1155–1166. doi:10.1007/s00170-010-2972-0

Das, M. C., Sarkar, B., & Ray, S. (2012). Decision making under conflicting environmental new MCDM method. *International Journal of Applied Decision Sciences*, *5*(2), 142–162. doi:10.1504/IJADS.2012.046505

Doniavi, A., Eskandarzade, M., Abdi, A., & Totonchi, A. (2008). Empirical Modelling of EDM Parameters Using Grey Relational Analysis. *Asian Journal of Scientific Research*, *1*(5), 502–509. doi:10.3923/ajsr.2008.502.509

Gaidhani, Y.B. & Kalamani, V.S. (2013). Abrasive water jet review and parameter selection by AHP method. *IOSR Journal of Mechanical and Civil Engineering, 8*(5), 1-6.

Gupta, T. V. K., Ramkumar, J., Tandon, P., & Vyas, N. S. (2012). Application of Grey Relational Analysis for Geometrical Characteristics in Abrasive Water Jet Mille Channels. *5th International & 26th All India Manufacturing Technology, Design and Research Conference, 393*, 1-6.

Hashish, M. (1984). A modeling study of metal cutting with abrasive water jets. *Journal of Engineering Materials and Technology*, *106*(1), 88–100. doi:10.1115/1.3225682

Hashish, M. (1991). Advances in composite machining with abrasive waterjets. *ASME Process Manufacturing Computer Materials*, *49*(27), 93–111.

Huo, L. A., Lan, J., & Wang, Z. (2011). New parametric prioritization methods for an analytical hierarchy process based on a pairwise comparison matrix. *Mathematical and Computer Modelling*, *54*(11-12), 2736–2749. doi:10.1016/j.mcm.2011.06.062

Hutyrová, Z., Ščučka, J., Hlaváček, H. P., & Zeleňák, M. (2015). Turning of wood plastic composites by water jet and abrasive water jet. *International Journal of Advanced Manufacturing Technology*. doi:10.1007/s00170-015-7831-6

Jagadish, B. S., & Ray, A. (2015). Prediction and optimization of process parameters of green composites in AWJM process using response surface methodology. *International Journal of Advanced Manufacturing Technology*. doi:10.1007/s00170-015-8281-x

Jagadish. Bhowmik, S., & Ray, A. (2015). Prediction of surface roughness quality of green abrasive water jet machining: A soft computing approach. *Journal of Intelligent Manufacturing*. doi:10.1007/s10845-015-1169-7

Kao, J. Y., Tsao, C. C., Wang, S. S., & Hsu, C. Y. (2010). Optimization of the EDM parameters on machining Ti–6Al–4V with multiple quality characteristics. *International Journal of Advanced Manufacturing Technology*, *47*(1-4), 395–402. doi:10.1007/s00170-009-2208-3

Karande, P., & Chakraborty, S. (2012). Application of multi-objective optimization on the basis of ratio analysis (MOORA) method for materials selection. *Materials & Design*, *37*(1), 317–324. doi:10.1016/j.matdes.2012.01.013

Lin, J. L., & Lin, C. L. (2002). The use of the orthogonal array with gray relational analysis to optimize the electrical discharge machining process with multiple performance characteristics. *International Journal of Machine Tools & Manufacture*, *42*(2), 237–244. doi:10.1016/S0890-6955(01)00107-9

Lin, Y., & Lee, H. (2009). Optimization of machining parameters using magnetic-force-assisted EDM based on gray relational Analysis. *International Journal of Advanced Manufacturing Technology*, *42*(11-12), 1052–1064. doi:10.1007/s00170-008-1662-7

Liu, F., Zhang, H., & Cheng, X. H. (1999). A decision-making framework model for green manufacturing and the case study. *Journal of Mechanical Engineering*, *35*(5), 11–15.

Machining Data Center. (1980). Cincinnati, OH: Machining Data Handbook, Inc.

Markarian, J. (2002). Additive developments aid growth in wood-plastic composites. *Plastics Additives & Compounding*, *4*(11), 18–21. doi:10.1016/S1464-391X(02)80174-X

Mehul, A. R., & Chirag, P. P. (2013). Parametric Optimization of Magnetic Abrasive Water Jet Machining of AISI 52100 Steel using Grey Relational Analysis. *International Journal of Engineering Research and Applications*, *3*(4), 527–530.

Melngk, S. A., & Smith, R. T. (1996). *Green Manufacturing*. Dearborn, MI: Society of Manufacturing Engineers.

Mhamunkar, M. M., & Rau, N. (2016). Optimization of Process Parameter of CNC Abrasive Water Jet Machine for Titanium Ti 6Al 4V material. *International Journal of Advanced Research in Science, Engineering, and Technology.*, *3*(3), 1640–1646.

Patel, K. M., Pandey, P. M., & Rao, P. V. (2010). Optimization of process parameters for multi-performance characteristics in EDM of Al2O3 ceramic composite. *International Journal of Advanced Manufacturing Technology*, *47*(9-12), 1137–1147. doi:10.1007/s00170-009-2249-7

Patel, S. R., & Shaikh, A. A. (2013). A Review on Machining of Fiber Reinforced Plastic using Abrasive Waterjet. *International Journal of Innovative Technology & Adaptive Management*, *1*(3), 10–18.

Prasad, K., & Chakraborty, S. (2012). Application of multi-objective optimization on the basis of ratio analysis (MOORA) method for materials selection. *Materials & Design*, *37*, 317–324. doi:10.1016/j.matdes.2012.01.013

Pritchard, G. (2004). Two technologies merge wood-plastic composites. *Plastics Additives & Compounding*, *6*(4), 18–21. doi:10.1016/S1464-391X(04)00234-X

Rao, R. R., & Kalyankar, V. D. (2014). Optimization of Modern Machining Processes Using Advanced Optimization Techniques: A Review. *International Journal of Advanced Manufacturing Technology*, *73*(5-8), 1159–1188. doi:10.1007/s00170-014-5894-4

Roy, B. (2005). Paradigms and challenges. In J. Figueira, S. Greco, & M. Ehrgott (Eds.), *multiple criteria decision analysis. State of the art surveys*. New York, NY: Springer. doi:10.1007/0-387-23081-5_1

Saaty, T. L. (1989). Decision making with the analytic hierarchy process. *Journal de Chirurgie*, *1*(1), 83–98.

Sarkar, A., Panja, S. C., Das, D., & Sarkar, B. (2015). Developing an efficient decision support system for non-traditional machine selection: An application of MOORA and MOOSRA. *Journal of Production & Manufacturing Research*, *3*(1), 324–342. doi:10.1080/21693277.2014.895688

Sheng, P., & Srinivasan, M. (1995). Multi-objective process planning in environmentally conscious manufacturing: A feature-based approach. *CIRP Annals - Manufacturing Technology, 44*(1), 433-437.

Shokrani, A., Dhokia, V., & Newman, S. T. (2012). Environmentally conscious machining of difficult-to-machine materials with regard to cutting fluids. *International Journal of Machine Tools & Manufacture*, *57*, 83–101. doi:10.1016/j.ijmachtools.2012.02.002

Singaravel, B., Selvaraj, T., & Vinod, S. (2016). Multi-objective optimization of turning parameters using the combined MOORA and Entropy method. *Transactions of the Canadian Society for Mechanical Engineering*, *40*(1), 101–111.

Singh, S., & Verma, M. (2012). Parametric Optimization of Electric Discharge Drill Machine Using Taguchi Approach. *Journal of Engineering and Applied Sciences (Asian Research Publishing Network)*, *1*(3), 39–47.

Sivapirakasam, S. P., Mathew, J., & Surianarayanan, M. (2011). Multi-attribute decision making for green electrical discharge machining. *Expert Systems with Applications*, *38*(7), 8370–8374. doi:10.1016/j.eswa.2011.01.026

Tang, L., & Du, Y. T. (2014). Experimental study on green electrical discharge machining in the tap water of Ti–6Al–4V and parameters optimization. *International Journal of Advanced Manufacturing Technology*, *70*(1-4), 469–475. doi:10.1007/s00170-013-5274-5

Tzeng, G. H., & Huang, J. J. (2011). *Multiple attribute decision making: methods and applications.* CRC Press.

Vedansh, C. (2015). Multi-Response Optimization of Process Parameters of Abrasive Water Jet Machining for Stainless Steel AISI 304 Using VIKOR Approach Coupled with Signal to Noise Ratio Methodology. *Journal of Advanced Manufacturing Systems*, *14*(2), 107–121. doi:10.1142/S0219686715500080

Weller, E.J. (1984). *Non-traditional machining processes*. SME Dearborn.

Yadav, S. K., & Patel, S. K. (2013). *Optimization of green electro-discharge machining using VIKOR* (Thesis). Dept of Mech Eng., NIT, Rourkela, India.

Yeo, S. H., Neo, K. G., & Tan, H. C. (1998). Assessment of health hazard in the production of printed paper packages. *International Journal of Advanced Manufacturing Technology*, *14*(5), 376–384. doi:10.1007/BF01178918

Yeo, S. H., & New, A. K. (1999). A method for green process planning in EDM. *International Journal of Advanced Manufacturing Technology*, *15*(4), 287–291. doi:10.1007/s001700050068

Zavadskas, E. K., & Turskis, Z. (2010). A new additive ratio assessment (ARAS) method in multicriteria decision-making. *Technological and Economic Development of Economy*, *16*(2), 159–172. doi:10.3846/tede.2010.10

Zavadskas, E. K., Turskis, Z., Antucheviciene, J., & Zakarevicius, A. (2012). Optimization of weighted aggregated sum product assessment. *Elektronika ir Elektrotechnika*, *122*(6), 3–6. doi:10.5755/j01.eee.122.6.1810

Zavadskas, E. K., Turskis, Z., & Kildiene, S. (2014). State of art surveys of overviews on MCDM/MADM methods. *Technological and Economic Development of Economy*, *20*(1), 165–179. doi:10.3846/20294913.2014.892037

Chapter 13
Tribological Performance Optimization of Electroless Nickel Coatings Under Lubricated Condition

Santanu Duari
Jadavpur University, India

Arkadeb Mukhopadhyay
Jadavpur University, India

Tapan Kumar Barman
Jadavpur University, India

Prasanta Sahoo
Jadavpur University, India

ABSTRACT

The present chapter aims to determine optimal tribo-testing condition for minimum coefficient of friction and wear depth of electroless Ni-P, Ni-P-W and Ni-P-Cu coatings under lubrication using grey relational analysis. Electroless Ni-P, Ni-P-W and Ni-P-Cu coatings are deposited on AISI 1040 steel substrates. They are heat treated at suitable temperatures to improve their hardness. Coating characterization is done using scanning electron microscope, energy dispersive X-Ray analysis and X-Ray diffraction techniques. Typical nodulated surface morphology is observed in the scanning electron micrographs of all the three coatings. Phase transformation on heat treating the deposits is captured through the use of X-Ray diffraction technique. Vicker's microhardness of the coatings in their as-deposited and heat treated condition is determined. Ni-P-W coatings are seen to exhibit the highest microhardness. Friction and wear tests under lubricated condition are carried out following Taguchi's experimental design principle. Finally, the predominating wear mechanism of the coatings is discussed.

DOI: 10.4018/978-1-5225-3401-3.ch013

INTRODUCTION

Surface modification techniques via the application of coatings have received considerable attention in a quest to achieve enhanced physical, mechanical, electrical and tribological properties of industrial products. Amongst several techniques, deposition of nickel alloy and composite coatings by the electroless method has been widely accepted and investigated by researchers due to their excellent mechanical and tribological properties (Sahoo & Das, 2011; Sudagar et al., 2013; Loto, 2016). A key feature of the coatings is the deposit uniformity i.e. intricate parts receive equable deposition like the exposed ones. Moreover, the use of electricity is not involved allowing deposition over different substrates with proper surface activation (Mukhopadhyay et al., 2016a; 2017a). This led to improvement in properties of magnesium alloys, glass/carbon fiber reinforced composites, etc. by the deposition of EN coatings (Correa et al., 2017; Balaraju et al., 2016; Zuleta et al., 2017). Therefore, electroless deposited nickel coatings find wide usage in electrical, electronics, chemical, aerospace, automobile, printing and textiles industries (Agarwala & Agarwala, 2003; Gadhari & Sahoo, 2016a).

Electroless nickel (EN) coatings are broadly classified as pure nickel, alloy and composite coatings. Pure nickel (99%) is obtained from hydrazine baths and has been discontinued lately due to the associated hazards. EN alloy coatings are primarily obtained from sodium hypophosphite and sodium borohydride based baths leading to the formation of Ni-P (Hur et al., 1990; Sahoo, 2009) and Ni-B (Das and Sahoo, 2011; Duari et al., 2016; Mukhopadhyay et al., 2016b; 2016c; 2017b; 2017c) coatings respectively. Electroless Ni-P coatings have excellent anti-wear, anti-friction and anti-corrosion properties (Staia et al., 1996; Sahoo, 2009; Mukhopadhyay et al., 2016d; 2017a; Panja & Sahoo, 2014; 2015; Panja et al., 2016). Ni-P coatings are also being investigated as potential candidates for high temperature applications (Kundu et al., 2016; Masoumi et al., 2012; Ghaderi et al., 2016). The tribological properties are enhanced on heat treatment and the incorporation of composites like Al_2O_3 (Gadhari & Sahoo, 2016b), TiO_2 (Gadhari & Sahoo, 2016c), SiC (Franco et al., 2016), MoS_2 (Li et al., 2013), WS_2 (Sivandipoor & Ashrafizadeh, 2012), PTFE (Der Ger et al., 2003), SiO_2 (Sadeghzadeh-Attar et al., 2016), diamond (Reddy et al., 2000), etc. Nitriding improves the adhesion of the coatings with the steel substrate (Soares et al., 2017). Duplex and multi-layered coatings are also observed to improve microhardness, corrosion resistance and tribological behavior of EN coatings (Song et al., 2017; Vitry & Bonin, 2017; Liu et al., 2016). Usually, the deposits in their as-plated condition may be a mixture of amorphous and nanocrystalline phases or completely amorphous depending on the amount of P. The crystallinity of the coatings increases with an increase in the P content while the microhardness decreases (Czagány et al., 2017). Heat treatment leads to the precipitation of crystalline nickel and its phosphides. These hard phases improve the microhardness and wear resistance. The wear mechanisms reported for Ni-P alloy and its composite variants are adhesive, abrasive, spalling, fatigue, etc. depending on the test conditions.

Additional elements may be co-deposited to form ternary/ poly alloy EN coatings such as Ni-P-W (Palaniappa & Seshadri, 2008; Mukhopadhyay et al., 2016e; 2016f; Duari et al., 2017), Ni-P-Cu (Liu & Zhao, 2004), Ni-Sn-P (Popoola et al., 2016), Ni-Zn-P (Veeraraghavan et al., 2004), Ni-Co-P (Sankara Narayanan et al., 2003), Ni-Mo-P (Liu et al., 2016), etc. Amongst the wide range of alloy and poly alloy coatings, electroless Ni-P-W and Ni-P-Cu have enhanced hardness. Other advantages include wear resistance, anti-corrosion properties and thermal reliability in comparison with the binary Ni-P alloy (Palaniappa & Seshadri, 2008; Tien & Duh, 2004; Sahoo & Roy, 2017; Krasteva et al., 1994). Several investigations have been carried out to improve the tribological properties and corrosion resistance of Ni-P-W and Ni-P-Cu coatings. The use of Taguchi's experimental design technique along with grey relational analysis has been successfully demonstrated in optimizing the coating bath parameters to obtain enhanced anti-wear, anti-friction and anti-corrosion properties of Ni-P-W and Ni-P-Cu coatings (Roy & Sahoo, 2012; 2014; Sahoo et al., 2014).Electroless Ni-P, Ni-P-Cu, Ni-P-PTFE and Ni-P-CNT coatings show excellent tribological behavior under water/oil lubrication (Xu et al., 2014; Der Ger et al., 2003; Chen et al., 2002). The natural lubricious properties are particularly enhanced on incorporation of PTFE to binary Ni-P deposits and they show excellent tribological behavior under boundary lubrication regime (Ramalho & Miranda, 2007). Engine cylinders may be protected from the corrosive attack of bio-lubricants by the application of Ni-P-Cu coatings (Xu et al., 2014).

From extensive literature review on EN coatings, it may be concluded that several studies report the tribological characteristics of the alloy and composite coatings mostly under dry condition. There have been several instances where the optimization of Ni-P coatings and its variants for minimum friction and wear has been carried out for dry non-lubricated sliding condition (Sahoo, 2009; Sahoo & Pal, 2007; Sahoo & Das, 2015). Ni-P binary and ternary alloy coatings may also prove to be potential candidates for applications involving tribological contact under lubricated condition. Some investigations regarding the tribological behavior of electroless Ni-P coating and its variants under lubricated condition have been also carried out. But it is noticeable that studies concerning the optimization of the tribological behavior of EN coatings under lubrication are scanty. Therefore, design optimization of EN coatings under lubricated condition is necessitated owing to its extensive usage in industries. In the present chapter, friction and wear characteristics of autocatalytically deposited Ni-P, Ni-P-W and Ni-P-Cu are considered under lubricated condition. Coating deposition is done on AISI 1040 steel specimens. They are heat treated at suitable temperatures to enhance their microhardness. Coating characterization is done using scanning electron microscope (SEM), energy dispersive X-Ray (EDX) analysis and X-Ray diffraction (XRD) technique. Their tribological behavior is investigated under oil lubrication. Friction and wear tests are carried out by varying tribological test parameters namely the applied normal load, sliding speed and sliding duration at three different levels. Wear depth and coefficient of friction (COF) are considered to be the responses. Optimal setting of the tribo-test parameters for simultaneous minimization of wear depth and COF of the deposits under oil lubrication is obtained from grey relational analysis. Optimized test conditions may prove to be beneficial for exploring the true potential of EN coatings under lubrication. Statistical analysis of variance (ANOVA) is carried out to determine the significant parameters affecting the friction and wear characteristics of the coatings. A confirmation test is performed to determine the improvement in tribological behavior of the coatings compared to an initial test run. Finally the wear mechanism is obtained for the coatings from SEM micrographs of worn specimens.

EXPERIMENTAL DETAILS AND FUNDAMENTAL CONSIDERATIONS

Electroless Nickel Deposition

For the deposition of EN coatings, AISI 1040 steel specimens are chosen with high precision and of N5 (R_a=0.4μm) roughness grade. It has been reported in several studies that the EN coatings follow the substrate roughness. Therefore, it is necessary to select the specimens having lower deviations ($\pm 1\%$) in surface roughness which would otherwise affect the tribological behavior of the deposits. The specimens are initially cleaned in detergent water followed by rinsing in running water. The specimens are again rinsed thoroughly using deionized water following which acetone degreasing is carried out. After degreasing, the specimens are given a pickling/activation treatment in HCl (50%). Finally, post activation, the specimens are again rinsed in deionized water and immersed immediately into the electroless bath.

The electroless coating bath formulation and its deposition condition are selected after several trial and error methods and from literature review. The chemicals used to obtain the electroless bath for Ni-P, Ni-P-W and Ni-P-Cu coatings are enlisted in Table 1 while the deposition conditions are enlisted in Table 2. In all the three cases sodium hypophosphite is used as the reducing agent whereas, appropriate source of nickel ions, stabilizers, complexing agents, etc. are marked in Table 1. The coated specimens are heat treated appropriately in a box furnace to achieve enhanced microhardness. Electroless Ni-P and Ni-P-W coatings are heat treated at 400°C for 1h. Electroless Ni-P-Cu coatings are heat treated at 500°C for 1h. The heat treatment temperatures are selected based on literature review to obtain enhanced hardness of the coatings. There are several controversies regarding the deposition mechanism of EN acid/alkaline bath coatings (Loto, 2016). But the electrochemical and atomic hydrogen mechanisms are widely accepted. The electrochemical mechanism may be explained by the following reactions as mentioned by Sahoo & Das (2011):

$$H_2PO_2^- + H_2O \rightarrow H_2PO_3^- + 2H + 2e^- \tag{1}$$

$$Ni^{2+} + 2e^- \rightarrow Ni \tag{2}$$

$$2H^+ + 2e^- \rightarrow H_2 \tag{3}$$

$$H_2PO_2^- + 2H^+ + e^- \rightarrow P + 2H_2O \tag{4}$$

Table 1. Coating bath formulation

Composition	Values (g/L)		
	Ni-P	Ni-P-W	Ni-P-Cu
Nickel sulphate	30	20	30
Nickel chloride	30	×	×
Sodium hypophosphite	10	20	15
Sodium succinate	12	×	×
Sodium citrate	×	35	15
Ammonium sulphate	×	30	×
Lactic acid	×	5	×
Sodium tungstate	×	25	×
Copper sulphate	×	×	0.5

On the other hand, in atomic hydrogen mechanism, due to the catalytic dehydrogenation of hypophosphite molecule adsorbed at the surface, release of atomic hydrogen takes place (Sahoo & Das, 2011). Further, this nascent hydrogen reduces nickel at the catalyst surface (Loto, 2016). The atomic hydrogen mechanism may be represented as follows (Sahoo & Das, 2011):

$$H_2PO_2^- + H_2O \rightarrow HPO_3^{--} + H^+ + 2H_{ads} \tag{5}$$

$$2H_{ads} + Ni^{++} \rightarrow Ni + 2H^+ \tag{6}$$

$$H_2PO_2 + H_{ads} \rightarrow H_2O + OH^- + P \tag{7}$$

$$\left(H_2PO_2\right)^{2-} + H_2O \rightarrow H^+ + \left(HPO_3\right)^{2-} + H_2 \tag{8}$$

The nascent or adsorbed hydrogen also reduces small amounts of the hypophosphite to water, hydroxyl ion and P. Nevertheless several mechanisms have been proposed for the electroless deposition technique and some of them are summarized by Loto (2016) in his work.

Coating Characterization

To ensure proper deposition of the coatings to the substrate it is necessary to observe its microstructure and composition. Moreover, the coating microstructure and composition affect its tribological behavior. Therefore, coating characterization is an essential part after its deposition. In the present chapter, coating characteristics are determined using different techniques such as:

- Scanning electron microscope (SEM) to study the surface morphology (Make: JEOL, JSM 6360).
- Energy dispersive X-Ray (EDX) analysis to determine the coating composition.
- X-Ray diffraction (XRD) technique to study the phase transformations (Make: Rigaku, Miniflex employing Cu Kα radiation).

Table 2. Deposition conditions

Deposition Condition	Values		
	Ni-P	Ni-P-W	Ni-P-Cu
pH	4.5	7 to 8	9.5
Temperature	80 ± 2°C	90 ± 2°C	85 ± 2°C
Duration of coating	3 hrs		
Bath volume (ml)	200 ml		

SEM and XRD are carried out for Ni-P, Ni-P-W and Ni-P-Cu coatings in as-deposited as well as post heat treatment. EDX analysis of the coatings is carried out only in their as-deposited condition.

Microhardness Measurement

Hardness of a material is a measure of its ability to resist permanent deformation/shape change on the application of a constant compressive force. Nowadays, indentation hardness techniques are gaining immense importance in industries since they are simple, fast and relatively cheap material quality control method (Broitman, 2017). The indentation hardness methods may be classified further as macro-indentation, micro-indentation and nano-indentation tests depending on the ranges of compressive load applied. For EN coatings, it has been observed that there is a close relationship between its hardness and wear resistance. Therefore, the present chapter reports the Vicker's microhardness measurements for the EN coatings in their as-deposited and heat treated condition. In this method, indentation is carried out by a diamond indenter in the form of a pyramid with a square base and an apex angle of 136°. The Vicker's hardness is calculated as follows:

$$HV = 1854.4\frac{F}{d^2} \tag{9}$$

where '*F*' is the applied force (in gf) and '*d*' is the average length of diagonal. Considering the coating thickness, the Vicker's microhardness of the deposits in as-deposited condition and after heat treatment are measured at an indentation load of 100gf, dwell time of 15s and an indentation speed of 25μm/s. Average of six microhardness measurements is considered. The measurements are carried out following ASTM standard E384-16 on a UHL microhardness tester (VMHTMOT, Technische Mikroskopie).

Tribological Tests

The tribological tests of Ni-P coatings are carried out under lubrication on a block-on-roller multi-tribotester. In this configuration, the heat treated Ni-P coatings (deposited on AISI 1040 steel blocks of dimension 20×20×8 mm) are held stationary against a hard chromium coated (~65HRc) roller (Ø50×20mm) counterface of EN8 specification. Dead weights are placed on a loading pan attached to a 1:5 loading lever to apply normal load. The roller speed and the sliding duration can be controlled via a computer. Wear is indicated in terms of displacement in 'μm' and is captured using a linear voltage resistance transducer (LVDT) having an accuracy of 1±1% of measured value. The frictional force is measured using a beam type load cell of 1000N capacity and an accuracy of 1±1% of measured value. The wear depth and COF are indicated on a computer in real time during the tests. The tribo-testing of electroless Ni-P-W and Ni-P-Cu coatings has been carried out on a pin-on-disc type configuration test setup. The coated specimens in the form of 'pins' (deposited on AISI 1040 steel of dimension Ø6×30mm) are held stationary against a rotating counterface disc (Ø165×8mm) of EN 31 specification (hardness ~62HRc). Normal load is applied via a 1:1 loading lever by placing dead weights on a loading pan attached to the lever. The wear depth and COF are measured by a LVDT and load cell respectively of accuracy 1±1% of measured value. The sliding duration and sliding speed are changed from a controller attached to the tribo-tester. Track diameter is kept fixed at 80mm.

It may be noted here that the wear is expressed as wear depth i.e. in terms of displacement. Since the counterface material in consideration is very hard, it is expected that wear of only the coatings take place. Therefore, the parameter wear depth may be considered to represent the actual wear loss of the coatings. The lubricant considered in the present work is SERVO PRIDE 40, commercially available engine oil manufactured by Indian Oil. This lubricant is essentially meant for heavy earth moving vehicles. Since EN coatings find wide usage in automotive industries, engine oil is considered while carrying out friction and wear tests under lubrication.

Planning of Experiments

The best way to capture the system response is by carrying out experiments. A system/process generally comprises of some input/s and output/s which are affected by controllable as well as uncontrollable parameters. The purpose of carrying out experiments is to observe how the system responds with the varying controllable parameters or to determine how the controllable parameters should be set such that the effect of the uncontrollable factors is minimized. But considering the present industrial demands and cost effectivity, it is difficult to vary one factor at a time on a trial and error basis and observe the system responses. Therefore, a scientific approach to experimentation is necessary which would reduce the variability in output and extract maximum information about a system by carrying out minimum number of trials. This approach is known as design of experiments (DOE). The use of DOE may prove to be highly effective in cases of product and process optimization, study the effect of multiple factors simultaneously on a response, influence of parameters on the system under consideration, etc. The first step is the selection of response variables and process parameters. The present chapter considers wear depth and COF as the responses. Tribological behavior of EN coatings is seen to be affected by the coating bath composition. The friction and wear characteristics also depend on the tribological test parameters. Therefore, tribological test parameters such as the load, speed and duration of test are selected as the process parameters. The tribological test parameters along with their values are considered at three different levels as laid down in Table 3. All other experimental parameters are kept constant.

At this point, it may be noted here that the values of tribological test parameters that have been selected for Ni-P coatings do not match with the values set for Ni-P-W and Ni-P-Cu coatings. Moreover, the experimental setup for Ni-P is a block-on-roller arrangement while that of Ni-P-W and Ni-P-Cu deposits is a pin-on-disc machine. Therefore it may seem that the tests are not standardized for all the

Table 3. Process parameters along with their levels

Coating	Design factors	Unit	Levels		
			1	2	3
Ni-P	Load (L)	N	50	75	100
	Speed (S)	rpm	60	70	80
	Time (T)	min	5	10	15
Ni-P-W & Ni-P-Cu	Load (L)	N	10	20	30
	Speed (S)	rpm	60	80	100
	Time (T)	min	5	10	15

three coating variants. But since the present chapter considers only statistical analysis of tribological test results and the application of DOE to optimize the responses, such standardization may not be necessary. Moreover, the test parameters in all the three cases are changed in an incremental manner and are equally spaced. Therefore, the statistical significance of the test parameters in controlling the responses may be determined.

Now it can be seen that the variation of process parameters is a case of 3 factors being varied at 3 levels. Several design techniques may be adopted such as full factorial, response surface methodology (RSM), Taguchi's design technique, etc. The present chapter reports the use of Taguchi's orthogonal array (OA) design methodology. The Taguchi OA has been widely employed by researchers and in industries to optimize performance characteristics (Roy & Sahoo, 2012; 2014; Das & Sahoo, 2011; Mukhopadhyay et al., 2016b; 2016d; 2016f). The Taguchi's OA have proven capabilities in optimizing complex engineering and industrial processes from minimum number of trials. They also have capabilities of designing a robust system and to reduce the cost related to evaluation of responses with varying process parameters. Hence, the Taguchi's experimental design technique has been adopted in the present work. The selection of an OA is based upon the degrees of freedom (DOF) of an experiment. For an individual factor, the DOF assigned is 2 while interaction of two factors is assigned a DOF 4 (2×2). Therefore, three individual factors constitutes of 6 DOF while the interaction has 12 DOF (4×3). Now, in the present case, three factors and their interactions constitute a total DOF 18 (6+12). According to the Taguchi method, an OA having DOF higher than or equal to that of the whole experiment needs to be chosen. Therefore, Taguchi's L_{27} OA satisfies the requirement which has 26 DOF.

Grey Relational Analysis

Taguchi's S/N ratio method, which is a logarithmic transformation of mean squared deviation, has been applied in several problems. But it is considered to be an offline quality control method and is more suitable for single response optimization problems. But industrial complexities involve several responses to be dealt simultaneously. Moreover, the individual responses may have different objectives i.e. one response may require minimization while others need to be maximized. In such cases, the multiple responses may be converted to a single response such that further analysis on this multiple performance index (MPI) can be carried out. To cater into multi-variate and multi-response system optimization, grey relational analysis (GRA) has been introduced by Deng (1989). GRA is useful in dealing with incomplete, imprecise and uncertain information. The GRA is based on minimization of absolute distance of sequences from an ideal referential sequence. In the grey system, complete information is considered to be white while no information is black. Any level of information in between i.e. incomplete information may be termed as grey.

The first step in GRA is normalization of sequences between 0 and 1. Different responses may have different range of values and units. To convert them to comparable sequences, they are normalized. Now normalization may be carried out following lower-the-better, higher-the-better or nominal-the-best quality characteristics depending upon the desired characteristics of the responses. They may be expressed as follows:

$$Lower-the-better: x_i^*(k) = \frac{\max x_i(k) - x_i(k)}{\max x_i(k) - \min x_i(k)} \tag{10}$$

$$Higher - the - better : x_i^*(k) = \frac{x_i(k) - \min x_i(k)}{\max x_i(k) - \min x_i(k)} \tag{11}$$

$$Nominal - the - best : x_i^*(k) = 1 - \frac{\left|x_i(k) - x_t(k)\right|}{\max x_i(k) - \min x_i(k)} \tag{12}$$

where $x_i^*(k)$ is the value after normalization, while $\min x_i(k)$ and $\max x_i(k)$ are the smallest and largest values of the kth response. The target of $x_i(k)$ is $x_t(k)$ in Equation 12.

After normalization, the next step is calculation of grey relational coefficient (GRC). The GRC is a measure of the deviation of experimental sequences from the ideal sequence or rather the correlation between desired and actual experimental data. A higher GRC close to 1 is an indication of better system performance. The GRC may be calculated as:

$$\xi_i(k) = \frac{\Delta_{\min} + \zeta\Delta_{\max}}{\Delta_{0i}(k) + \zeta\Delta_{\max}} \tag{13}$$

where Δ_{0i} is the absolute difference value between $x_0(k)$ and $x_i^*(k)$ may be termed as the deviation sequence. The values $\Delta_{\min}$ and $\Delta_{\max}$ are the minimum and maximum values of absolute difference of all the comparing sequences. ζ is known as the distinguishing coefficient and controls the value of $\Delta_{\max}$ when it gets too large. Lower the value of ζ the higher is its distinguishing ability. Usually, $\zeta \in \left\{0,1\right\}$, though its value is taken to be 0.5 due to its moderating distinguishing effects and stability of outcomes (Das & Sahoo, 2011).

The last step is the generation of MPI which is the grey relational grade (GRG) in this work. The GRG is obtained by finding the average of the GRC of the responses. Thus, multiple responses are converted to a single grade which may be further considered for analysis. A higher value of GRG indicates that the result is closer to near optimum value. It may be expressed as:

$$\gamma_i = \frac{1}{n}\sum_{i=1}^{n}\xi_i(k) \tag{14}$$

where n is the number of experiments.

Confirmation Tests

After carrying out GRA, a validation test is carried out to ascertain if there has been any improvement in the grade predicted by GRA compared to the initial test condition. The initial test condition considered is the combination of test parameters at their mid-levels i.e. L2S2T2. A similar procedure has been adopted in this chapter. Moreover, the predicted GRG and experimentally obtained grade are compared to see if they are in agreement or not. The GRG at the optimum combination may be predicted as follows:

$$\hat{\gamma} = \gamma_m + \sum_{i=1}^{o}\left(\gamma_i - \gamma_m\right) \quad (15)$$

where $\hat{\gamma}$ is the predicted value of GRG, γ_m is the mean value of all the grades, γ_i is the mean grade at the optimal level for a process parameter and o is the number of process parameters which is three in this case.

Analysis of Variance

A mathematical method known as the analysis of variance (ANOVA) may be employed for the determination of the statistical influence of process parameters in controlling the responses. This is achieved by separating the total variability of the GRG into contributions of the individual process parameters and error. The ANOVA table generally consists of the DOF, sum of squares (SS), mean square (MS), F-ratio and the contribution percentage of the system parameters under consideration. The total sum of squared deviations constitutes of SS due to the process parameters and the error. The F-ratio also known as the variance ratio determines the significant effect of a process parameter on the response. It is the ratio of regression MS and the mean squared error. At a confidence level α, a parameter may be considered significant if the calculated F-ratio is higher than the tabulated values.

RESULTS AND DISCUSSION

Surface Morphology of the Coatings

The surface morphology of the coatings is investigated using SEM. The SEM micrographs of the coatings in their as-deposited condition are presented in Figure 1. Electroless Ni-P, Ni-P-W and Ni-P-Cu coatings exhibit a nodular and dense surface morphology. The surface appears to be free of any porosity. In Figure 1(b) bright particles are seen to be scattered over certain parts which may be due to the overgrowth of nickel. For Ni-P-Cu coatings it is observed that the nodules form clusters in Figure 1(c). The size of nodules seems to be higher for Ni-P-W coatings followed by Ni-P and Ni-P-Cu. Electroless Ni-P and Ni-P-W coatings exhibit a smooth surface finish, though the roughness of Ni-P-Cu is higher as visible in the SEM micrograph due to the formation of clusters of the nodules. Smooth surface finish in Ni-P-Cu coatings has been observed by Liu et al. (2015) as the copper concentration in the electroless solution increases. The lower copper concentration may have led to a slightly higher surface roughness of Ni-P-Cu coating in the present work. Balaraju et al. (2009) has observed the surface morphology of Ni-P-W coatings for various concentration of sodium tungstate. It has been concluded that incorporation of tungsten results in a cloudy appearance. As the concentration is increased scattered small nodules are seen along with agglomeration of big nodules. Due to the higher concentration of sodium tungstate in the present work, the scattered particles are observed in Figure 1(b) for Ni-P-W coatings. Thus, the observations made in the present work corroborates well with the literature. Due to the compact nodulated structure, the as-deposited EN coatings exhibit high resistance to corrosion. The absence of pores or grain boundaries in coatings is an indication of the deposits being amorphous in their as-deposited condition.

Figure 1. As-deposited EN coatings: (a) Ni-P; (b) Ni-P-W; and (c) Ni-P-Cu

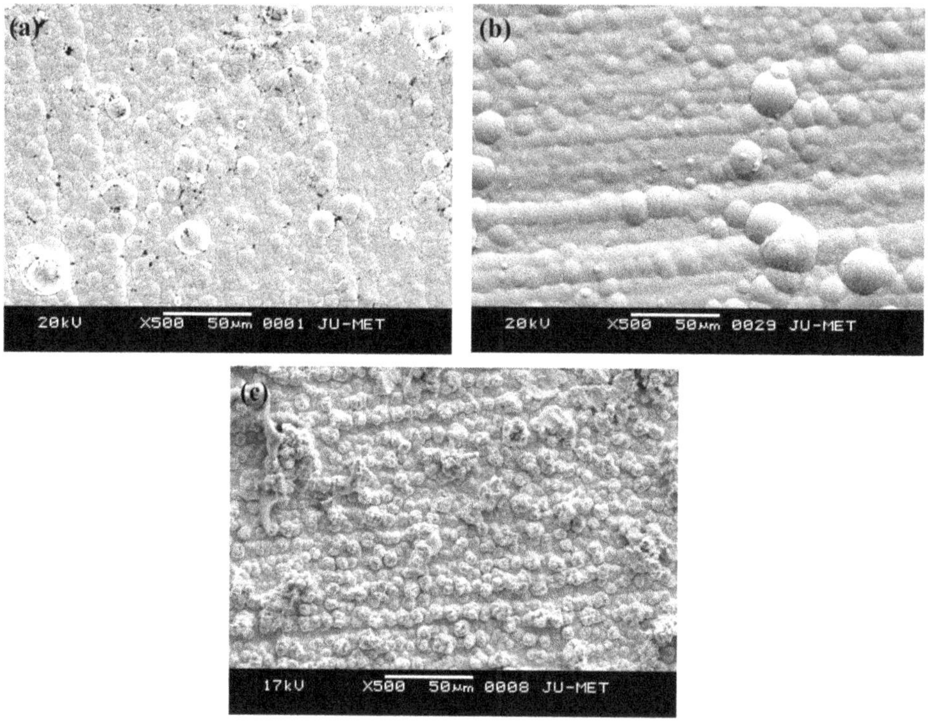

The SEM micrograph of the EN coatings after heat treatment is presented in Figure 2. On heat treatment, the size of the nodules increases for Ni-P (Figure 2(a)) and Ni-P-Cu (Figure 2(c)) coatings. Specific grain boundaries are observed for all the three variants. This is an indication of crystallization of the deposits taking place post heat treatment. The Ni-P coatings (Figure 2(a)) turn bluish at places due to the formation of nickel oxides. Pores are observed in case of Ni-P coatings in Figure 2(b). In case of Ni-P-W coatings some amount of volume contraction has been observed and the nodules seem to be compact. Such volume contractions may be observed in Ni-P-W coatings due to the precipitation of the phosphides out of the supersaturated solid solution of Ni as matrix containing P and W. Therefore, it may be concluded from the surface morphology that phase transformation of the deposits occur on heat treatment. Such behavior of EN coatings is in line with the literature (Sahoo & Das, 2011; Sahoo, 2009; Sudagar et al., 2013; Mukhopadhyay et al., 2016a; 2016d; 2017a; Balaraju et al., 2009).

Composition Analysis

The composition of electroless Ni-P, Ni-P-W and Ni-P-Cu coatings in terms of their respective weight percentages of elements deposited is determined using EDX and is listed in Table 4. In electroless Ni-P coating, the percentage of P deposited is around 10.8%. Electroless Ni-P coatings may be classified as low P (3-5%), medium P (6-9%) and high P (10-14%). The microhardness as well as crystallinity of the coatings depends on the weight percentage of P. Generally, the low and medium P coatings exhibit an X-ray amorphous/nanocrystalline structure. The Ni-P coatings with high P exhibit amorphous structure. In the present work, the Ni-P coating lies in the high P range and is therefore expected to be amorphous.

Figure 2. Heat treated EN coatings: (a) Ni-P; (b) Ni-P-W; and (c) Ni-P-Cu

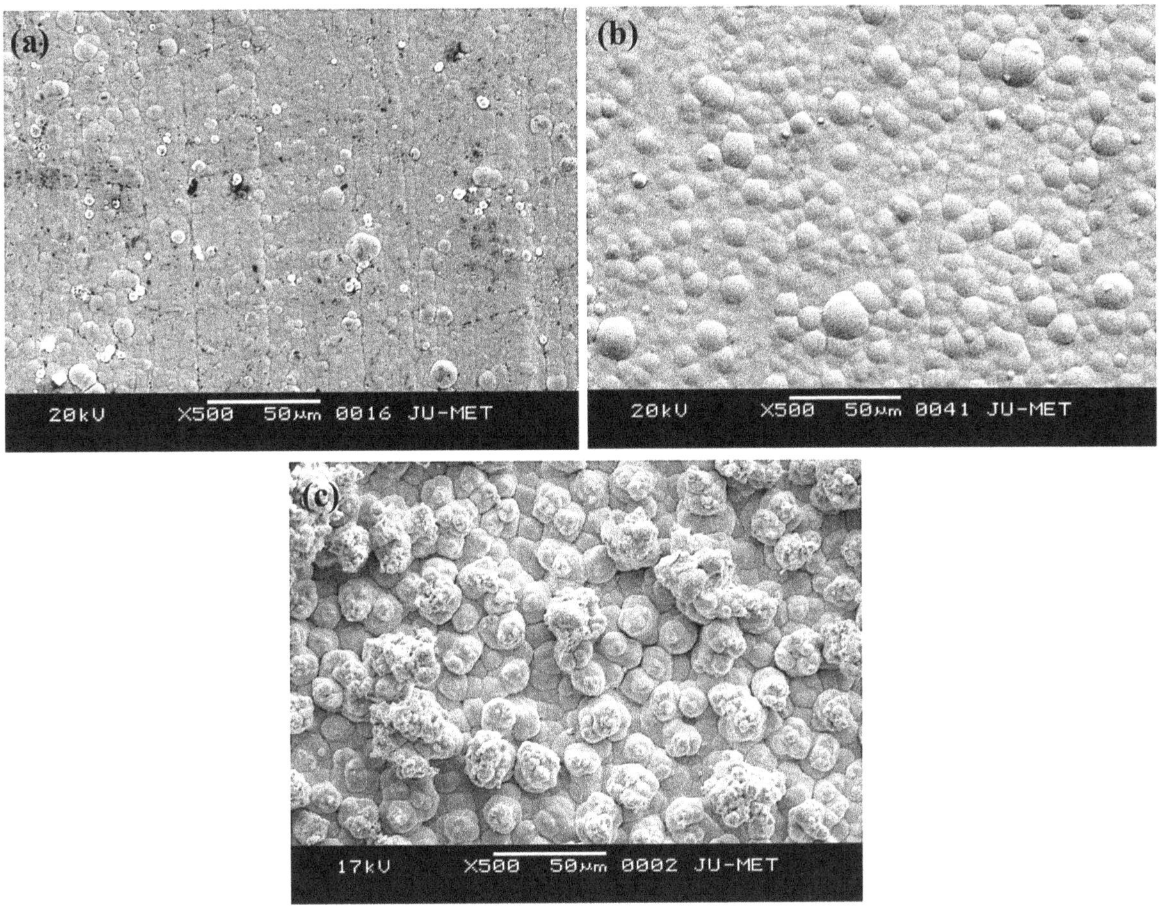

Table 4. Composition of EN coatings in their as-deposited condition

EN Coatings	%Ni	%P	%W	%Cu
Ni-P	89.2	10.8	-	-
Ni-P-W	87.8	8	4.2	-
Ni-P-Cu	85.38	10.62	-	4

This is in agreement with the SEM image observed in Figure 1(a). In case of Ni-P-W and Ni-P-Cu coatings, the percentage of W and Cu deposited is 4.2% and 4% respectively. Incorporation of W leads to a decrease in P content and a slight decrease in Ni too. While in case of Ni-P-Cu coatings, a decrease in nickel content is mostly observed.

Phase Transformation Analysis

The structural state of the coatings is determined using XRD technique. To understand certain physiological behavior and explain the same, it is necessary to study the induced crystalline phases. XRD plots of the EN coatings in their as-deposited condition are presented in Figure 3. Electroless Ni-P coatings exhibit a broad peak with superimposed short peaks in Figure 3(a). This means the coatings are amorphous in their as-deposited condition. Similar result has been inferred from the SEM (Figure 1 (a)) and EDX analysis and the XRD results consolidate the same. In general, a broad peak is observed in X-Ray diffractograms due to disorder in arrangement of atoms. The crystallinity of the coatings is determined by the rate of segregation of the Ni and P atoms (Palaniappa & Seshadri, 2008). Compared to Ni, the rate of diffusion of P is lower. To achieve segregation of Ni and P, a higher amount of P atoms need to be moved from a given area per unit time (Palaniappa & Seshadri, 2008). Therefore, the coatings lying in the high P range show amorphous characteristics. Similar observations have been made by other researchers and the present observation is in line with the same (Sahoo, 2009; Panja & Sahoo, 2014; 2015; Panja et al., 2016; Mukhopadhyay et al., 2016d; 2017a; Sahoo & Roy, 2017; Hur et al., 1990).

The XRD plot of as-deposited Ni-P-W coating is illustrated in Figure 3(b). The reflection from Ni (111) is observed to be sharper. This occurs due to a reduction in the P content as observed in Table 4. It has been observed by Balaraju et al. (2009) that with an increase in W, the reflection from Ni (111) becomes sharper. Therefore, as-deposited Ni-P-W exhibit amorphous/nanocrystalline phases. The reflection from Ni (111) is comparatively broader in Figure 3(c). The as-deposited Ni-P-Cu coatings are therefore observed to be amorphous though, the diffractogram (Figure 3(c)) seems to be sharper compared to that of Ni-P coatings (Figure 3(a)). Similar observation has been made by Balaraju and Rajam (2005). The peak of the as-deposited Ni-P-Cu coating may appear slight sharper due the overlapping of Fe (110) peak of the substrate. Nevertheless, the coating is thought to be amorphous (Yu et al., 2001).

Figure 3. XRD results of coatings in their as-deposited condition: (a) Ni-P; (b) Ni-P-W; and (c) Ni-P-Cu

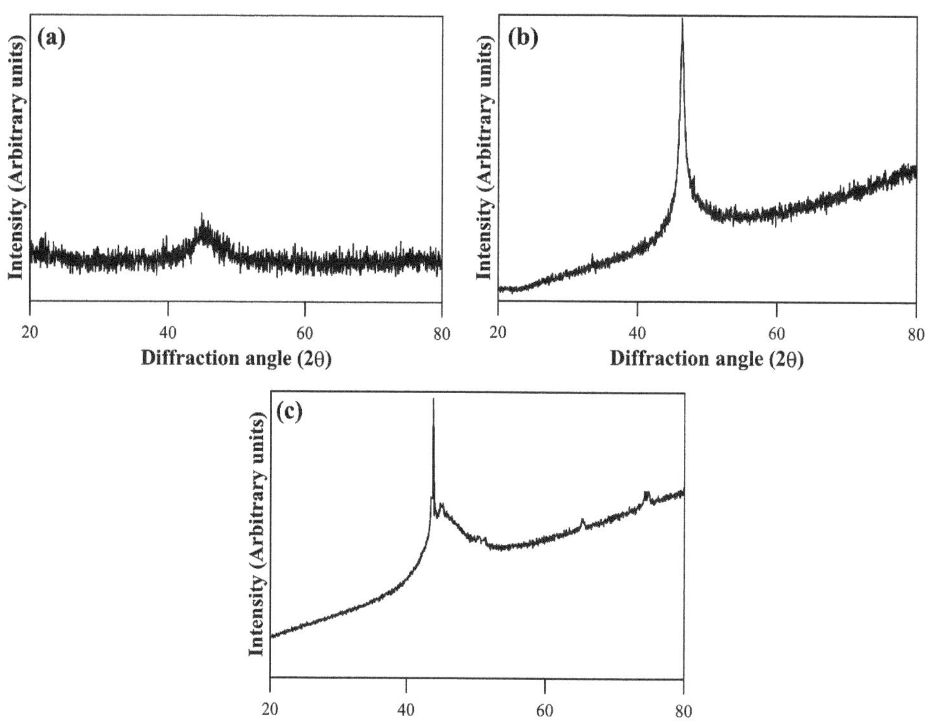

Figure 4. XRD results of coatings in their heat treated condition: (a) Ni-P; (b) Ni-P-W; and (c) Ni-P-Cu

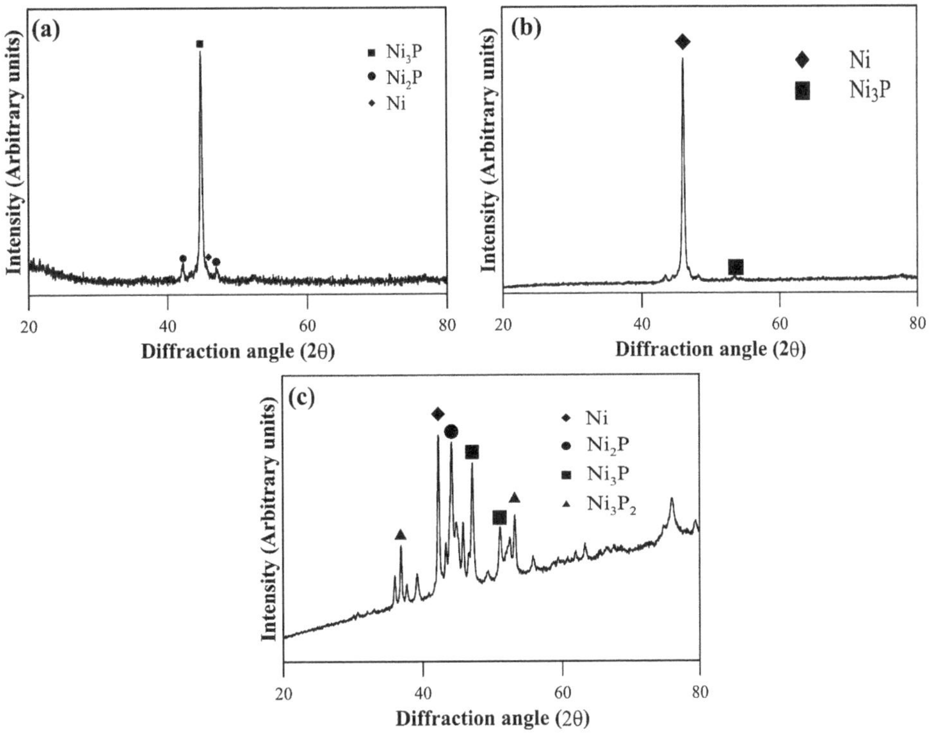

On heat treatment electroless Ni-P coatings become crystalline with the precipitation of crytalline Ni_3P, Ni_2P, Ni as can be seen in Figure 4(a). In case of electroless Ni-P-W coatings precipitation of crystalline Ni (111) and Ni_3P takes place (Figure 4(b)). The precipitation of crystalline Ni, Ni_3P and Ni_2P is also observed for electroless Ni-P-Cu coating on heat treatment along with a meta-stable phase of Ni_3P_2. Therefore, it is concluded that EN binary and ternary alloy coatings are either amorphous or a mixture of amorphous and nanocrystalline phases which depend on the P content of the coatings in as-deposited condition. On heat treatment, the coatings become crystalline with the precipitation of crystalline nickel and its phosphides. The observations are seen to be in agreement with the literature (Balaraju et al., 2009; Balaraju & Rajam, 2005; Yu et al., 2001).

The surface morphology, composition and phases of EN coatings depend on the bath composition and deposition condition. The amount of nickel source, reducing agent, stabilizers, accelerators, complexing agents and the third alloying element source is determined by the catalytic activity and electrochemical potential of the process (Sahoo & Roy, 2017). Therefore, for different EN deposition process, the chemical composition as well as the deposition condition varies. Nevertheless, from literature review and by trial and error process, an effort has been made to keep the content of W and Cu deposition almost same as can be seen in Table 4. Thus, to a certain extent, the effect of co-deposition of the third alloying element to the binary Ni-P alloy on the coating characteristics is revealed. But, a more standardized electroless bath composition needs to be formulated where the chemical composition and deposition condition are kept almost similar which would enable the true comparison of the coating characteristics.

Microhardness of the Coatings

Microhardness of the EN coatings pre and post heat treatment is measured using Vicker's micro-indentation technique. The microhardness results are presented in Figure 5. Electroless Ni-P, Ni-P-W and Ni-P-Cu coatings are represented as C1, C2 and C3 respectively in Figure 5. Ni-P coatings in as-deposited state exhibit a microhardness of around 530HV_{100}. As already mentioned, the phase structure and the microhardness depend on the content of P. Higher P content results in lower microhardness. The microhardness of the as-deposited Ni-P coatings is similar to the results obtained by other researchers for similar percentage of P (Panja & Sahoo, 2014). Heat treatment results in an increase in the microhardness of Ni-P coatings to around 750HV_{100}. This marked improvement in microhardness of Ni-P coatings may be attributed to the precipitation of hard phosphides phases on heat treatment. The microhardness of electroless Ni-P-W coating in its as-plated state is around 650HV_{100}. This is quite higher compared to Ni-P coating. The improvement in microhardness of Ni-P coatings on alloying with W is due to the solid solution strengthening phenomenon. For electroless Ni-P-W coatings, strengthening caused due to W in solid solution form in Ni matrix has been reported by Wu et al. (2004). Balaraju and Rajam (2005) also observed a similar phenomenon. Heat treatment results in a remarkable improvement in the microhardness (~1011HV_{100}) of Ni-P-W coatings as can be seen in Figure 5. A high hardness of heat treated electroless Ni-P-W coatings is attributed to the formation of a solid solution of Ni(W) on annealing as well as the precipitation of stable Ni_3P phase (Figure 4b). Electroless Ni-P-Cu coatings have a microhardness of around 570HV_{100} in as-deposited condition while on heat treatment it increases to almost 770HV_{100}. The microhardness of Ni-P-Cu coatings is mostly similar to that of Ni-P coatings and a significant change is not observed on the incorporation of Cu to the binary Ni-P alloy. Therefore, the microhardness of Ni-P-W coatings is observed to be the highest amongst the three EN variants investigated in this chapter.

Optimization of Tribological Behavior Using Grey Relational Analysis

The tribological test results of electroless Ni-P, Ni-P-W and Ni-P-Cu coatings are given in Table 5 in their appropriate sequence as laid down in Taguchi's L_{27} OA. It can be clearly seen in Table 5 that the experimental sequence which yields the lowest value of wear depth may not give the lowest COF and vice versa. Therefore, the use of an efficient technique like GRA is necessary for optimization. Taguchi's S/N ratio technique is meant primarily for the optimization of single response problems. Therefore, the complexity of multi-response optimization is overcome by the use of GRA in this chapter. It has also been used widely for the optimization of properties of EN coatings. Though, some recent studies have reported the use of genetic algorithm (Mukhopadhyay et al., 2016g; 2017a) and a modified version of the GRA i.e. a hybrid grey fuzzy logic (Mukhopadhyay et al., 2015; 2016b; 2016d) for the optimization of tribological behavior of EN coatings, the use of GRA is still suggested due to its simplicity and quick decision making capabilities. The present chapter therefore reports the use of GRA for the optimization of tribological behavior of electroless Ni-P, Ni-P-W and Ni-P-Cu coatings.

The first step in GRA is the grey relational generation which involves the normalization of experimental data. Normalization may be carried out following the different quality characteristics appropriate for a response. Wear depth and COF of the coatings are to be reduced i.e. minimized. Therefore, lower-the-better characteristic is adopted to normalize the wear depth and COF. The results are normalized using Equation 10 and enlisted in Table 6. After normalization, the sequences become comparable since their values range between 0 and 1.

Figure 5. Microhardness of electroless nickel coatings

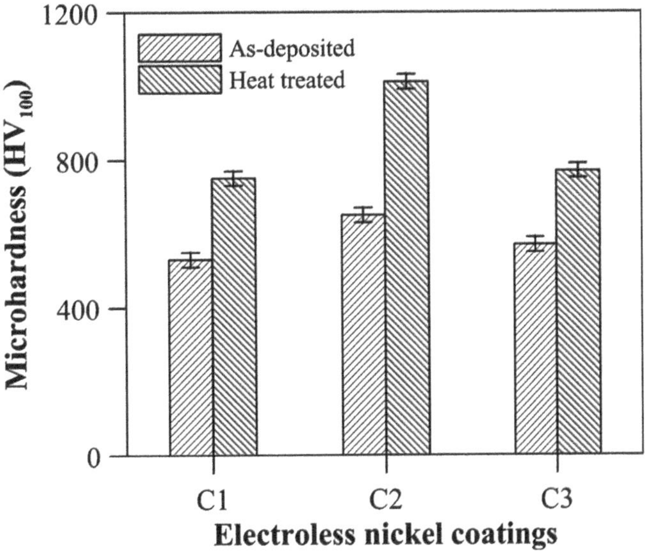

After normalization, the GRC is computed. The GRC is calculated based on Equation 13 and are presented in Table 7. It indicates the deviation of a sequence from the ideal best i.e. 1. A higher value of GRC for a sequence is an indication of the better system quality pertaining only to the response under consideration. Now since there are two responses, GRA suggests the conversion of them to a single MPI by calculating the average GRC of the responses as mentioned in Equation 14 which is the GRG. The GRG of all the experiments is given in Table 8 along with their order for Ni-P, Ni-P-W and Ni-P-Cu coatings. Thus, the multi-response problem is converted to a single grade i.e. the GRG. A higher value of GRG is an indication of near optimal condition or better system performance.

After computation of the GRG, further analysis is carried out using it. The present experimental design under consideration is an OA. Therefore, the effect of each tribological test parameter at different levels may be easily separated. For an instance, the effect of applied normal load at levels 1, 2 and 3 may be calculated by finding the average of the GRG at those levels. This means, the effect of load at level 1 is calculated by finding the average of the first nine experiments. Similarly, the effect of load at level 2 and level 3 is computed from the averages of experiments 10-18 and 19-27 respectively. Similar calculations are also done for the speed and time. This is represented in Table 9 as the response table for means of GRG. Additionally, the response table also consists of ranks based on delta values (Table 9). Delta values are calculated by finding the difference between the highest and lowest values in a column. The rank signifies the contribution of the process parameter in controlling the response or rather the relative magnitude of effects. In case of Ni-P coatings, rank 1 is assigned to load. This indicates load is the most significant parameter that controls the tribological behavior of Ni-P coatings under lubrication followed by speed and time which have been assigned rank 2 and 3 respectively. Similarly, for Ni-P-W coatings, speed is assigned rank 1 followed by load and time as observed in Table 9. In case of electroless Ni-P-Cu coating, the most significant influence is from load followed by speed and time as indicated by delta statistics in Table 9. The average grade of all the 27 experiments for Ni-P, Ni-P-W and Ni-P-Cu coatings is also indicated in Table 9.

Table 5. Results of the tribological tests carried out under lubricated condition

Exp. No.	Wear Depth (μm)			COF		
	Ni-P	Ni-P-W	Ni-P-Cu	Ni-P	Ni-P-W	Ni-P-Cu
1	11.84	4.398	11.329	0.084	0.098	0.038
2	12.87	5.114	13.075	0.082	0.110	0.044
3	13.95	5.559	13.548	0.078	0.119	0.039
4	11.93	4.265	12.782	0.067	0.080	0.01
5	12.40	6.370	14.051	0.067	0.089	0.009
6	13.85	9.552	14.062	0.065	0.115	0.008
7	12.21	7.803	13.607	0.055	0.033	0.006
8	13.30	9.859	14.836	0.052	0.036	0.010
9	13.34	11.204	15.178	0.052	0.049	0.010
10	15.14	7.982	14.757	0.053	0.039	0.122
11	16.67	8.795	16.775	0.051	0.048	0.124
12	17.86	9.439	17.64	0.048	0.047	0.122
13	15.89	9.723	16.592	0.047	0.097	0.072
14	16.97	11.414	18.643	0.047	0.108	0.078
15	17.16	11.674	20.384	0.046	0.113	0.084
16	15.67	8.890	19.526	0.043	0.118	0.038
17	16.53	12.646	21.468	0.042	0.153	0.038
18	16.62	13.921	22.216	0.043	0.165	0.039
19	17.50	6.534	20.978	0.041	0.040	0.158
20	18.33	9.012	22.983	0.036	0.043	0.149
21	18.82	10.084	24.422	0.037	0.041	0.150
22	17.34	8.837	21.791	0.039	0.106	0.137
23	17.56	11.299	24.832	0.035	0.113	0.144
24	18.22	12.783	25.920	0.032	0.114	0.147
25	16.55	11.814	24.052	0.028	0.052	0.134
26	17.53	14.476	25.996	0.024	0.063	0.132
27	18.31	15.520	26.998	0.024	0.066	0.131

Corresponding to the response table for means of GRG (Table 9), the main effects plot is obtained and presented in Figure 6. From the main effects plot, the optimum setting of test parameters for Ni-P coatings predicted is L1S3T1 which means an applied normal load of 50N, speed of 80rpm and time duration of 5min leads to enhanced tribological behavior under lubrication. Electroless Ni-P-W coatings show optimal tribological behavior for a parametric combination of L1S1T1 i.e. the lowest level of all the test parameters (applied normal load of 10N, sliding speed of 60rpm for a sliding duration of 5min) as can be concluded from Figure 6. Again, for electroless Ni-P-Cu coatings, the optimal parametric setting predicted by GRA is L1S3T1 viz. for a load of 10N, speed of 100rpm and a sliding duration of 5min (Figure 6). Though, the slope of main effects plot (Figure 6) and the delta statistics presented

Table 6. Normalized values of experimental data

Exp. No.	Normalized Wear Depth			Normalized COF		
	Ni-P	Ni-P-W	Ni-P-Cu	Ni-P	Ni-P-W	Ni-P-Cu
1	1.000	0.988	1.000	0.000	0.508	0.789
2	0.852	0.925	0.889	0.033	0.417	0.750
3	0.698	0.885	0.858	0.100	0.348	0.783
4	0.987	1.000	0.907	0.283	0.644	0.974
5	0.920	0.813	0.826	0.283	0.576	0.980
6	0.712	0.530	0.826	0.317	0.379	0.987
7	0.947	0.686	0.855	0.483	1.000	1.000
8	0.791	0.503	0.776	0.533	0.977	0.974
9	0.785	0.383	0.754	0.533	0.879	0.974
10	0.527	0.670	0.781	0.517	0.955	0.237
11	0.308	0.598	0.652	0.550	0.886	0.224
12	0.138	0.540	0.597	0.600	0.894	0.237
13	0.420	0.515	0.664	0.617	0.515	0.566
14	0.265	0.365	0.533	0.617	0.432	0.526
15	0.238	0.342	0.422	0.633	0.394	0.487
16	0.451	0.589	0.477	0.683	0.356	0.789
17	0.328	0.255	0.353	0.700	0.091	0.789
18	0.315	0.142	0.305	0.683	0.000	0.783
19	0.189	0.798	0.384	0.717	0.947	0.000
20	0.070	0.578	0.256	0.800	0.924	0.059
21	0.000	0.483	0.164	0.783	0.939	0.053
22	0.212	0.594	0.332	0.750	0.447	0.138
23	0.181	0.375	0.138	0.817	0.394	0.092
24	0.086	0.243	0.069	0.867	0.386	0.072
25	0.325	0.329	0.188	0.933	0.856	0.158
26	0.185	0.093	0.064	1.000	0.773	0.171
27	0.073	0.000	0.000	1.000	0.750	0.178

in the response table (Table 9) denote to a certain extent the influential parameters in controlling the tribological behavior of the coatings, actual statistical significance may be understood from ANOVA.

Statistical Significance of the Tribological Test Parameters

The statistical significance of the test parameters and their interactions along with their contribution to the tribological behavior of EN coatings under lubrication is analyzed from ANOVA and is carried out in MINITAB software. Table 10 enlists the ANOVA results for Ni-P deposits. At a confidence level of 99%, the highest contribution is observed from load followed by speed. The interaction of load and time

Table 7. Grey relational coefficient of wear depth and COF of EN coatings

Exp. No.	GRC Wear Depth			GRC COF		
	Ni-P	Ni-P-W	Ni-P-Cu	Ni-P	Ni-P-W	Ni-P-Cu
1	1.000	0.977	1.000	0.333	0.504	0.704
2	0.772	0.869	0.818	0.341	0.462	0.667
3	0.623	0.813	0.779	0.357	0.434	0.697
4	0.975	1.000	0.844	0.411	0.584	0.950
5	0.862	0.728	0.742	0.411	0.541	0.962
6	0.635	0.516	0.741	0.423	0.446	0.974
7	0.904	0.614	0.775	0.492	1.000	1.000
8	0.705	0.501	0.691	0.517	0.957	0.950
9	0.699	0.448	0.671	0.517	0.805	0.950
10	0.514	0.602	0.696	0.508	0.917	0.396
11	0.419	0.554	0.590	0.526	0.815	0.392
12	0.367	0.521	0.554	0.556	0.825	0.396
13	0.463	0.508	0.598	0.566	0.508	0.535
14	0.405	0.440	0.517	0.566	0.468	0.514
15	0.396	0.432	0.464	0.577	0.452	0.494
16	0.477	0.549	0.489	0.612	0.437	0.704
17	0.427	0.402	0.436	0.625	0.355	0.704
18	0.422	0.368	0.418	0.612	0.333	0.697
19	0.381	0.713	0.448	0.638	0.904	0.333
20	0.350	0.542	0.402	0.714	0.868	0.347
21	0.333	0.492	0.374	0.698	0.892	0.345
22	0.388	0.552	0.428	0.667	0.475	0.367
23	0.379	0.444	0.367	0.732	0.452	0.355
24	0.354	0.398	0.349	0.789	0.449	0.350
25	0.426	0.427	0.381	0.882	0.776	0.373
26	0.380	0.355	0.348	1.000	0.688	0.376
27	0.350	0.333	0.333	1.000	0.667	0.378

also has some influence with a contribution percentage of 15.58%. According to F-statistics, time and the interaction of load and speed is also significant.

The results of ANOVA of Ni-P-W coatings are presented in Table 11. The F-statistics indicate that load, speed, time and the interaction between load and speed are influential in controlling the tribological behavior of Ni-P-W coatings under lubrication. The highest contribution is from sliding speed with 34.94% followed by interaction of load and speed (21.68%). The contribution from load is slightly lower than the interaction parameter. Sliding duration also has some significant contribution in the process. Similarly, the results of ANOVA for Ni-P-Cu coatings is obtained and laid down in Table 12. It can be deduced that applied normal load is the most significant factor with 95.69% contribution to the tribologi-

Table 8. Grey relational grade and their order for EN coatings

Exp. No.	Ni-P		Ni-P-W		Ni-P-Cu	
	Grade	Order	Grade	Order	Grade	Order
1	0.667	5	0.740	5	0.852	5
2	0.557	11	0.665	11	0.742	8
3	0.490	23	0.624	14	0.738	9
4	0.693	2	0.792	3	0.897	1
5	0.637	7	0.634	12	0.852	4
6	0.529	15	0.481	21	0.858	3
7	0.698	1	0.807	2	0.887	2
8	0.611	8	0.729	6	0.820	6
9	0.608	9	0.626	13	0.810	7
10	0.511	21	0.759	4	0.546	14
11	0.473	26	0.684	9	0.491	16
12	0.462	27	0.673	10	0.475	18
13	0.515	20	0.508	18	0.567	12
14	0.486	25	0.454	22	0.515	15
15	0.487	24	0.442	24	0.479	17
16	0.545	13	0.493	20	0.596	10
17	0.526	17	0.378	26	0.570	11
18	0.517	18	0.351	27	0.558	13
19	0.510	22	0.808	1	0.391	20
20	0.532	14	0.705	7	0.375	22
21	0.516	19	0.692	8	0.360	25
22	0.528	16	0.513	17	0.398	19
23	0.556	12	0.448	23	0.361	24
24	0.572	10	0.423	25	0.350	27
25	0.654	6	0.602	15	0.377	21
26	0.690	3	0.521	16	0.362	23
27	0.675	4	0.500	19	0.356	26

cal behavior. F-statistics reveal that speed, time and the interaction of load vs. speed are also significant. But their percentage contributions are quite low.

It may be seen in Table 10 that for electroless Ni-P coatings, load is the most significant parameter (Duari et al., 2014). Similar results have been also obtained for Ni-P-Cu coatings (Table 12). But for Ni-P-W coatings, the contribution from sliding speed is highly significant (Table 11). Such differences may arise due to the different microstructure of the coatings and wear mechanisms that occur at microscopic levels during the wear process.

Table 9. Response table for means of grey relational grade

	Ni-P			Ni-P-W			Ni-P-Cu		
Level	**L**	**S**	**T**	**L**	**S**	**T**	**L**	**S**	**T**
1	0.609	0.524	0.591	0.678	0.706	0.669	0.829	0.552	0.612
2	0.502	0.555	0.563	0.527	0.522	0.580	0.533	0.586	0.565
3	0.581	0.614	0.539	0.579	0.556	0.535	0.370	0.593	0.554
Delta	0.107	0.090	0.052	0.151	0.184	0.135	0.459	0.041	0.059
Rank	1	2	3	2	1	3	1	3	2
	Mean grade = 0.564			*Mean grade = 0.595*			*Mean grade = 0.577*		

Figure 6. Main effects plot for means of grey relational grade

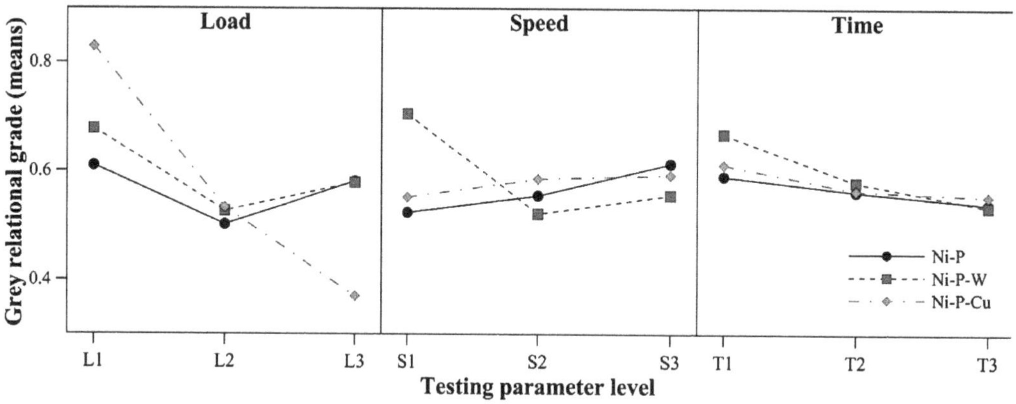

Table 10. ANOVA results of Ni-P coatings

Source	DOF	Seq SS	Adj MS	F-Ratio	% Contribution
L	2	0.0560	0.0280	74.31*	38.45
S	2	0.0374	0.0187	49.58*	25.65
T	2	0.0120	0.0060	15.94*	8.25
L × S	4	0.0131	0.0033	8.68*	8.99
L × T	4	0.0227	0.0057	15.05*	15.58
S × T	4	0.0015	0.0004	0.97	1.01
Error	8	0.0030	0.0004		2.07
Total	26	0.1457			100.00

*Significant at a confidence level of 99%.

Table 11. ANOVA results of Ni-P-W coatings

Source	DOF	Seq SS	Adj MS	F-Ratio	% Contribution
L	2	0.1053	0.0527	41.99*	21.39
S	2	0.1720	0.0860	68.59*	34.94
T	2	0.0844	0.0422	33.64*	17.14
L × S	4	0.1068	0.0267	21.28*	21.68
L × T	4	0.0117	0.0029	2.33	2.37
S × T	4	0.0022	0.0006	0.44	0.45
Error	8	0.0100	0.0013		2.04
Total	26	0.4924			100.00

*Significant at a confidence level of 99%.

Table 12. ANOVA results of Ni-P-Cu coatings

Source	DOF	Seq SS	Adj MS	F-Ratio	% Contribution
L	2	0.9733	0.4867	1971.84*	95.69
S	2	0.0086	0.0043	17.5*	0.85
T	2	0.0173	0.0086	34.96*	1.70
L × S	4	0.0128	0.0032	12.99*	1.26
L × T	4	0.0025	0.0006	2.55	0.25
S × T	4	0.0006	0.0002	0.67	0.06
Error	8	0.0019	0.0002		0.19
Total	26	1.0172			

*Significant at a confidence level of 99%.

Confirmation Test Results

To ascertain the improvement in results due to the application of GRA, a confirmation test is carried out. Comparisons are made with the initial test condition (which is the mid level combination of test parameters i.e. L2S2T2). The predicted GRG is calculated using Equation 15. Confirmation test results for electroless Ni-P, Ni-P-W and Ni-P-Cu coatings are laid down in Table 13. For electroless Ni-P coatings, the improvement in GRG at optimal condition compared to the initial test run is around 43.62% which is quite noteworthy. Moreover, the predicted grade agrees well with the experimentally obtained ones. A good reduction in the wear depth of Ni-P coatings is noticed due to optimization of process parameters using GRA. But a slight increase in the COF takes place from 0.047 to 0.055. Now since this is a case of multi-objective optimization, the goal is to ascertain the combination of process parameters which will yield low wear as well as COF simultaneously. This simultaneous minimization of both characteristics together results in some amount of compromise in the system performance at optimal condition.

Confirmation test results of Ni-P-W coating indicate an improvement in 62.98% in the GRG at optimal condition (Table 13). Moreover, the wear depth reduces from 11.414μm to 4.398μm and the COF reduces from 0.108 to 0.098. Therefore, the use of GRA has improved the tribological behavior of the coatings

Table 13. Confirmation test results of EN coatings

		Initial Parameter	Optimal Parameter	
			Predicted	**Experimental**
Ni-P	Level	L2S2T2	L1S3T1	L1S3T1
	COF	0.047		0.055
	Wear depth (μm)	16.968		12.213
	Grade	0.486	0.686	0.698
	Improvement of GRG = 0.212 (43.62%)			
Ni-P-W	Level	L2S2T2	L1S1T1	L1S1T1
	COF	0.108		0.098
	Wear depth (μm)	11.414		4.398
	Grade	0.454	0.863	0.740
	Improvement of GRG = 0.286 (62.98%)			
Ni-P-Cu	Level	L2S2T2	L1S3T1	L1S3T1
	COF	0.078		0.006
	Wear depth (μm)	18.643		13.607
	Grade	0.515	0.880	0.887
	Improvement of GRG = 0.372 (72.23%)			

significantly under lubrication. The GRG predicted by Equation 15 and experimentally obtained grade corroborate well with each other.

The confirmation test results of Ni-P-Cu coatings are laid down in Table 13. A high improvement of 72.23% is observed at the optimized condition. The wear depth is seen to reduce from 18.643μm to 13.607μm when the initial test results are compared with the optimal ones. The COF also improves significantly from 0.078 to 0.006 at the optimum condition.

Therefore, the confirmation test results in Table 13 indicate that the use of Taguchi's experimental design technique along with GRA has been applied successfully to optimize the tribological behavior of the coatings under oil lubricated condition. A significant improvement in the GRG is achieved for all the three coatings.

Analysis of Wear Mechanism

The wear mechanism of Ni-P, Ni-P-W and Ni-P-Cu coatings under lubricated sliding condition is ascertained by examining the worn specimens under SEM. The SEM images of the worn surfaces are presented in Figure 7. The worn surface of Ni-P coating (Figure 7(a)) is mainly characterized by ductile failure along with high degree of plasticity in the sliding direction. Fine parallel scratch marks may be observed. This is an indication of micro-cutting taking place. This may have occurred due to the occasional entrapment of debris particle. Pits and prows may not be observed. Coating delamination in the form of patches is also not observed. Therefore, abrasive wear is the predominant mechanism under lubricated condition for Ni-P coating. Under dry sliding condition, a mixture of abrasive and adhesive mode has been observed by Sahoo and Pal (2007). The use of lubrication prevents adhesion of the coating and counterface material.

Figure 7. Worn specimen at highest level of test parameters: (a) Ni-P; (b) Ni-P-W; and (c) Ni-P-Cu

The worn surface of electroless Ni-P-W coating can be seen in Figure 7(b). The surface is characterized by micro-cutting and mainly surface smoothening. It seems that the top of the nodules have been abraded. Long parallel scratch marks in the direction of sliding are not observed. Generally, the long parallel grooves are produced due to wear debris cutting through the material by displacing it along the side of the wear tracks. This results in three body abrasion. In this case, it seems two body abrasion is predominant i.e. the hard counterface abrades the coating material, though, this kind of surface smoothening effect leads to enhanced tribological behavior compared to dry sliding condition (Mukhopadhyay et al. 2016g). Under dry condition, heat treated electroless Ni-P-W coatings mostly exhibit abrasive wear with wear debris playing an important role (Mukhopadhyay et al., 2016e). But in case of lubricated condition, the flowing lubricant may have carried away the generated wear debris leading to two body abrasion.

The SEM micrograph of worn surface of Ni-P-Cu coatings may be observed in Figure 7(c). Most of the nodules seem to remain un-deformed. Some of the nodules have been fractured. Micro-cutting or micro-ploughing effects are not observed. Pits or prows are also absent. Therefore, the possibility of abrasive or adhesive wear is ruled out. During the sliding wear test, due to the repeated nature of loading, formation of sub-surface cracks may take place which eventually propagate resulting in fracture of the nodules. Moreover, the heat treatment may have resulted in the coatings becoming brittle though the hardness improves.

From the foregoing analysis, it is clear that the EN coatings do not undergo similar kind of deformation under lubricated condition. Ni-P coatings undergo abrasive wear with signs of plastic deformation. On the other hand, two body abrasions are concluded for Ni-P-W coating. Electroless Ni-P-Cu coatings exhibit failure due to fracture of the nodules. Therefore, it is necessary to select the appropriate coating depending on the service condition for which it is being chosen. Nevertheless, the wear depth and COF exhibited by the EN coatings under lubricated condition are quite low as observed in Table 5 which indicates its suitability for use in tribological contact applications under lubrication.

CONCLUSION

The present chapter discusses the autocatalytic deposition Ni-P, Ni-P-W and Ni-P-Cu coatings on AISI 1040 steel specimens and tribological characterization under lubricated condition. The deposited coatings are heat treated at suitable temperatures to improve their microhardness. Deposit characterization is undertaken using SEM, EDX and XRD. The heat treated coatings are subjected to tribological tests under lubrication. The use of GRA to optimize the tribological behavior is also reported. The following conclusions may be inferred from the investigation:

1. SEM micrographs of the coatings reveal the nodulated surface morphology of electroless Ni-P, Ni-P-W and Ni-P-Cu coatings. The size of nodules of Ni-P-W coating is comparatively higher than Ni-P or Ni-P-Cu coating. Ni-P-Cu coatings exhibit cluster of nodulated structures. On heat treatment the size of nodules increases and formations of specific grain boundaries are observed for Ni-P and Ni-P-Cu coatings. The nodules of Ni-P-W coating after heat treatment appear to be somewhat compact.
2. EDX analysis of the coatings in their as-deposited condition confirms the presence of respective elements in the coatings. Electroless Ni-P and Ni-P-Cu coatings lie in the high P range. Whereas, the P content decreases due to the co-deposition of tungsten.
3. XRD analysis reveals that electroless Ni-P and Ni-P-Cu coatings are amorphous in their as-deposited condition. Ni-P-W coating exhibit X-ray amorphous/nanocrystalline structure. Heat treatment results in the precipitation of crystalline nickel and its phosphides.
4. In as-deposited condition, electroless Ni-P-W coating has the highest hardness compared to Ni-P and Ni-P-Cu. Microhardness of all the three coating variants are seen to improve after heat treatment.
5. The use of Taguchi's OA along with GRA significantly improves the tribological behavior of the coatings under lubricated condition. GRA predicts the optimal parametric setting for Ni-P coatings to be 50N load, 80rpm speed and 5min sliding duration. For Ni-P-W coatings, optimal tribological behavior is obtained at 10N load, 60rpm speed and 5min sliding duration. GRA predicted optimal parametric condition for Ni-P-Cu coatings is 10N load, 100rpm speed and 5min sliding time.
6. Statistical ANOVA reveals that load is the most influential factor in controlling the tribological behavior of Ni-P and Ni-P-Cu coating under lubrication. While for Ni-P-W coating, speed has the highest contribution.
7. Confirmation test indicates a remarkable improvement in the GRG at optimal condition from which it is prominent that the design optimization is gainfully applied to achieve optimum tribological behavior of the coatings under lubrication.
8. The wear mechanism of Ni-P coating is seen to be predominantly abrasive with plastic deformation of the nodules and micro-cutting effects caused due to wear debris. Ni-P-W coatings also undergo abrasive wear. A surface smoothening effect is also observed for Ni-P-W coating. Ni-P-Cu coatings suffer fracture of the nodules which may be due to the repeated nature of loading and the coatings turning brittle due to heat treatment.

The present chapter may prove to be beneficial for selection of EN coatings where tribological contact under lubrication is involved. The statistical results and different characterization results will aid suitably to the design and applicability of the coatings to various mating parts where friction and wear under lubricated condition occurs. Future research may be carried out by considering bio-lubricants which are increasingly replacing petroleum based lubricants. The flow rate of the lubricant may be also considered as a parameter to achieve minimum quantity lubrication which may result in a more economic system. Moreover, advanced optimization techniques such as the swarm based or genetic algorithms are increasingly becoming popular since they allow evaluation of a global optima or a set of pareto-optimal front which enables suitable selection of parameters. Such optimization techniques may be implemented to further enhance the tribological properties under lubrication.

ACKNOWLEDGMENT

The authors gratefully acknowledge the financial support of University Grants Commission (UGC), Govt. of India through Major Research Project vide Ref. No. F No. 41-984/2012 (SR) dated 25.07.2012.

REFERENCES

Agarwala, R. C., & Agarwala, V. (2003). Electroless alloy/composite coatings: A review. *Sadhana*, *28*(3-4), 475–493. doi:10.1007/BF02706445

Balaraju, J. N., Kalavati, , & Rajam, K. S. (2009). Surface morphology and structure of electroless ternary NiWP deposits with various W and P contents. *Journal of Alloys and Compounds*, *486*(1-2), 468–473. doi:10.1016/j.jallcom.2009.06.173

Balaraju, J. N., Radhakrishnan, P., Ezhilselvi, V., Kumar, A. A., Chen, Z., & Surendran, K. P. (2016). Studies on electroless nickel polyalloy coatings over carbon fibers/CFRP composites. *Surface and Coatings Technology*, *302*, 389–397. doi:10.1016/j.surfcoat.2016.06.040

Balaraju, J. N., & Rajam, K. S. (2005). Electroless deposition of Ni-Cu-P, Ni-W-P and Ni-W-Cu-P alloys. *Surface and Coatings Technology*, *195*(2-3), 154–161. doi:10.1016/j.surfcoat.2004.07.068

Broitman, E. (2017). Indentation hardnessmeasurements at macro-, micro-, and nanoscale: A critical overview. *Tribology Letters, 65*, 18. DOI: 10.1007/s11249-016-0805-5

Chen, W. X., Tu, J. P., Gan, H. Y., Xu, Z. D., Wang, Q. G., Lee, J. Y., & Zhang, X. B. et al. (2002). Electroless preparation and tribological properties of Ni-P-Carbon nanotube composite coatings under lubricated condition. *Surface and Coatings Technology*, *160*(1), 68–73. doi:10.1016/S0257-8972(02)00408-5

Correa, E., Mejía, J. F., Castaño, J. G., Echeverría, F., & Gómez, M. A. (2017). Tribological characterization of electroless Ni–B coatings formed on commercial purity magnesium. *Journal of Tribology*, *139*(5), 051302. doi:10.1115/1.4036169

Czagány, M., Baumli, P., & Kaptay, G. (2017). The influence of the phosphorous content and heat treatment on the nano-micro-structure, thickness and micro-hardness of electroless Ni-P coatings on steel. *Applied Surface Science*, *423*, 160–169. doi:10.1016/j.apsusc.2017.06.168

Das, S. K., & Sahoo, P. (2011). Tribological characteristics of electroless Ni–B coating and optimization of coating parameters using Taguchi based grey relational analysis. *Materials & Design*, *32*(4), 2228–2238. doi:10.1016/j.matdes.2010.11.028

Deng, J. (1989). Introduction to grey system. *Journal of Grey System*, *1*(1), 1–24.

Der Ger, M., Hou, K. H., Wang, L. M., & Hwang, B. J. (2003). The friction and wear of Ni–P–PTFE composite deposits under water lubrication. *Materials Chemistry and Physics*, *77*(3), 755–764. doi:10.1016/S0254-0584(02)00153-0

Duari, S., Barman, T. K., & Sahoo, P. (2014). Study of tribological properties of Ni-P coatings under lubrication. *IOSR Journal of Mechanical and Civil Engineering*, *6*, 22–26.

Duari, S., Mukhopadhyay, A., Barman, T. K., & Sahoo, P. (2016). Optimization of wear performance of electroless Ni-B coating under lubrication. *International Journal of Engineering and Technologies*, *7*, 94–103. doi:10.18052/www.scipress.com/IJET.7.94

Duari, S., Mukhopadhyay, A., Barman, T. K., & Sahoo, P. (2017). Study of wear and friction of chemically deposited Ni-P-W coating under dry and lubricated condition. *Surfaces and Interfaces*, *6*, 177–189. doi:10.1016/j.surfin.2017.01.009

Franco, M., Sha, W., Aldic, G., Malinov, S., & Çimenoğlu, H. (2016). Effect of reinforcement and heat treatment on elevated temperature sliding of electroless Ni–P/SiC composite coatings. *Tribology International*, *97*, 265–271. doi:10.1016/j.triboint.2016.01.047

Gadhari, P., & Sahoo, P. (2016a). Electroless nickel-phosphorus composite coatings: A review. *International Journal of Manufacturing, Materials, and Mechanical Engineering*, *6*(1), 14–50. doi:10.4018/IJMMME.2016010102

Gadhari, P., & Sahoo, P. (2016b). *Effect of annealing temperature and alumina particles on mechanical and tribological properties of Ni-P-Al_2O_3 composite coatings*. Silicon. doi:10.1007/s12633-016-9452-6

Gadhari, P., & Sahoo, P. (2016c). Effect of TiO_2 particles on micro-hardness, corrosion, wear and friction of Ni–P–TiO_2 composite coatings at different annealing temperatures. *Surface Review and Letters*, *23*, 1550082. DOI: 10.1142/S0218625X15500821

Ghaderi, M., Rezagholizadeh, M., Vaghefi, S. M., & Heidary, A. (2016). Investigation of high temperature wear resistance of electroless nickel coating with different contents of phosphorous. *Protection of Metals and Physical Chemistry of Surfaces*, *52*(3), 538–542. doi:10.1134/S2070205116030126

Hur, K. H., Jeong, J. H., & Lee, D. N. (1990). Microstructures and crystallization of electroless Ni-P deposits. *Journal of Materials Science*, *25*(5), 2573–2584. doi:10.1007/BF00638061

Krasteva, N., Fotty, V., & Armyanov, S. (1994). Thermal stability of Ni-P and Ni-Cu-P amorphous alloys. *Journal of the Electrochemical Society*, *141*(10), 2864–2867. doi:10.1149/1.2059246

Kundu, S., Das, S. K., & Sahoo, P. (2016). *Tribological behaviour of electroless Ni-P deposits under elevated temperature*. Silicon. doi:10.1007/s12633-016-9450-8

Li, Z., Wang, J., Lu, J., & Meng, J. (2013). Tribological characteristics of electroless Ni–P–MoS_2 composite coatings at elevated temperatures. *Applied Surface Science*, *264*, 516–521. doi:10.1016/j.apsusc.2012.10.055

Liu, H., Lv, Y. Y., Liu, Z., & Thompson, G. E. (2016). Dry sliding wear behaviour and structural characteristics of laser-annealed electroless Ni–P/Ni–Mo–P duplex coatings. *Tribology International*, *103*, 343–351. doi:10.1016/j.triboint.2016.07.014

Liu, J., Wang, X., Tian, Z., Yuan, M., & Ma, X. (2015). Effect of copper content on the properties of electroless Ni-Cu-P coatings prepared on magnesium alloys. *Applied Surface Science*, *356*, 289–293. doi:10.1016/j.apsusc.2015.08.072

Liu, Y., & Zhao, Q. (2004). Study of electroless Ni–Cu–P coatings and their anti-corrosion properties. *Applied Surface Science*, *228*(1), 57–62. doi:10.1016/j.apsusc.2003.12.031

Loto, C. A. (2016). Electroless nickel plating–A review. *Silicon*, *8*(2), 177–186. doi:10.1007/s12633-015-9367-7

Masoumi, F., Ghasemi, H. R., Ziaei, A. A., & Shahriari, D. (2012). Tribological characterization of electroless Ni–10% P coatings at elevated test temperature under dry conditions. *International Journal of Advanced Manufacturing Technology*, *62*(9-12), 1063–1070. doi:10.1007/s00170-011-3832-2

Mukhopadhyay, A., Barman, T. K., & Sahoo, P. (2016c). Effect of operating temperature on tribological behavior of as-plated Ni-B coating deposited by electroless method. *Tribology Transactions*. DOI: doi: 10.1080/10402004.2016.1271929

Mukhopadhyay, A., Barman, T. K., & Sahoo, P. (2017b). Tribological behavior of sodium borohydride reduced electroless nickel alloy coatings at room and elevated temperatures. *Surface and Coatings Technology*, *321*, 464–476. doi:10.1016/j.surfcoat.2017.05.015

Mukhopadhyay, A., Barman, T. K., & Sahoo, P. (2017c). Effects of heat treatment on tribological behavior of electroless Ni–B coating at elevated temperatures. *Surface Review and Letters*, 1850014. doi:10.1142/S0218625X18500142

Mukhopadhyay, A., Duari, S., Barman, T. K., & Sahoo, P. (2015). Application of grey fuzzy logic for the optimization of tribological performance of electroless Ni-B coating. *Journal of Manufacturing Technology Research*, *7*(1/2), 1–24.

Mukhopadhyay, A., Duari, S., Barman, T. K., & Sahoo, P. (2016a). Wear analysis of electroless Ni-P coating under lubricated condition using fuzzy logic. *Portugaliae Electrochimica Acta*, *34*(1), 63–83. doi:10.4152/pea.201601063

Mukhopadhyay, A., Duari, S., Barman, T. K., & Sahoo, P. (2016b). Tribological performance optimization of electroless Ni–B coating under lubricated condition using hybrid grey fuzzy logic. *Journal of the Institution of Engineers (India): Series D, 97*(2), 215-231.

Mukhopadhyay, A., Duari, S., Barman, T. K., & Sahoo, P. (2016d). Optimization of friction and wear properties of electroless Ni–P coatings under lubrication using grey fuzzy logic. *Journal of the Institution of Engineers (India): Series D*. DOI: 10.1007/s40033-016-0133-9

Mukhopadhyay, A., Duari, S., Barman, T. K., & Sahoo, P. (2016e). Investigation of wear behavior of electroless Ni-P-W Coating under dry and lubricated conditions using RSM and fuzzy logic. *Portugaliae Electrochimica Acta*, *34*(4), 231–255. doi:10.4152/pea.201604231

Mukhopadhyay, A., Duari, S., Barman, T. K., & Sahoo, P. (2016f). Wear behavior of electroless Ni-P-W coating under lubricated condition-a Taguchi based approach. IOP Conference Series: Materials Science and Engineering, 149, 012004.

Mukhopadhyay, A., Duari, S., Barman, T. K., & Sahoo, P. (2016g). Optimization of wear behavior of electroless Ni-P-W coating under dry and lubricated conditions using genetic algorithm (GA). *Jurnal Tribologi*, *11*, 24–48.

Mukhopadhyay, A., Duari, S., Barman, T. K., & Sahoo, P. (2017a). Tribology of electroless Ni-P coating under lubricated condition: An RSM and GA approach. *International Journal of Surface Engineering and Interdisciplinary Materials Science*, *5*(1), 37–57. doi:10.4018/IJSEIMS.2017010103

Palaniappa, M., & Seshadri, S. K. (2008). Friction and wear behavior of electroless Ni–P and Ni–W–P alloy coatings. *Wear*, *265*(5), 735–740. doi:10.1016/j.wear.2008.01.002

Panja, B., Das, S. K., & Sahoo, P. (2016). Tribological behavior of electroless Ni–P coatings in various corrosive environments. *Surface Review and Letters*, *23*, 1650040. DOI: 10.1142/S0218625X16500402

Panja, B., & Sahoo, P. (2014). Taguchi-Grey based optimization of friction and wear of electroless Ni-P coatings in acidic environment. *International Journal of Surface Engineering and Interdisciplinary Materials Science*, *2*(2), 53–69. doi:10.4018/IJSEIMS.2014070104

Panja, B., & Sahoo, P. (2015). Tribological behavior of electroless Ni-P coatings in alkaline environment and optimization of coating parameters using Taguchi based grey relational analysis. *Indian Journal of Engineering and Materials Sciences*, *22*, 503–512.

Popoola, A. P. I., Loto, C. A., Osifuye, C. O., Aigbodion, V. S., & Popoola, O. M. (2016). Corrosion and wear properties of Ni-Sn-P ternary deposits on mild steel via electroless method. *Alexandria Engineering Journal*, *55*(3), 2901–2908. doi:10.1016/j.aej.2016.06.018

Ramalho, A., & Miranda, J. C. (2007). Tribological characterization of electroless NiP coatings lubricated with biolubricants. *Wear*, *263*(1), 592–597. doi:10.1016/j.wear.2006.10.027

Reddy, V. V. N., Ramamoorthy, B., & Nair, P. K. (2000). A study on the wear resistance of electroless Ni–P/Diamond composite coatings. *Wear*, *239*(1), 111–116. doi:10.1016/S0043-1648(00)00330-6

Roy, S., & Sahoo, P. (2012). Corrosion study of electroless Ni-P-W coatings using electrochemical impedance spectroscopy. *Portugaliae Electrochimica Acta*, *30*(3), 203–220. doi:10.4152/pea.201203203

Roy, S., & Sahoo, P. (2014). Parametric optimization of corrosion and wear of electroless Ni-P-Cu coating using grey relational coefficient coupled with weighted principal component analysis. International Journal of Mechanical and Materials Engineering, 9(1).

Sadeghzadeh-Attar, A., AyubiKia, G., & Ehteshamzadeh, M. (2016). Improvement in tribological behavior of novel sol-enhanced electroless Ni-P-SiO_2 nanocomposite coatings. *Surface and Coatings Technology, 307*, 837–848. doi:10.1016/j.surfcoat.2016.10.026

Sahoo, P. (2009). Wear behaviour of electroless Ni–P coatings and optimization of process parameters using Taguchi method. *Materials & Design, 30*(4), 1341–1349. doi:10.1016/j.matdes.2008.06.031

Sahoo, P., & Das, S. K. (2011). Tribology of electroless nickel coatings – A review. *Materials & Design, 32*(4), 1760–1775. doi:10.1016/j.matdes.2010.11.013

Sahoo, P., & Das, S. K. (2015). Wear performance optimization for electroless Ni-P Coating. *International Journal of Surface Engineering and Interdisciplinary Materials Science, 3*(2), 1–17. doi:10.4018/IJSEIMS.2015070101

Sahoo, P., & Pal, S. K. (2007). Tribological performance optimization of electroless Ni–P coatings using the Taguchi method and grey relational analysis. *Tribology Letters, 28*(2), 191–201. doi:10.1007/s11249-007-9264-3

Sahoo, P., & Roy, S. (2017). Tribological behavior of electroless Ni-P, Ni-P-W and Ni-P-Cu coatings: A comparison. *International Journal of Surface Engineering and Interdisciplinary Materials Science, 5*(1), 1–15. doi:10.4018/IJSEIMS.2017010101

Sahoo, P., Roy, S., & Davim, J. P. (2014). Design and selection of chemically deposited Ni-P-W coatings for optimum tribological behavior. In L. Santo & J. P. Davim (Eds.), *Surface Engineering Techniques and Applications: Research Advancements* (pp. 45–72). Engineering Science Reference (an imprint of IGI Global). doi:10.4018/978-1-4666-5141-8.ch002

Sankara Narayanan, T. S. N., Selvakumar, S., & Stephen, A. (2003). Electroless Ni–Co–P ternary alloy deposits: Preparation and characteristics. *Surface and Coatings Technology, 172*(2), 298–307. doi:10.1016/S0257-8972(03)00315-3

Sivandipoor, I., & Ashrafizadeh, F. (2012). Synthesis and tribological behaviour of electroless Ni-P-WS_2 composite coatings. *Applied Surface Science, 263*, 314–319. doi:10.1016/j.apsusc.2012.09.051

Soares, M. E., Soares, P., Souza, P. R., Souza, R. M., & Torres, R. D. (2017). The effect of nitriding on adhesion and mechanical properties of electroless Ni–P coating on AISI 4140 steel. *Surface Engineering, 33*(2), 116–121. doi:10.1080/02670844.2016.1148831

Song, G. S., Sun, S., Wang, Z. C., Luo, C. Z., & Pan, C. X. (2017). Synthesis and characterization of electroless Ni–P/Ni–Mo–P duplex coating with different thickness combinations. *Acta Metallurgica Sinica. English Letters*. doi:10.1007/s40195-017-0603-6

Staia, M. H., Castillo, E. J., Puchi, E. S., Lewis, B., & Hintermann, H. E. (1996). Wear performance and mechanism of electroless Ni-P coating. *Surface and Coatings Technology, 86*, 598–602. doi:10.1016/S0257-8972(96)03086-1

Sudagar, J., Lian, J., & Sha, W. (2013). Electroless nickel, alloy, composite and nano coatings–A critical review. *Journal of Alloys and Compounds, 571*, 183–204. doi:10.1016/j.jallcom.2013.03.107

Tien, S. K., & Duh, J. G. (2004). Thermal reliability of electroless Ni–P–W coating during the aging treatment. *Thin Solid Films*, *469*, 268–273. doi:10.1016/j.tsf.2004.08.179

Veeraraghavan, B., Kim, H., & Popov, B. (2004). Optimization of electroless Ni–Zn–P deposition process: Experimental study and mathematical modeling. *Electrochimica Acta*, *49*(19), 3143–3154. doi:10.1016/j.electacta.2004.01.035

Vitry, V., & Bonin, L. (2017). Formation and characterization of multilayers borohydride and hypophosphite reduced electroless nickel deposits. *Electrochimica Acta*, *243*, 7–17. doi:10.1016/j.electacta.2017.04.152

Wu, F. B., Tien, S. K., Chen, W. Y., & Duh, J. G. (2004). Microstructure evaluation and strengthening mechanism of Ni-P-W alloy coatings. *Surface and Coatings Technology*, *177-178*, 312–316. doi:10.1016/j.surfcoat.2003.09.010

Xu, Y., Zheng, X., Hu, X., Yin, Y., & Lei, T. (2014). Preparation of the electroless Ni–P and Ni–Cu–P coatings on engine cylinder and their tribological behaviors under bio-oil lubricated conditions. *Surface and Coatings Technology*, *258*, 790–796. doi:10.1016/j.surfcoat.2014.07.079

Yu, H. S., Luo, S. F., & Wang, Y. R. (2001). A comparative study on the crystallization behavior of electroless Ni-P and Ni-Cu-P deposits. *Surface and Coatings Technology*, *148*(2-3), 143–148. doi:10.1016/S0257-8972(01)01345-7

Zuleta, A. A., Correa, E., Castaño, J. G., Echeverría, F., Baron-Wiecheć, A., Skeldon, P., & Thompson, G. E. (2017). Study of the formation of alkaline electroless Ni-P coating on magnesium and AZ31B magnesium alloy. *Surface and Coatings Technology*, *321*, 309–320. doi:10.1016/j.surfcoat.2017.04.059

Chapter 14
Multi-Characteristic Optimization in Wire Electrical Discharge Machining of Inconel-625 by Using Taguchi-Grey Relational Analysis (GRA) Approach:
Optimization of an Existing Component/Product for Better Quality at a Lower Cost

Pradeep Kumar Karsh
National Institute of Technology Silchar, India

Hari Singh
National Institute of Technology Kurukshetra, India

ABSTRACT

Inconel 625 is a nickel-chromium based super-alloy which is mostly used in high end applications due to its excellent chemical and fabrication properties combined with high strength and outstanding corrosion resistance. Wire EDM is most commonly the used process for machining this material and gives high accuracy and precision in the machined parts. In this paper, an attempt is made to undertake an experimental investigation to optimize the input variables like pulse on time, pulse off time, peak current, and servo voltage in Wire EDM for achieving simultaneous optimization of cutting rate, surface roughness, dimensional deviation, and wire wear ratio. Taguchi's L9 orthogonal array with one replication has been used as an experimental design and then analysis of variance (ANOVA) has been applied to determine the significant parameters and their contributions on response variables. Taguchi method has been combined with grey relational analysis for multi-characteristic optimization. Finally, results are validated with the help of confirmation experiments.

DOI: 10.4018/978-1-5225-3401-3.ch014

INTRODUCTION

Today's manufacturing industries promote the use of modern and sophisticated materials like ceramics, super alloys, composites etc. for their manufacturing base (Singh, 2012). These materials provide greater mechanical strength without increasing the weight of overall component. However, the major problem with these materials the type of machining they require for their processing as they cannot be economically machined by conventional tools available (Rao and Kalyankar, 2014). The material used in this study is a nickel-chromium based super alloy Inconel 625. This material is an excellent choice for sea water applications, aerospace field, chemical processing industries and nuclear energy sector due to its ultimate properties (Yan et al., 2013).

In this research work, the material is machined with the help of WEDM which is presently the most versatile machine tool available for machining of electrically conductive materials. WEDM is a non - convectional machining process used in aerospace, die making industries, etc. due to high accuracy and precision work obtained. This machining process is capable to machine two dimensional and three dimensional intricate shape and design due to very low material removal rate (Tosun et al., 2003). When the suitable voltage is applied, discharge occurs between the wire and work material where a suitable gape is maintained between them. Dielectric like deionized water is applied in the gape which acts as insulation resistance. Due to sparking process temperature reaches about 8300 – 11000 °C. The microchips formed in WEDM are carried away by the dielectric fluid and which also cools the workpiece. Wire is fed continuously maintaining the gape of 0.025 to 0.05mm between the wire and work (Sharma et al., 2013). There are many types of wire like brass wire, copper wire and coated wire etc. work as cutting tool in WEDM so that manufacturing of costly tool is avoided (Maher et al., 2015). WEDM is thermo- electric process in which electrical energy is converted into heat by which spark is produced. This spark melts and vaporizes the work piece in the form of small burr which is carried away by the dielectric fluid. In this process, electrode and work piece are separated by small distance, so mechanical stress is negligible. Generally deionized water is applied as a dielectric fluid which helps in producing spark and maintaining gape between work material and wire electrode (Alias et al., 2012). The wire movement is controlled by the Computerized Numerical Control, so that the accuracy obtained is very high. Taper cutting is also possible by the WEDM with the help of CNC controller. It is capable to machine all electrically conductive materials irrespective of their toughness and hardness. WEDM is mostly used in automotive, aerospace, mould, tool and die making, medical, optical, dental, and jewelry industries (Ramakrishnan and Karunamoorthy, 2008). Figure 1 represents the WEDM process.

BACKGROUND

Hewidy et al. (2005) applied response surface methodology (RSM) to modal and optimize the input variables for machining of Inconel in Wire EDM machining. They conclude that as peak current (IP) rises material removal rate (MRR) also increase. This increase is, however, diminished after 7 A. Puri and Bhattacharyya (2005) also used RSM to model white layer depth in WEDM. Central composite rotatable design of second order was applied to study four process variables. Manna and Bhattacharyya (2006) applied Taguchi and Gauss elimination method to optimize the process parameters of Al/Sic-

Figure 1. Representation of WEDM process

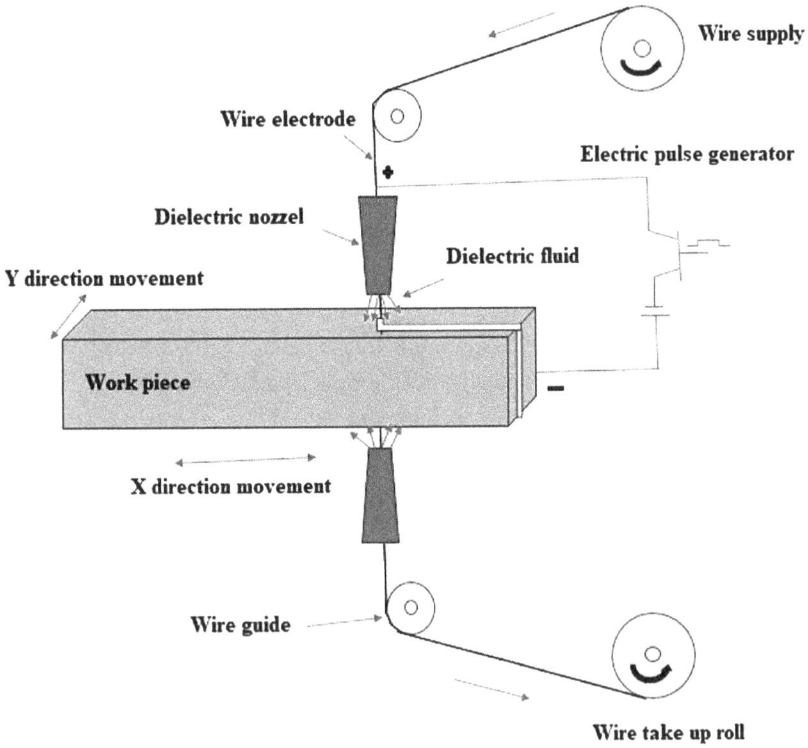

MMC. Open gape voltage and pulse on time (T_{on}) were found the significant factors for MRR. Newton et al. (2009) tried to find the parameters which affect the development of re-cast layer in Inconel 718. They concluded that average thickness of re-cast layer increased when energy per spark, current pulse duration and peak discharge current were increased. They found that average thickness of re-cast layer vary from 5 to 9 μm. Wire diameter and spark cycle time played no major role in the formation of recast layer. Ramakrishnan and Karunamoorthy (2006) applied robust design of experiment for achieving multi response optimization. The study revealed that T_{on} and current intensity are significant factors in affecting the various responses. Singh and Garg (2009) have attempted to find the effect of different parameters on the MRR in WEDM for hot die steel (H-11). They found that MRR is directly proportional to T_{on} and IP, while inversely proportional to pulse off time (T_{off}) and servo voltage (SV). Hargrove and Ding developed finite element method (FEM) to model temperature distribution in workpiece under different conditions of cutting parameters. Shandilya et al. (2012) have optimized the input variables of SiC_p/6061 Al metal matrix composite by using RSM. They considered SV, T_{on}, T_{off}, and wire feed rate (WF) as input parameters, and cutting width (kerf) as output parameter. Rajyalakshmi and Ramaiah (2013) used Taguchi – grey analysis to find the optimum cutting condition for Inconel 825 in wire EDM. They combined the orthogonal array design of experiment with GRA. It was found that MRR increases from 119.625 to126.85mm^3/min and surface roughness decrease from 1.68 to 1.44 μm. Goswami and Kumar (2014) found the relationship between the different process parameters like T_{on}, T_{off}, IP, SV,

WF, and wire tension (WT) with responses like MRR, surface integrity, and WWR of Nimonic 80A alloy. Kumar and Agarwal (2012) applied multi-objective genetic algorithm to optimize the machining parameters for maximizing MRR and surface finish. A mathematical model was then optimized based on Non-dominated Sorting Genetic Algorithm-II to obtain a Pareto-optimal solution set. Baig and Venkaiah (2015) optimized the process parameters of wire EDM for Hastelloy C276. MRR and Kerf width were taken as response variables. Atzeni et al. (2015) used zinc coated brass wire for cutting of Inconel 718 to optimize the process parameters in WEDM and modify the surface and sub surface. SEM and EDS were used to evaluate possible variation in the surface chemical composition. They found that as nominal energy per length (E – value) increases, surface roughness also increases. Yuan et al. (2008) used multi-objective optimization based on Gaussian process regression (GPR) to optimize high speed wire EDM process. Mean current, T_{on} and T_{off} were taken as input parameters, and SR and MRR as output parameters. They compared different types of models like BPNN, SVR and RVM with GPR, and conclude that GPR is the better technique in terms of accuracy, reliability and probabilistic variance. Tosun and Cogun (2013) investigated AISI 4140 steel using regression analysis and concluded that the WWR rises with increase in T_{on} and open circuit voltage while it decline with increase in wire speed and dielectric fluid pressure. Chiang and Chang (2006) machined Al_2O_3 particle reinforced material (6061 alloy) on WEDM and investigated the effects of various input parameters on surface roughness and surface removal rate. They applied Taguchi's L18 mixed OA and Grey Relational Analysis design technique to optimize the results. The main findings of their study were that the arc on-time and T_{on}, T_{off} and SV are significant factor of CR and SR. Nain et al. (2017) applied ANN model to determine the surface roughness, waviness and MRR of UdimetL605. Unune and Mali (2017) used low-frequency vibration workpiece to improve the performance of WEDM. Nayak and Mahapatra (2016) optimized process parameters of WEDM for taper turning of deep cryo-treated Inconel 718. Chalisgaonkar and Kumar (2015) used multiple user preference to multiple optimization and modelling of titanium.

Multi-characteristics optimization of Inconel 625 by using GRA method is still a topic to be explored. The effect of machining parameters on Inconel 625 has not been fully explored using WEDM with brass wire as electrode. In the present study, single and multi-characteristic optimization of process parameters of WEDM are carried out. In the literature study, it is found that the effect of machining parameters on Inconel 625 has not been fully explored using WEDM with brass wire as electrode. In this paper, Taguchi and GRA approach are successfully applied on optimization of process parameters.

EXPERIMENTAL SETUP AND WORK MATERIAL

In present study, experiment are carried out on a wire-cut EDM machine of Electronica Machine Tools Ltd. The pictorial view of the WEDM is shown in Figure 2. Inconel 625 is taken as work material of size 80*80*14 mm plate. Inconel is a nickel based super alloy which is used for high temperature application due to its high melting point and good mechanical properties. Table 1 and Table 2 give the chemical properties of Inconel 625. After machining slots are produced in a plate as shown in Figure 3. The pictorial view of the surface roughness tester is shown in Figure 4. Brass wire with 0.25 mm is used as electrode. The other parameters for machining process are kept constant.

Figure 2. Pictorial view of WEDM
Courtesy: NIT Kurukshetra.

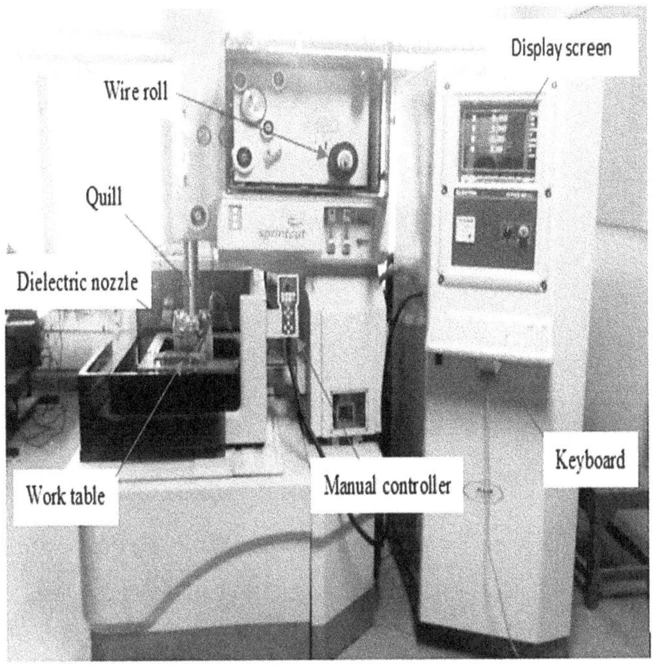

Table 1. Mechanical properties of Inconel 625

Properties	Density	Melting Point	Coefficient of Expansion	Modulus of Rigidity
Value	8.44 g/cm^3	1290 – 1350 °C	12.8μm/m°C	79 KN/mm^2

Table 2. Composition of Inconel 625

Elements	Nickel	Chromium	Molybdenum	Columbium + Tantalum	Iron
% value	58%	21 – 24%	8 – 10%	3.45 – 4.25%	5%

METHODOLOGIES

Taguchi Method

Taguchi method is applied to investigate the simultaneous effect of the selected process parameters on the mean, and variance of process responses. This method involve orthogonal arrays for experimental design to assign the parameters to various columns of the arrays. In Taguchi method, instead of testing all possible combinations it only tests certain pairs of combinations. Taguchi method involves the following steps.

Figure 3. Slots produced in work piece

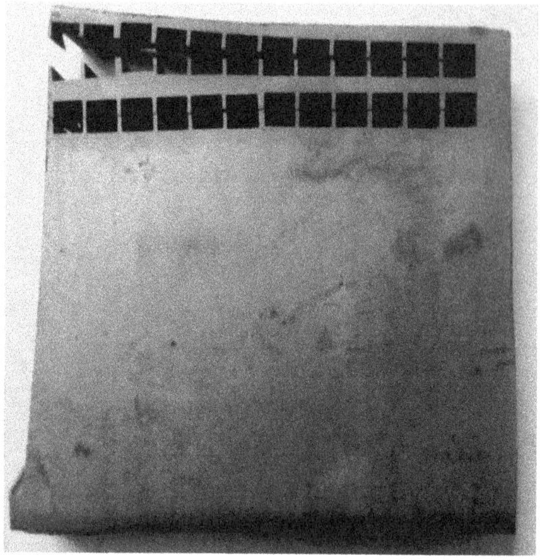

Figure 4. Surface roughness tester
Courtesy: NIT Kurukshetra.

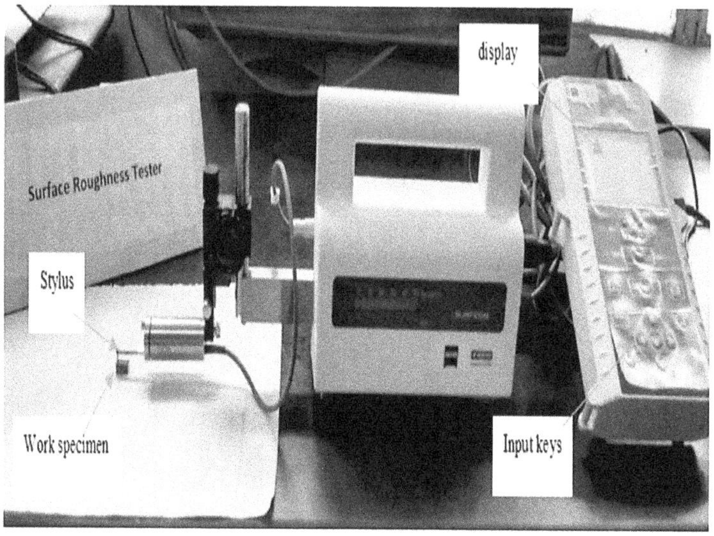

Selection of Parameters and Levels

L9 orthogonal array (OA) with one replication based on Taguchi method is used to design the experiment. In this study L9 OA is used because it is capable to find effect of all four parameters individually, and interaction between the parameters is not considered. All four factors have three levels i.e. each factor have two degree of freedom (DOF) as shown in Table 3. In L9 OA with one replication, there are 17 degree of freedom, so there is sufficient DOF for errors. So L9 OA is sufficient to find the effect of all four process parameters. By increasing number of experiments, interaction can be included, but this is a

Table 3. Parameters and their levels

Factors	Parameters	Levels		
		L1	L2	L3
A	Pulse on Time T_{on} (μs)	105	115	125
B	Pulse off Time T_{off} (μs)	35	45	55
C	Peak current IP (A)	120	140	160
D	Servo voltage SV (V)	40	60	80

wastage of resources for the less amount of accuracy in experiment. The selection of process parameters and their levels is based upon some pilot study and review of literature.

Experimental Data

The experimental data for cutting rate (CR), surface roughness (SR), dimensional deviation (DD), and wire wear ratio (WWR) are given in Table 4. Experiments were performed using Taguchi experimental design methodology and each experiment was repeated twice for obtaining S/N values.

GREY RELATIONAL ANALYSIS (GRA) ARROACH

GRA is applied for optimization of different responses simultaneously. In this study, an attempt is made to optimize the CR, SR, DD and WWR simultaneously. In GRA method all responses are converted into a single objective. In GRA method, degree of approximation is measured by using the grey relationship grade (Siddiquee et al., 2012). Steps involved in applying GRA approach are given below.

Table 4. Experimental data

Exp. No.	T_{on}	T_{off}	IP	SV	CR_1	CR_2	SR_1	SR_2	DD_1	DD_2	WWR_1	WWR_2
1	105	35	120	40	1.015	0.968	3.3630	3.164	2.10	2.08	0.2160	0.2300
2	105	45	140	60	0.566	0.550	1.5750	1.757	1.76	1.80	0.1593	0.1699
3	105	55	160	80	0.331	0.353	0.7650	0.813	1.72	1.68	0.1540	0.1240
4	115	35	140	80	1.243	1.282	5.0440	5.280	1.82	1.78	0.1880	0.1930
5	115	45	160	40	1.860	2.020	5.8940	6.066	2.03	2.07	0.2055	0.2330
6	115	55	120	60	0.745	0.730	3.8800	3.820	1.67	1.65	0.1600	0.1500
7	125	35	160	60	2.756	3.118	7.2330	7.001	1.74	1.78	0.3450	0.3300
8	125	45	120	80	1.538	1.614	6.2860	5.981	1.92	1.88	0.2150	0.2140
9	125	55	140	40	1.989	1.847	6.6575	6.736	2.06	1.98	0.1970	0.1890

Calculation of S/N Ratios

The formulae used for finding the signal to noise (S/N) ratio using Taguchi method for different quality characteristics are given in Equations 1 & 2. The calculated value is shown in Table 5.

$$\left(S / N\right)_{HB} = -10 * \log\left(\sum\left(\frac{1}{Y^2}\right) / n\right) \tag{1}$$

$$\left(S / N\right)_{LB} = -10 * \log\left(\sum\left(Y^2\right) / n\right) \tag{2}$$

where, n is the number of repetitions, and Y is the experimental value.

Equation (1) is used for higher the better characteristic like cutting rate and equation (2) is used for lower the better characteristic like SR to calculate S/N ratio.

Data Preprocessing of S/N Ratios

In data preprocessing, original sequence is transformed into comparable sequence. In this step, a linear normalization of the experimental results in the range of 0 and 1 has been performed. This process is also called the grey relational generating. The experimental values are normalized by using Equations (3) & (4). Data preprocessing of S/N ratios of experimental values for the performance characteristics are shown in Table 6.

$$Z_i(k) = \frac{\left[Z_i^{(0)}(m) - \min_{(i)} Z_i^{(0)}(m)\right]}{\left[\max_{(i)} Z_i^{(0)}(m) - \min_{(i)} Z_i^{(0)}(m)\right]} \tag{3}$$

Table 5. Mean and S/N ratio of experimental results

Experiment No.	Mean CR	S/N CR	Mean SR	S/N SR	Mean DD	S/N DD	Mean WWR	S/N WWR
1	0.9915	-0.0813	3.2635	-10.2770	2.09	-6.400	0.2230	13.0290
2	0.5580	-5.0699	1.6600	-4.4460	1.78	-5.008	0.1646	15.6660
3	0.3420	-9.3329	0.7890	2.0544	1.70	-4.609	0.1390	17.0890
4	1.2625	2.0215	5.1620	-14.2580	1.80	-5.105	0.1905	14.4010
5	1.9400	5.7338	5.9800	-15.5340	2.05	-6.023	0.2190	13.1640
6	0.7370	-2.6460	3.8500	-11.7094	1.66	-4.400	0.1550	16.1880
7	2.9370	9.3080	7.1200	-17.0470	1.76	-4.910	0.3375	9.4322
8	1.5760	3.9435	6.1330	-15.7560	1.90	-5.575	0.2145	13.3710
9	1.9180	5.6391	6.6965	-16.5170	2.02	-6.108	0.1930	14.2860

Table 6. Data preprocessing of S/N ratios of experimental results for performance characteristics

Exp. No.	CR	SR	DD	WWR
Reference sequence Comparability sequence	1.00	1.00	1.00	1.00
1	0.4963	0.6455	1.000	0.5302
2	0.2286	0.3403	0.304	0.1858
3	0.000	0.000	0.1045	0.000
4	0.6091	0.8539	0.3525	0.351
5	0.8082	0.9207	0.8115	0.5126
6	0.3580	0.7205	0.000	0.117
7	1.000	1.000	0.255	1.000
8	0.7122	0.9324	0.5875	0.4855
9	0.8031	0.9722	0.854	0.366

$$Z_i(k) = \frac{\left[\max_{(i)} Z_i^{(0)}(m) - Z_i^{(0)}(m)\right]}{\left[\max_{(i)} Z_i^{(0)}(m) - \min_{(i)} Z_i^{(0)}(m)\right]} \tag{4}$$

where $Z_i(m)$ is sequence after data preprocessing, $Z_i^{(0)}(m)$ is original sequence where $i = 1, 2, 3, \ldots.m$ and $m = 1, 2, 3, \ldots.n$, $\min_{(i)} Z_i^{(0)}(m)$ is the minimum value of original sequence and $\max_{(i)} Z_i^{(0)}(m)$ is the maximum value of original sequence.

Calculation of Grey Relational Coefficient (GRC)

Before calculation of GRC value, it is required to calculate deviation sequence by using Equation (5). Calculated dimensional values are reported in Table 7.

Table 7. Deviation sequence

Deviation Sequence	Δ_{01} (1)	Δ_{01} (2)	Δ_{01} (3)	Δ_{01} (4)
1	0.5037	0.3545	0.000	0.4698
2	0.7714	0.6597	0.696	0.8142
3	1.000	1.000	0.8955	1.000
4	0.3909	0.1461	0.6475	0.649
5	0.1918	0.0793	0.1885	0.4874
6	0.642	0.2795	1.000	0.883
7	0.000	0.00	0.745	0.000
8	0.2878	0.0676	0.4125	0.5145
9	0.1969	0.0278	0.146	0.634

$$\Delta_{0i}(m) = \left|Z_0^*(m) - Z_i^0(m)\right| \tag{5}$$

where, Δ_{0i} (m) is the dimensional sequence.

For example Δ_{01} can be calculated as follows:

$$\Delta_{01}(1) = \left|1 - 0.4963\right| = 0.5037 \tag{6}$$

$$\Delta_{01}(2) = \left|1 - 0.6455\right| = 0.3545 \tag{7}$$

Same calculation is done for other values of $i = 1 - 9$ and $m = 1 - 4$. The results are listed in Table 7.

After calculation of dimensional deviation, GRC values are determine using Eq. (10) and the results are tabulated in Table 8.

$$\Delta_{max} = \Delta_{03}(1) = \Delta_{03}(2) = \Delta_{06}(3) = \Delta_{03}(4) = 1.000 \tag{8}$$

$$\Delta_{min} = \Delta_{07}(1) = \Delta_{07}(2) = \Delta_{01}(3) = \Delta_{07}(4) = 0.000 \tag{9}$$

$$\xi_i(m) = \frac{\Delta \min + \xi * \Delta \max}{\Delta_{0i}(m) + \xi * \Delta \max} \tag{10}$$

Table 8. GRC value

Exp. No.	GRC Value ξ_i (k)				GRG	Rank
	CR ξ_i (1)	SR ξ_i (2)	DD ξ_i (3)	WWR ξ_i (4)		
1	0.4981	0.5851	1.000	0.5155	0.6496	4
2	0.3932	0.4311	0.4180	0.3804	0.4056	8
3	0.333	0.3333	0.3582	0.3333	0.3393	9
4	0.5612	0.7738	0.4357	0.4351	0.5514	6
5	0.7227	0.8631	0.7262	0.5063	0.7045	3
6	0.4370	0.6414	0.3333	0.3615	0.4432	7
7	1.000	1.000	0.4016	1.000	0.8504	1
8	0.6346	0.8809	0.5479	0.4928	0.6390	5
9	0.7174	0.9473	0.7739	0.4409	0.7198	2

$$0 < \xi_i(m) \leq 1$$

where, $\xi_i(m)$ is the GRC value, $\Delta_{0i}(m)$ is the deviation sequence and ξ is the differentiating coefficient that varies from 0 to 1; in this study it is taken as 0.5. For Example, GRC value for $\Delta_{0i}(1)$ is calculated as

$$\xi_1(1) = \frac{0.00 + 0.5 * 1}{0.5037 + 0.5 * 1} = 0.4981 \quad (11)$$

Similar procedure is applied for $i = 1$ to 9, and the results are summarized in Table 9.

After calculation of GRC value, the Grey Relational Grade (GRG) is determined by averaging all the GRC values in a particular experimental run. For example

$$\left(GRC\right)_1 = \left(0.4981 + 0.5851 + 1 + 0.5155\right) / 4 = 0.6496 \quad (12)$$

In this study, all the responses have same weightage. Next, ranking is done according to the respective GRG values - the highest GRG value is given the first rank and the lowest GRG value is given the last rank. It is clear from the Table 8 that experiment number 7 has the highest GRG value and thus it is given the first rank, and so on. Therefore, the experiment number 7 provides the best condition for multiple characteristics optimization. Larger the value of GRG, better the multiple response.

RESULTS AND DISCUSSION

Analysis of Results by Using Taguchi Method and ANOVA

Effect of Input Variables on Cutting Rate

It is observed from main effects plot (Figure 5) as T_{on} and IP rises, CR increases whereas, it decline with increase in T_{off} and SV. This is due to the fact that the discharge energy boosted with the rising in T_{on} and IP, leading to a faster cutting rate. Analysis of variance (ANOVA) was performed to determine the significant parameters affecting the cutting rate. Analysis of variance (ANOVA) for cutting rate is reported in Table 9. Cutting rate is the "higher the better" type of quality characteristic. The Figure 5 reveals that T_{on} at third level (A_3), T_{off} at first level (B_1), IP at third level (C_3) and SV at first level (D_1) correspond to the maximum value of CR. At the optimal levels of significant variables, CR obtained its optimum value. In ANOVA table P (%) is the percentage of significance for an individual variable under consideration.

$$\mu = \overline{A_3} + \overline{B_1} + \overline{C_3} + \overline{D_1} - 3\overline{T} = 2.143 + 1.730 + 1.739 + 1.616 - (3 * 1.362) = 3.142 \quad (13)$$

Figure 5. Main effects plot for mean (cutting rate)

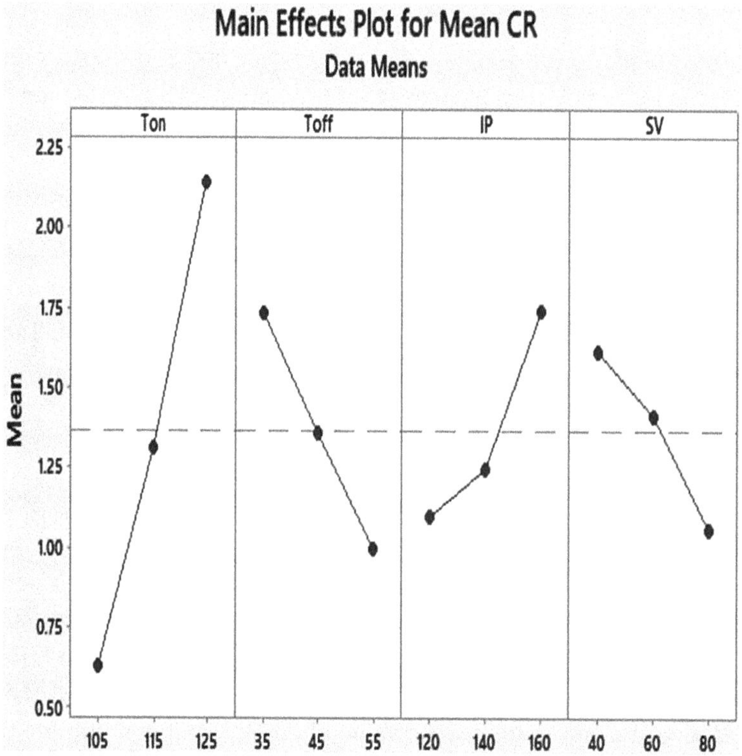

Table 9. ANOVA table for cutting rate (raw data)

Source	Degree of Freedom	Standard Deviation	Variance	F Ratio	P (%)
A	2	6.896	3.448	456.08	63.41
B	2	1.609	0.8045	106.55	14.79
C	2	1.348	0.6741	89.284	12.39
D	2	0.954	0.477	63.178	8.77
Error	9	0.068	0.00755		0.62
Total	17				

$$CI_{POP} = \pm\left[F_{\alpha}\left(n, f_e\right) * V_e / n_{eff}\right]^{1/2} = \pm 0.106 \tag{14}$$

where, $F_{\alpha}\left(n, f_e\right) = 3.01$, $V_e = 7.55 * 10^{-3}$, $n_{eff} = 18/(1+8) = 2$

Optimum value = 3.142 $\pm$0.106

Effect of Input Variables on Surface Roughness

Figure 6 shows that surface roughness increases rapidly when T_{on} increases; it decreases with increasing T_{off}. It is also evident that SR slightly varies with the IP. But SR first decreases rapidly and then slowly with increase in servo voltage. Analysis of variance (ANOVA) for surface roughness is reported in Table 10. The plot reveals that T_{on} at first level (A_1), T_{off} at third level (B_3), IP at first level (C_1) and SV at third level (D_3) provide minimum value of surface roughness. At the optimal levels of significant variables, SR obtained its optimum value.

Figure 6. Main effects plot for mean (surface roughness)

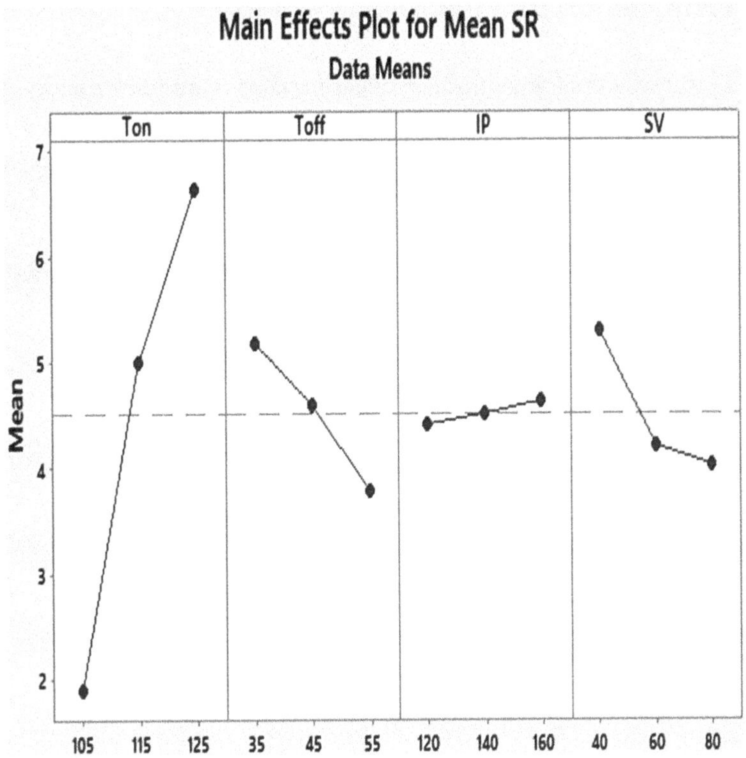

Table 10. ANOVA table for surface roughness (raw data)

Source	Degree of Freedom	Standard Deviation	Variance	F Ratio	P (%)
A	2	69.55	34.7755	1792.55	85.229
B	2	5.95	2.975	153.35	7.29
C	2	0.1324	0.0662	3.412	0.1622
D	2	5.7959	2.897	149.329	7.102
Error	9	0.1747	0.0194		
Total	17				

$$\mu = \overline{A_1} + \overline{B_3} + \overline{C_1} + \overline{D_3} - 3\overline{T} = 1.904 + 3.778 + 4.4155 + 4.028 - (3 * 4.5175) = 0.573 \quad (15)$$

$$CI_{POP} = \pm\left[F_\alpha\left(n, f_e\right) * V_e / n_{eff}\right]^{1/2} = \pm 0.170 \quad (16)$$

where, $F_\alpha\left(n, f_e\right) = 3.01$, $V_e = 0.0194$, $n_{eff} = 18 / (1+8) = 2$

Optimum value = 0.573 ±0.170

Effect of Input Variables on Dimensional Deviation

As shown in Figure 7, dimensional deviation first declines and, then rises rapidly when T_{on} increases. With increasing T_{off}, DD increases first and then decreases. It is also evident that DD decreases with increase in IP. Analysis of variance (ANOVA) for dimensional deviation is given in Table 11. Dimensional Deviation is the "lower the better" type quality characteristic, so it is clear from the main effects plot that the T_{on} at second level (A_2), T_{off} at third level (B_3), IP at third level (C_3) and, SV at second level (D_2) result the minimum value of dimensional deviation. At the optimal level of significant variables, DD achieved its optimum value.

Figure 7. Main effects plot for Mean (dimensional deviation)

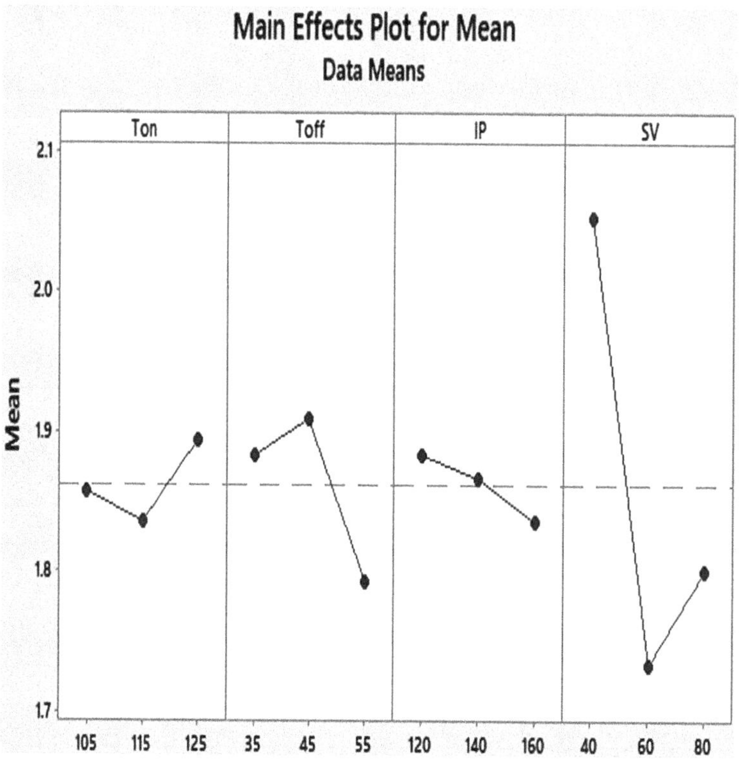

$$\mu = \overline{A}_2 + \overline{B}_3 + \overline{C}_3 + \overline{D}_2 - 3\overline{T} = 1.8636 + 1.793 + 1.836 + 1.733 - (3 * 41.862) = 0.0838 \quad (17)$$

$$CI_{POP} = \pm\left[F_{\alpha}\left(n, f_e\right) * V_e / n_{eff}\right]^{1/2} = \pm 0.0327 \quad (18)$$

where, $F_{\alpha}\left(n, f_e\right) = 3.01$, $V_e = 7.11 * 10^{-4}$, $n_{eff} = 18 / (1 + 8) = 2$

Optimum value = 1.6127 ± 0.0327

Effect of Input Variables on Wire Wear Ratio

Figure 8 shows the main effects plot for WWR. It is evident from the plot, when T_{on} increases, WWR also increases. This is due to the fact that discharge energy also increases with increase in T_{on} which leads to more wear on wire. The wire wear ratio decreases with increasing T_{off}. It is also evident that wire wear ratio first decreases and then increases with increase in IP. But with increase in SV, WWR first increases and then decreases. ANOVA for wire wear is given in Table 12. WWR is the "lower the better" type quality characteristic, so T_{on} at first level (A_1), T_{off} at third level (B_3), IP at second level C_2) and SV at third level (D_3) correspond to minimum value of wire wear ratio as seen from the plot. At the optimal levels of significant variables, WWR gets its optimum value.

$$\mu = \overline{A}_1 + \overline{B}_3 + \overline{C}_2 + \overline{D}_3 - 3\overline{T} = 0.175 + 0.162 + 0.183 + 0.181 - (3 * 0.204) = 0.089 \quad (19)$$

$$CI_{POP} = \pm\left[F_{\alpha}\left(n, f_e\right) * V_e / n_{eff}\right]^{1/2} = \pm 0.009 \quad (20)$$

where, $F_{\alpha}\left(n, f_e\right) = 3.01$, $V_e = 5.722 * 10^{-5}$, $n_{eff} = 18 / (1 + 8) = 2$

Optimum value = 0.089 ± 0.00927

Table 11. ANOVA table for DD (raw data)

Source	Degree of Freedom	Standard Deviation	Variance	F Ratio	P (%)
T_{on} A	2	0.0106	0.0053	7.454	2.56
T_{off} B	2	0.0455	0.02275	31.99	11.02
IP C	2	0.0074	0.0037	5.20	1.793
SV D	2	0.3427	0.1713	240.928	83.05
Error	9	0.0064	0.000711		1.577
Total	17				

Figure 8. Main effects plot for mean (wire wear ratio)

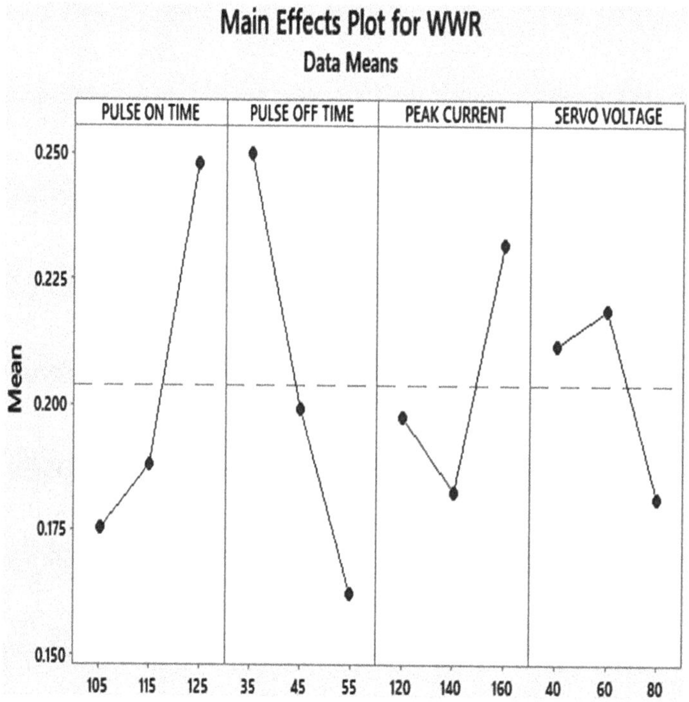

Table 12. ANOVA table for WWR (raw data)

Source	Degree of Freedom (DOF)	Standard Deviation	Variance	F Ratio	P (%)
A	2	0.0182	0.0091	159.00	32.85
B	2	0.0237	0.0118	206.22	42.77
C	2	0.00825	0.00412	70.124	14.48
D	2	0.00496	0.00248	43.34	8.953
Error	9	0.00051	0.000057		0.947
Total	17				

Effect of Input Variable on Material Removal Rate

Effect of pulse on time, pulse off time, peak current and servo voltage are shown in Figure 9, 10, 11 and 12. From the Figure 9, it is clear that with increase in pulse on time, material removal rate (MRR) is also increased. This is due to the increase in discharge energy. Whereas with increase in pulse off time, MRR is decrease due to decrease in discharge energy as shown in Figure 10. MRR is raised with increase in peak current as shown in Figure 11, whereas it decrease with increase in servo voltage as shown in Figure 12.

Figure 9. Variation of MRR with pulse on time

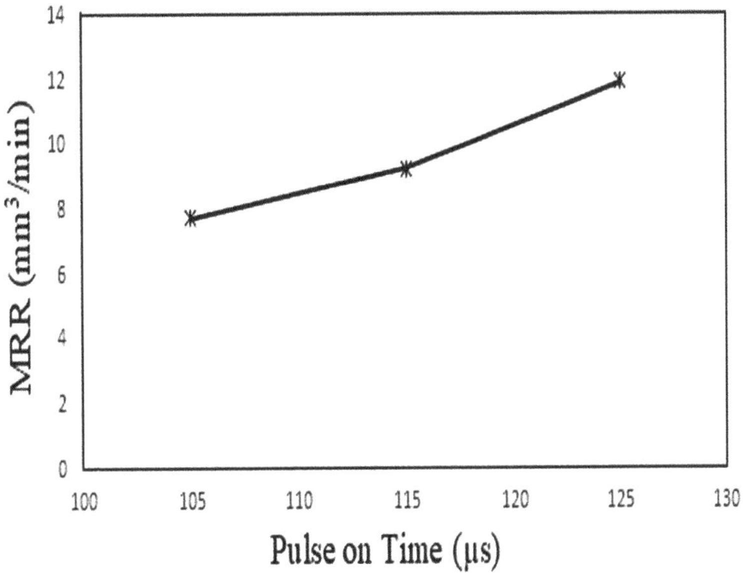

Figure 10. Variation of MRR with pulse off time

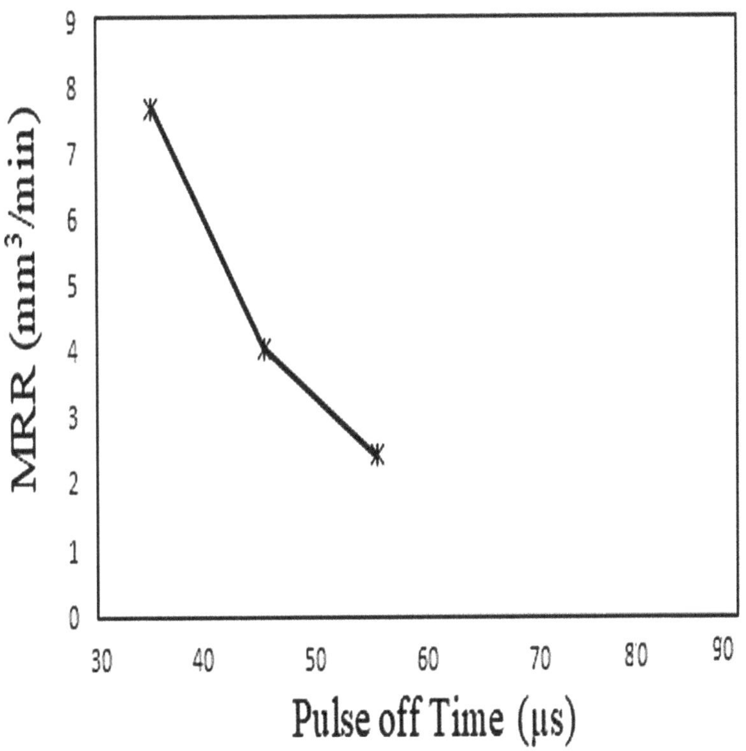

Figure 11. Variation of MRR with peak current

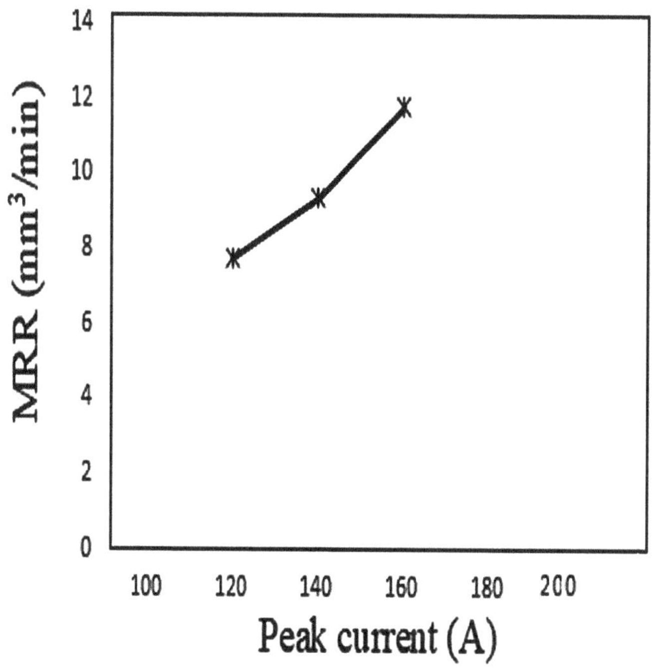

Figure 12. Variation of MRR with servo voltage

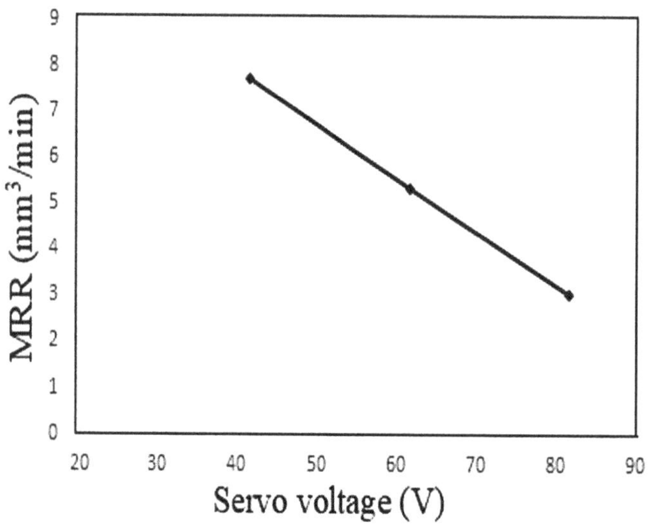

Analysis of Experimental Results Using GRA and ANOVA

The effects of input variables on GRG value is presented in Figure 13. The optimal setting for maximum GRG value i.e. better multiple response is $A_3B_1C_3D_1$. ANOVA is applied to find the most significant factors. The un-pooled and pooled ANOVA results are given in Tables 13 and 14 respectively. The

Figure 13. Effects of process parameters on grey relational grade

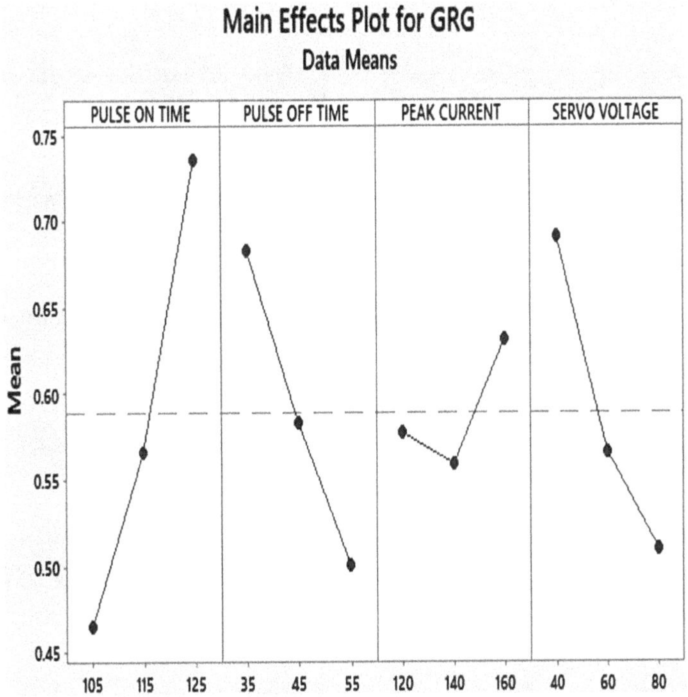

Table 13. Un-pooled ANOVA for GRG

Source	Degree of Freedom	Standard Deviation	Variance	P (%)
Pulse on time A	2	0.1129	0.0564	50.491
Pulse off time B	2	0.0504	0.0252	22.54
Peak current C	2	0.0085	0.0042	3.801
Servo voltage D	2	0.0516	0.0258	23.076
Error	0	0.0002		
Total	8			

Table 14. Pooled ANOVA for GRG

Source	DOF	Standard Deviation	Variance	F Ratio	SS′	P (%)
Pulse on time A	2	0.1129	0.0564	12.965	0.1042	46.601
Pulse off time B	2	0.0504	0.0252	5.793	0.0417	18.649
Servo voltage D	2	0.0516	0.0258	5.931	0.0429	19.186
Error	2	0.0087	0.00435			15.564
Total	8					

Table 15. Effect of parameters on GRG value

Parameters	Level 1	Level 2	Level 3
Pulse on time (T_{on})	0.4648	0.5663	**0.7364**
Pulse off time (T_{off})	**0.6838**	0.5830	0.5007
Peak current (IP)	0.5772	0.5589	0.6314
Servo voltage (SV)	**0.6913**	0.5664	0.5099

Optimal value of GRG = 0.9331; the optimal levels are indicated by bold type.

factor C (peak current) is insignificant factor which has least percentage of contribution, so it has been pooled. From the pooled ANOVA Table 14, it is found that T_{on} has maximum percentage contribution (46.601%) followed by SV and T_{off}.

Validation of Results

Confirmation experiment was conducted to validate the results. The machining parameters were set as $A_3 B_1 C_1 D_1$ for conducting confirmation experiment. The predicted values of all responses by Taguchi method and results obtained from confirmation experiment are reported in Table 16.

CONCLUSION

From the study following conclusions can be drawn:

1. T_{on} is the most important factor (63.41% contribution) in affecting the cutting rate, and cutting rate rises with rise in T_{on}.
2. Again, T_{on} is the major influencing factor (contributing 85.23%) in controlling the surface roughness.
3. For the dimensional deviation, SV is most significant factor (contributing 83.05%).
4. For the wire wear ratio, T_{on} is most significant factor (contributing 42.77%) and, WWR decline with rise in pulse off time.
5. Grey Relational Analysis in conjunction with Taguchi Methods has predicted that the optimum cutting condition for multi-response optimization is $A_3B_1C_3D_1$ corresponding to the highest GRG value.

Table 16. Results of confirmation test

Machining Characteristics	Optimal Setting of Parameters	Predicted Value by Taguchi Method	Results Obtained From Confirmation Experiment by GRA $A_3 B_1 C_1 D_1$
Cutting rate	$A_3 B_1 C_3 D_1$	3.142 ± 0.106	3.221 mm/min
Surface roughness	$A_1 B_3 C_1 D_3$	0.573 ± 0.170	0.66 μm
Dimensional deviation	$A_3 B_3 C_3 D_2$	1.612 ± 0.0327	1.512%
Wire wear ratio	$A_1 B_3 C_2 D_3$	0.089 ± 0.00927	0.102

6. The pooled ANOVA for GRG reveals that T_{on} is most significant factor (46.60%) on the GRG value followed by SV and T_{off}.

SCOPE FOR FUTURE WORK

In this paper, an attempt has carried out to investigate and optimize the different process parameters for machining of Inconel 625 in wire EDM. There are some suggestions for future work:

1. The effects of different input variables on wire breakage, recast layer thickness, and overcut should be investigated.
2. There are many mathematical models for optimization, by which experimental cost and time can be reduced.

REFERENCES

Alias, A., Abdullah, B., & Abbas, N. M. (2012). Influence of machine feed rate in WEDM of Titanium ti-6al-4v with constant current (6a) using Brass wire. *Procedia Engineering*, *41*, 1806–1811. doi:10.1016/j.proeng.2012.07.387

Atzeni, Bassoli, Gatto, Iuliano, Minetola, & Salmi. (2015). Surface and sub surface evaluation in coated wire electric discharge machining of Inconel alloy 718. *Procedia CIRP*, *33*, 389-394.

Baig & Venkaiah. (2014). Parametric optimization of WEDM for HastelloyC276, using GRA method. *International Journal of Engineering Development and Research*, *1*(2), 1-7.

Chalisgaonkar & Kumar. (2015). Multi-response optimization and modeling of trim cut WEDM operation of commercially pure titanium (CPTi) considering multiple user's preferences. *Engineering Science and Technology, an International Journal*, *18*, 125-134.

Chiang, K., & Chang, F. (2006). Optimization of the WEDM process of particle-reinforced material with multiple performance characteristics using grey relation analysis. *Journal of Materials Processing Technology*, *180*(1-3), 96–110. doi:10.1016/j.jmatprotec.2006.05.008

Goswami & Kumar. (2014). Investigation of surface integrity, material removal rate and wire wear ratio for WEDM of Nimonic 80A alloy using GRA and Taguchi method. *Engineering Science and Technology, an International Journal*, *17*, 173-184.

Hewidy, M. S., El-Taweel, T. A., & El-Safty, M. F. (2005). Modelling the machining parameters of wire electrical discharge machining of Inconel 601 using RSM. *Journal of Materials Processing Technology*, *169*(2), 328–336. doi:10.1016/j.jmatprotec.2005.04.078

Newton, Melkote, Watkins, Trejo, & Reister. (2009). Investigation of the effect of process parameters on the formation and characteristics of recast layer in wire-EDM of Inconel 718. *Materials Science and Engineering A*, (513–514), 208–215.

Keith Hargrove, S., & Ding, D. (2007). Determining cutting parameters in wire EDM based on workpiece surface temperature distribution. *International Journal of Advanced Manufacturing Technology*, *34*(3-4), 295–299. doi:10.1007/s00170-006-0609-0

Kumar, K., & Agarwal, S. (2012). Multi-objective parametric optimization on machining with wire electric discharge machining. *International Journal of Advanced Manufacturing Technology*, *62*(5-8), 617–633. doi:10.1007/s00170-011-3833-1

Maher, I., Sarhan, A. A. D., & Hamdi, M. (2015). Review of improvements in wire electrode properties for longer working time and utilization in wire EDM machining. *International Journal of Advanced Manufacturing Technology*, *76*(1-4), 329–351. doi:10.1007/s00170-014-6243-3

Manna, A., & Bhattacharyya, B. (2006). Taguchi and Gauss elimination method: A dual response approach for parametric optimization of CNC wire cut EDM of PRAlSiCMMC. *International Journal of Advanced Manufacturing Technology*, *28*(1-2), 67–75. doi:10.1007/s00170-004-2331-0

Nain, S. S., Garg, D., & Kumar, S. (2017). Prediction of the Performance Characteristics of WEDM on Udimet-L605 Using Different Modelling Techniques. *Materials Today: Proceedings*, *4*(2), 546–556. doi:10.1016/j.matpr.2017.01.056

Nayak & Mahapatra. (2016). Optimization of WEDM process parameters using deep cryo-treated Inconel 718 as work material. *Engineering Science and Technology, an International Journal*, *19*, 161–170.

Pragya Shandilya, P. K. (2012). Parametric optimization during wire electrical discharge machining using response surface methodology. *Procedia Engineering*, *38*, 2371–2377. doi:10.1016/j.proeng.2012.06.283

Puri, A. B., & Bhattacharyya, B. (2005). Modeling and analysis of white layer depth in a wire-cut EDM process through response surface methodology. *International Journal of Advanced Manufacturing Technology*, *25*(3-4), 301–307. doi:10.1007/s00170-003-2045-8

Rajyalakshmi, G., & Venkata Ramaiah, P. (2013). Multiple process parameter optimization of wire electrical discharge machining on Inconel 825 using Taguchi grey relational analysis. *International Journal of Advanced Manufacturing Technology*, *69*(5-8), 1249–1262. doi:10.1007/s00170-013-5081-z

Ramakrishnan, R., & Karunamoorthy, L. (2006). Multi response optimization of wire EDM operations using robust design of experiments. *International Journal of Advanced Manufacturing Technology*, *29*(1-2), 105–112. doi:10.1007/s00170-004-2496-6

Ramakrishnan, R., & Karunamoorthy, L. (2008). Modeling and multi-response optimization of Inconel 718 on machining of CNC WEDM process. *Journal of Materials Processing Technology*, *207*(1-3), 343–349. doi:10.1016/j.jmatprotec.2008.06.040

Sharma, N., Khanna, R., Gupta, R. D., & Sharma, R. (2013). Modeling and multi response optimization on WEDM for HSLA by RSM. *International Journal of Advanced Manufacturing Technology*, *67*(9-12), 2269–2281. doi:10.1007/s00170-012-4648-4

Siddiquee, A. N., Khan, Z. A., & Mallick, Z. (2010). Grey relational analysis coupled with principal component analysis for optimization design of the process parameters in in-feed centreless cylindrical grinding. *International Journal of Advanced Manufacturing Technology*, *46*(9-12), 983–992. doi:10.1007/s00170-009-2159-8

Singh, H., & Garg, R. (2009). Effects of process parameters on material removal rate in WEDM. *Journal of Achievements in Materials and Manufacturing Engineering of Achievements in Materials and Manufacturing Engineering*, *32*, 70–74.

Singh, S. (2012). Optimization of machining characteristics in electric discharge machining of 6061Al/Al2O3p/20P composites by grey relational analysis. *International Journal of Advanced Manufacturing Technology*, *63*(9-12), 1191–1202. doi:10.1007/s00170-012-3984-8

Tosun, N., & Cogun, C. (2013). An investigation on wire wear in WEDM. *Journal of Materials Processing Technology*, *134*(3), 273–278. doi:10.1016/S0924-0136(02)01045-2

Tosun, N., Cogun, C., & Pihtili, H. (2003). The Effect of Cutting Parameters on Wire Crater Sizes in Wire EDM. *International Journal of Advanced Manufacturing Technology*, *21*(10-11), 857–865. doi:10.1007/s00170-002-1404-1

Unune & Mali. (2017). Experimental investigation on low-frequency vibration assisted micro-WEDM of Inconel 718. *Engineering Science and Technology, an International Journal*, *20*, 222–231.

Venkata Rao, R., & Kalyankar, V. D. (2014). Optimization of modern machining processes using advanced optimization techniques: A review. *International Journal of Advanced Manufacturing Technology*, *73*(5-8), 1159–1188. doi:10.1007/s00170-014-5894-4

Yan, M.-T., Fang, G.-R., & Liu, Y.-T. (2013). An experimental study on micro wire-EDM of polycrystalline diamond using a novel pulse generator. *International Journal of Advanced Manufacturing Technology*, *66*(9-12), 1633–1640. doi:10.1007/s00170-012-4446-z

Yuan, J., Wang, K., Yu, T., & Fang, M. (2008). Reliable multi-objective optimization of high-speed WEDM process based on Gaussian process regression. *International Journal of Machine Tools & Manufacture*, *48*(1), 47–60. doi:10.1016/j.ijmachtools.2007.07.011

Compilation of References

Abbas, N. M., Solomon, D. G., & Bahari, M. F. (2007). A review of current research trends in electrical discharge machining (EDM). *International Journal of Machine Tools & Manufacture*, *47*(7-8), 1214–1228. doi:10.1016/j.ijmachtools.2006.08.026

Abdin, E. S., & Xu, W. (1998). Control Design and Dynamic Performance Analysis of a Wind Turbine-Induction Generator Unit. IEEE, 1198 – 1202.

Abella, R. J., Daschbach, J. M., & McNichols, R. J. (1994, April). Reverse Engineering Industrial Applications. *Computers & Industrial Engineering*, *26*(2), 381–385. doi:10.1016/0360-8352(94)90071-X

Abrate, S. (1997). Machining of composite materials. In P. K. Mallick (Ed.), *Composites engineering handbook*. New York: Marcel Dekker.

Acar, B. S. (1966). A new model for the design process. *Proceedings of the Institute of Mechanical Engineers, E*(210), 135-138.

Adepu, K., & Kolli, M. (2014). Effect of Additives Added in Dielectric Fluid on Electrical Discharge Machining of Titanium Alloy. *International Conference onAdvanced Research & innovations in Mechanical, Material science, Industrial Engineering & Management*, 247-250.

Advani, S. G. (1994). Resin Transfer Moulding Flow Phenomena in Polymeric Composites. In S. G. Advani (Ed.), *Flow and Rheology in Polymer Composite Manufacturing* (pp. 465–515). Newark, NJ: Elsevier Science.

AfsanehDorri. (2015). Mechanical and tribological properties of self-lubricating metal matrix nano composites reinforced by CNTs. *Composites. Part B, Engineering*, *77*, 402–420. doi:10.1016/j.compositesb.2015.03.014

Agarwala, R. C., & Agarwala, V. (2003). Electroless alloy/composite coatings: A review. *Sadhana*, *28*(3-4), 475–493. doi:10.1007/BF02706445

Ahmed, P. X. (1998). Culture and climate for innovation. *European Journal of Innovation Management*, *1*(1), 30–43. doi:10.1108/14601069810199131

Akgun, A. E., Byrne, J., Keskin, H., & Lynn, G. S. (2005). Knowledge networks in NPD projects: A transactive memory perspective. *Information & Management*, *42*, 1105–1120. doi:10.1016/j.im.2005.01.001

Akos, M., & Imre, T. (2013). The Importance Of Friction Stir Welding Tool. *Production Processes and Systems.*, *6*(1), 25–34.

Al-Ahmari, A. M. A. (2007). Predictive machinability models for a selected hard material in turning operations. *Journal of Materials procesing. Technology (Elmsford, N.Y.)*, *190*, 305–311.

Ala-Kleme, S., Kivikytö-Reponen, P., Liimatainen, J., Hellman, J., & Hannula, S. P. (2006). Abrasive wear properties of metal matrix composites produced by hot isostatic pressing. *Proc. Estonian Acad. Sci. Eng*, *12*(4), 445–454.

Alias, A., Abdullah, B., & Abbas, N. M. (2012). Influence of machine feed rate in WEDM of Titanium ti-6al-4v with constant current (6a) using Brass wire. *Procedia Engineering*, *41*, 1806–1811. doi:10.1016/j.proeng.2012.07.387

Al-Khudairi, O., & Ghasemnejad, H. (2015). To improve failure resistance in joint design of composite wind turbine blade materials. *Renewable Energy*, *81*, 936–951. doi:10.1016/j.renene.2015.04.015

Ansar, M., Xinwei, W., & Chouwei, Z. (2011). Modeling strategies of 3D woven composites: A review. *Composite Structures*, *93*(8), 1947–1963. doi:10.1016/j.compstruct.2011.03.010

Arewar, A. P., & Bhope, D. V. (2013). Stress analysis of axial flow fan impeller. *International Journal of Engineering Research and Applications*, *3*(1), 2086–2090.

Arpitha, G. R., Sanjay, M. R., Naik, L., & Yogesha, B. (2014). Mechanical Properties of Epoxy Based Hybrid Composites Reinforced with Sisal/SIC/Glass Fibers. *International Journal of Engineering Research and General Science*, *2*(5), 398–405.

Arslantürk, C. (2008). Correlation equations for optimum design of annular fins with temperature dependent thermal conductivity. *Heat and Mass Transfer*, *45*(4), 519–525. doi:10.1007/s00231-008-0446-9

Arslantürk, C. (2010). Analysis of thermal performance of annular fins with variable thermal conductivity by homotopy method. *Journal Of Thermal Science And Technology*, *30*(2), 1–7.

Asadi, P., Faraji, G., & Besharati, M. K. (2010). Producing of AZ91/SiC composite by friction stir processing (FSP). *International Journal of Advanced Manufacturing Technology*, *51*(1-4), 247–260. doi:10.1007/s00170-010-2600-z

Asadi, P., Givi, M. K. B., Abrinia, K., Taherishargh, M., & Salekrostam, R. (2011). Effects of SiC particle size and process parameters on the microstructure and hardness of AZ91/SiC composite layer fabricated by FSP. *Journal of Materials Engineering and Performance*, *20*(9), 1554–1562. doi:10.1007/s11665-011-9855-x

Ashik, K. P., & Sharma, R. S. (2015). A Review on Mechanical Properties of Natural Fiber Reinforced Hybrid Polymer Composites. *Journal of Minerals & Materials Characterization & Engineering*, *3*(05), 420–426. doi:10.4236/jmmce.2015.35044

Ashori, A. (2008). Wood–plastic composites as promising green-composites for automotive industries. *Bioresource Technology*, *99*(11), 4661–4667. doi:10.1016/j.biortech.2007.09.043 PMID:18068352

Atharifar, H., Lin, D. C., & Kovacevic, R. (2009). Numerical and experimental investigations on the loads carried by the tool during friction stir welding. *Journal of Materials Engineering and Performance*, *18*(4), 339–350. doi:10.1007/s11665-008-9298-1

Atherden, M. J. (2004). *Formula SAE Shock Absorber Design. Bachelor of Engineering*. University of Queensland.

Atilgan-Inan, E., Buyukkupcu, A., & Akinci, S. (2010). A content analysis of factors affecting NPD process. *Business and Economics Research Journal*, *1*(2), 87–100.

Attia, A. N. (2001). Surface metal matrix composites. *Materials & Design*, *22*(6), 451–457. doi:10.1016/S0261-3069(00)00081-9

Atzeni, Bassoli, Gatto, Iuliano, Minetola, & Salmi. (2015). Surface and sub surface evaluation in coated wire electric discharge machining of Inconel alloy 718. *Procedia CIRP*, *33*, 389-394.

Aziz, A., & Rahman, M. M. (2009). Thermal performance of a functionally graded radial fin. *International Journal of Thermophysics*, *30*(5), 1637–1648. doi:10.1007/s10765-009-0627-x

Badarinarayan, H., Yang, Q., & Zhua, S. (2009). Effect of tool geometry on static strength of friction stir spot-welded aluminum alloy. *Int. Mach. Tools Manuf.*, *49*(2), 142–148. doi:10.1016/j.ijmachtools.2008.09.004

Bader, M. G. (2002). Selection of composite materials and manufacturing routes for cost-effective performance. *Composites. Part A, Applied Science and Manufacturing*, *33*(7), 913–934. doi:10.1016/S1359-835X(02)00044-1

Bahadur, R., & Bar-cohen, A. (2005). Thermal Design and Optimization of Natural Convection Polymer Pin Fin Heat Sinks. *IEEE Transactions on Components and Packaging Technologies*, *28*(2), 238–246. doi:10.1109/TCAPT.2005.848498

Bahadur, R., & Bar-cohen, A. (2007). Orthotropic thermal conductivity effect on cylindrical pin fin heat transfer. *International Journal of Heat and Mass Transfer*, *50*(5-6), 1155–1162. doi:10.1016/j.ijheatmasstransfer.2006.04.025

Baig & Venkaiah. (2014). Parametric optimization of WEDM for HastelloyC276, using GRA method. *International Journal of Engineering Development and Research*, *1*(2), 1-7.

Balaraju, J. N., Kalavati, , & Rajam, K. S. (2009). Surface morphology and structure of electroless ternary NiWP deposits with various W and P contents. *Journal of Alloys and Compounds*, *486*(1-2), 468–473. doi:10.1016/j.jallcom.2009.06.173

Balaraju, J. N., Radhakrishnan, P., Ezhilselvi, V., Kumar, A. A., Chen, Z., & Surendran, K. P. (2016). Studies on electroless nickel polyalloy coatings over carbon fibers/CFRP composites. *Surface and Coatings Technology*, *302*, 389–397. doi:10.1016/j.surfcoat.2016.06.040

Balaraju, J. N., & Rajam, K. S. (2005). Electroless deposition of Ni-Cu-P, Ni-W-P and Ni-W-Cu-P alloys. *Surface and Coatings Technology*, *195*(2-3), 154–161. doi:10.1016/j.surfcoat.2004.07.068

Baldwin, C. Y., & Clark, K. B. (2000). *Design Rules: The Power of Modularity*. Cambridge, MA: MIT Press.

Banerjee, A., & Pohit, G. (2014). Crack Investigation of Rotating Cantilever Beam by Fractal Dimension Analysis. *Procedia Technology*, *14*, 188–195. doi:10.1016/j.protcy.2014.08.025

Banu, A., Ali, M. Y., & Rahman, M. A. (2014). Micro-Electro Discharge Machining of non-conductive zirconia ceramic: Investigation of MRR & recast layer hardness. *International Journal of Advanced Manufacturing Technology*, *75*(1-4), 257–267. doi:10.1007/s00170-014-6124-9

Baraskar, S. S., Banwait, S. S., & Laroiya, S. C. (2013). Multiobjective Optimization of Electrical Discharge Machining Process Using a Hybrid Method. *Materials and Manufacturing Processes*, *28*(4), 348–354. doi:10.1080/10426914.2012.700152

Barczak, G., Griffin, A., & Khan, K. B. (2009). Trends and drivers of success in NPD practices: Results of the 2003 PDMA best practice study. *Journal of Product Innovation Management*, *26*(1), 3–23. doi:10.1111/j.1540-5885.2009.00331.x

Barik, A. K., Dash, S. K., & Guha, A. (2014b). New correlations for prediction of air entrainment into an Infrared suppression (IRS) device. *Applied Ocean Research*, *47*, 303–312. doi:10.1016/j.apor.2014.06.007

Barik, A. K., Dash, S. K., & Guha, A. (2015). Experimental and numerical investigation of air entrainment into an infrared suppression device. *Applied Thermal Engineering*, *75*, 33–44. doi:10.1016/j.applthermaleng.2014.05.042

Barik, A. K., Dash, S. K., Patro, P., & Mohapatra, S. (2014a). Experimental and numerical investigation of air suction into a louvered funnel. *Applied Ocean Research*, *48*, 176–185. doi:10.1016/j.apor.2014.08.009

Beard, C and Easingwood, C (1996). New Product Launch: Marketing Action and launch tactics for high technology products. *Journal of Industrial Marketing Management, 25*, 87-103.

Becker, H. A., Hottel, H. C., & Williams, G. C. (1963). Mixing and flow in ducted turbulent jets. In *9th International Symposium on Combustion*. Pittsburg, PA: The Combustion Institute. doi:10.1016/B978-1-4832-2759-7.50009-1

Bembde, M. S., & Sawale, J. K. (2014). Review of Powder Mixed Dielectric on Material Removal Rate in EDM. *International Journal of Science and Research*, *12*(4), 821–824.

Benedict, G. F. (1987). *Non-traditional manufacturing processes*. New York: Marcel Decker Inc.

Berbon, P. B., Bingel, W. H., Mishra, R. S., Bampton, C. C., & Mahoney, M. W. (2001). Friction stir processing: A tool to homogenize nanocomposite aluminum alloys. *Scripta Materialia*, *44*(1), 61–66. doi:10.1016/S1359-6462(00)00578-9

Berwick, A. (1971). The Short Leg Brace For Ankle Foot Stability In Adults. *The Australian Journal of Physiotherapy*, *17*(3), 81–84. doi:10.1016/S0004-9514(14)61110-7 PMID:25025930

Besharati, M. K., & Asadi, P. (2014). *Advances in Friction Stir Welding and Processing*. Woodhead Publishing.

Bhaskaran, S. R., & Krishnan, V. (2009). Effort, Revenue and Cost Sharing Mechanisms for Collaborative NPD. *Management Science*, *55*(7), 1152–1169. doi:10.1287/mnsc.1090.1010

Bhope, D. V., & Padole, P. M. (2004). Experimental and theoretical analysis of stresses, noise and flow in centrifugal fan impeller. *Mechanism and Machine Theory*, *39*(12), 1257–1271. doi:10.1016/j.mechmachtheory.2004.05.015

Bhowmick, M., Mukhopadhyay, S., & Alagirusamy, R. (2012). Mechanical properties of natural fibre reinforced composites. *Textile Progress*, *44*(2), 85–140. doi:10.1080/00405167.2012.676800

Bhuyan, D., & Kumar, K. (in press). 3D Cad Modelling and Computational Fluid Analysis of Piston Valve of Twin Tube Shock Absorbers. *Materials Today: Proceedings*.

Bist, A., Saini, J. S., & Sharma, B. (2016). A review of tool wear prediction during friction stir welding of aluminium matrix composite. *Transactions of Nonferrous Metals Society of China*, *26*(8), 2003–2018. doi:10.1016/S1003-6326(16)64318-2

Boothroyd, G., & Winston, A. K. (1989). *Non-Conventional Machining Processes. In Fundamentals of Machining & Machine Tools* (p. 491). New York: Marcel Dekker Inc.

Booz, Allen, and Hamilton. (1991). *New Product Management*. Booz, Allen and Hamilton Inc.

Borges, J. E. (1994). A three-dimensional inverse method for turbo machinery: Part-1Theory. *ASME Trans J Turbo*, *112*(3), 346–354. doi:10.1115/1.2927666

Brauers, W. K. M., & Zavadskas, E. K. (2006). The MOORA method and its application to privatization in a transition economy. *Control and Cybernetics*, *35*(2), 445–469.

Bregman, D. J. J., van der Krogt, M. M., De Groot, V., Harlaar, J., Wisse, M., & Collins, S. H. (2011). The effect of ankle foot orthosis stiffness on the energy cost of walking: A simulation study. *Clinical Biomechanics (Bristol, Avon)*, *26*(9), 955–961. doi:10.1016/j.clinbiomech.2011.05.007 PMID:21723012

Broitman, E. (2017). Indentation hardnessmeasurements at macro-, micro-, and nanoscale: A critical overview. *Tribology Letters, 65*, 18. DOI: 10.1007/s11249-016-0805-5

Brown, T. (2008). Design Thinking. *Harvard Business Review*, *86*(6), 84–92. PMID:18605031

Bruschke, M. V., & Advani, S. G. (1990). A Finite Element/Control Volume Approach to Mould Filling in Anisotropic Porous Media. *Polymer Composites*, *11*(6), 398–405. doi:10.1002/pc.750110613

Bruschke, M. V., & Advani, S. G. (1994). A Numerical Approach to Model Non-Isothermal Viscous Flow through Fibrous Media with Free Surfaces. *International Journal for Numerical Methods in Fluids*, *19*(7), 575–603. doi:10.1002/fld.1650190704

Calhoun, R. D., Yalvac, S., Wetters, D. G., Wu, C. H., Wang, T. J., Tsai, J. S., & Lee, L. J. (1996). Mould Filling Analysis in Resin Transfer Moulding. *Polymer Composites*, *17*(2), 251–264. doi:10.1002/pc.10610

Campo, A., & Cui, J. (2008). Temperature/Heat Analysis of Annular Fins of Hyperbolic Profile Relying on the Simple Theory for Straight Fins of Uniform Profile. *Journal of Heat Transfer*, *130*(5), 0545011–0545014. doi:10.1115/1.2885162

Cao, L., Bai, Z., Huang, J., OuYang, H., & Wu, J. (2017). Ablation properties of Cf/C-SiC MoSi2 Composites: Effects of Hydrothermal penetration temperature. *Journal of Alloys and Compounds*, *703*, 45–55. doi:10.1016/j.jallcom.2017.01.316

Carbonell, P., & Rodriguez, A. I. (2006). The impact of market characteristics and innovation speed on perceptions of positional advantage and new product performance. *International Journal of Research in Marketing*, *23*(1), 1–12. doi:10.1016/j.ijresmar.2006.01.002

Chai, S. T., & Mason, W. H. (1996). Landing Gear Integration in Aircraft Conceptual Design. Virginia Polytechnic Institute and State University.

Chai, F., Zhang, D., & Li, Y. (2014). Effect of Thermal History on Microstructures and Mechanical Properties of AZ31 Magnesium Alloy Prepared by Friction Stir Processing. *Materials (Basel)*, *7*(3), 1573–1589. doi:10.3390/ma7031573 PMID:28788532

Chakraborty, S. (2011). Applications of the MOORA method for decision making in a manufacturing environment. *International Journal of Advanced Manufacturing Technology*, *54*(9-12), 1155–1166. doi:10.1007/s00170-010-2972-0

Chalisgaonkar & Kumar. (2015). Multi-response optimization and modeling of trim cut WEDM operation of commercially pure titanium (CPTi) considering multiple user's preferences. *Engineering Science and Technology, an International Journal*, *18*, 125-134.

Chandrashekar, M., Muralidhar, M., Murali Krishna, C., & Dixit, U. S. (2010). Application of soft computing techniques in machining performance prediction & optimization: A literature review. *International Journal of Advanced Manufacturing Technology*, *46*(5-8), 445–464. doi:10.1007/s00170-009-2104-x

Chang, C. I., Lee, C. J., Chuang, C. H., Pei, H. R., & Huang, J. C. (2007). On Mg-Al-Zn Intermetallic Alloys Made by Friction Stir Processing Containing Quasi-Crystals or Amorphous Phases. Trans Tech Publications.

Chang, C. I., Wang, Y. N., Pei, H. R., Lee, C. J., & Huang, J. C. (2006). On the Hardening of Friction Stir Processed Mg-AZ31 Based Composites with 5–20% Nano-ZrO2 and Nano-SiO2 Particles. *Materials Transactions*, *47*(12), 2942–2949. doi:10.2320/matertrans.47.2942

Chan, H. C. (2010). Linkage community based innovation and speed to market: The mediating role of the NPD process. *The International Journal of Organisational Innovation*, *2*(4), 49–60.

Charit, I., & Mishra, R. S. (2005). Low temperature super plasticity in a friction-stir-processed ultrafine grained Al–Zn–Mg–Sc alloy. *Acta Materialia*, *53*(15), 4211–4223. doi:10.1016/j.actamat.2005.05.021

Chaudhari, S. C., Yadav, C. O., & Damor, A. B. (2013). A comparative study of mix flow pump impeller cfd analysis and experimental data of submersible pump. *International Journal of Research in Engineering & Technology*, *1*(3), 57–64.

Chaudhury, P., Samantaray, S., & Sahu, S. (2017). Multi Response Optimization of Powder Additive Mixed Electrical Discharge Machining by Taguchi Analysis. *Materials Today: Proceedings*, *4*(2), 2231–2241. doi:10.1016/j.matpr.2017.02.070

Cheng, G. R. (1997). *Genetic Algorithm and Engineering Design*. New York: John Wiley.

Chen, W. X., Tu, J. P., Gan, H. Y., Xu, Z. D., Wang, Q. G., Lee, J. Y., & Zhang, X. B. et al. (2002). Electroless preparation and tribological properties of Ni-P-Carbon nanotube composite coatings under lubricated condition. *Surface and Coatings Technology*, *160*(1), 68–73. doi:10.1016/S0257-8972(02)00408-5

Chiang, K., & Chang, F. (2006). Optimization of the WEDM process of particle-reinforced material with multiple performance characteristics using grey relation analysis. *Journal of Materials Processing Technology*, *180*(1-3), 96–110. doi:10.1016/j.jmatprotec.2006.05.008

Chikhradze, N. M., Marquis, F. D., & Abashidze, G. S. (2015). Hybrid fiber and nanopowder reinforced composites for wind turbine blades. *Journal of Materials Research and Technology*, *4*(1), 60–67. doi:10.1016/j.jmrt.2015.01.002

Chikofsky, E. J., & Cross, J. H. (1990). Reverse Engineering and Design Recovery: A Taxonomy. *IEEE Software*, *7*(1), 13–17. doi:10.1109/52.43044

Choo, S.-H., Lee, S., & Kwon, S.-J. (1999). Effect of flux addition on the microstructure and hardness of TiC-reinforced ferrous surface composite layers fabricated by high-energy electron beam irradiation. *Metallurgical and Materials Transactions. A, Physical Metallurgy and Materials Science*, *30*, 1211. doi:10.1007/s11661-999-0271-x

Choun, Y. S., Park, J., & Choi, I. K. (2014, October). Effects of mechanical property variability in lead rubber bearings on the response of seismic isolation system for different ground motions. *Nuclear Engineering and Technology*, *46*(5), 605–618. doi:10.5516/NET.09.2014.718

Chowdhury, S. M., Chen, D. L., Bhole, S. D., & Cao, X. (2010). Effect of spin tool thread orientation on fatigue strength of friction stir welded AZ31B-H34 Mg butt joints. *Proc Eng.*, *2*(1), 825–833. doi:10.1016/j.proeng.2010.03.089

Chowdhury, S. M., Chen, D. L., Bhole, S. D., & Cao, X. (2010, August). Tensile properties of a friction stir welded magnesium alloy: Effect of spin tool thread orientation and weld pitch. *Materials Science and Engineering A*, *527*(21-22), 6064–6075. doi:10.1016/j.msea.2010.06.012

Coit, D. W., Jackson, B. T., & Smith, A. E. (1998). Static Neural Network process models: Considerations and case studies. *International Journal of Production Research*, *36*(11), 2953–2967. doi:10.1080/002075498192229

Colegrove, P. A., & Shercliff, H. R. (2004). Two-dimensional CFD modeling of flow round profiled FSW tooling. *Science and Technology of Welding and Joining*, *9*(6), 483–492. doi:10.1179/136217104225021832

Collier, M., Steel, R., Nelson, T., Sorensen, C., & Packer, S. (2003). Grade development of polycrystalline cubic boron nitride for friction stir processing of ferrous alloys. *J. Materials Science Forum, 426–432*, 3011–3016.

Cooper, R. G., & Edgett, S. J. (2003). Overcoming the crunch in resources for new product development. *Research Technology Management, 46*(3), 48-59.

Cooper, R. G., Edgett, S. J., & Kleinschmidt, E. J. (2002). Optimising the stage-gate process: What best practice companies do – II. *Research Technology Management, 45*(6), 43-51.

Cooper, R. G. (1999). From Experience: The invisible success factors in product innovation. *Journal of Product Innovation Management*, *16*(4), 333–351. doi:10.1016/S0737-6782(99)00005-3

Cooper, R. G. (2005). *Product Leadership: Pathways to Profitable Innovation*. Basic Books.

Cooper, R. G. (2008). The Stage-Gate® idea-to-launch process update, what's new and NexGen systems. *Journal of Product Innovation Management*, *25*(3), 213–232. doi:10.1111/j.1540-5885.2008.00296.x

Cooper, R. G. (2011). *Winning at New Products: Creating Value through Innovation* (4th ed.). Basic Books.

Cooper, R. G., & Kleinschmidt, E. J. (2007). Winning businesses in product development: Critical Success Factors. *Research Technology Management*, *50*(3), 52–66.

Cooper, R., Wooton, B., & Bruce, M. (1998). Requirements Capture: Theory and Practice. *Technovation*, *18*(8/9), 497–511. doi:10.1016/S0166-4972(98)00033-9

Correa, E., Mejía, J. F., Castaño, J. G., Echeverría, F., & Gómez, M. A. (2017). Tribological characterization of electroless Ni–B coatings formed on commercial purity magnesium. *Journal of Tribology*, *139*(5), 051302. doi:10.1115/1.4036169

Courtenay, J. C., Johns, M. A., Galembeck, F., Deneke, C., Lanzoni, E. M., Costa, C. A., & Sharma, R. I. (2017). Surface modified cellulose scaffolds for tissue engineering. *Cellulose (London, England)*, *24*(1), 253–267. doi:10.1007/s10570-016-1111-y

Cox, K., & Echtermeyer, A. (2012). Structural design and analysis of a 10MW wind turbine blade. *Energy Procedia*, *24*, 194–201. doi:10.1016/j.egypro.2012.06.101

Cus, F., & Zuperl, U. (2006). Approach to optimization of cutting conditions by using Artificial Neural network. *Journal of Materials Processing Technology*, *173*(3), 281–290. doi:10.1016/j.jmatprotec.2005.04.123

Czagány, M., Baumli, P., & Kaptay, G. (2017). The influence of the phosphorous content and heat treatment on the nano-micro-structure, thickness and micro-hardness of electroless Ni-P coatings on steel. *Applied Surface Science*, *423*, 160–169. doi:10.1016/j.apsusc.2017.06.168

Czop, P. et al.. (2011). A Computational Fluid Flow Analysis of a Disc Valve System. *Journal of KONES Powertrain and Transport*, *18*(1).

Danese, P., & Filippini, R. (2010). Modularity and the impact on new product development time performance: Investigating the moderating effects of supplier involvement and interfunctional integration. *International Journal of Operations & Production Management*, *30*(11), 1191–1209. doi:10.1108/01443571011087387

Darras, B. M., Khaishesh, M. K., Abu-Faraha, F. K., & Omar, M. A. (2007). Friction stir processing of commercial AZ31 magnesium alloy. *Journal of Materials Processing Technology*, *191*(1-3), 77–81. doi:10.1016/j.jmatprotec.2007.03.045

Darvishi, M. T., Khani, F., & Aziz, A. (2016). Numerical Investigation For a hyperbolic Annular Fin with Temperature dependent Thermal Conductivity. *Propulsion and Power Research*, *5*(1), 55–62. doi:10.1016/j.jppr.2016.01.005

Das, A., Roy, A. K., & Kumar, K. (2013). Design and Stress Analysis of a Mixed Flow Pump Impeller. *International Journal of Mechanical Engineering and Computer Applications*, *1*(5), 1–7.

Dash Wood, R. J., Grimes, R., Harrison, A. W., & Flower, H. M. (2001). High Strain Rate Super plastic Aluminium Alloys: The Way Forward. *Materials Science Forum*, *339*, 357–359.

Das, M. C., Sarkar, B., & Ray, S. (2012). Decision making under conflicting environmental new MCDM method. *International Journal of Applied Decision Sciences*, *5*(2), 142–162. doi:10.1504/IJADS.2012.046505

Das, S. K., & Sahoo, P. (2011). Tribological characteristics of electroless Ni–B coating and optimization of coating parameters using Taguchi based grey relational analysis. *Materials & Design*, *32*(4), 2228–2238. doi:10.1016/j.matdes.2010.11.028

Dayan, M., & Basarir, A. (2010). Antecedents and consequences of team reflexivity in NPD projects. *Journal of Business and Industrial Marketing*, *25*(1), 18–29. doi:10.1108/08858621011009128

De Giorgi, M., Scialpi, A., Panella, F. W., & De Filippis, L. A. C. (2009). Effect of shoulder geometry on residual stress and fatigue properties of AA6082 fsw joints. *Journal of Mechanical Science and Technology*, *23*(1), 26–35. doi:10.1007/s12206-008-1006-4

Debnath, K., Singh, I., Dvivedi, A., & Kumar, P. (2013). Natural Fibre-Reinforced Polymer Composites for Wind Turbine Blades: Challenges and Opportunities. In Recent Advances in Composite Materials for Wind Turbine Blades. The World Academic Publishing Co. Ltd.

Deb, S. K. (2004). Seismic base isolation – an overview. *Current Science*, *87*(10), 1426–1430.

Demaria, C., Ruiz, E., & Trochu, F. (2007). In-Plane Anistropic Permeability Characterisation of Deformed Woven Fabrics by Unidirectional Injection. Part II: Prediction Model and Numerical Simulations. *Polymer Composites*, *28*(6), 812–827. doi:10.1002/pc.20108

Deng, J. (1989). Introduction to grey system. *Journal of Grey System*, *1*(1), 1–24.

Deogonda, P., & Chalwa, V. N. (2013). Mechanical Property of Glass Fiber Reinforcement Epoxy Composites. *International Journal of Scientific Engineering and Research.*, *1*(4), 6–9.

Der Ger, M., Hou, K. H., Wang, L. M., & Hwang, B. J. (2003). The friction and wear of Ni–P–PTFE composite deposits under water lubrication. *Materials Chemistry and Physics*, *77*(3), 755–764. doi:10.1016/S0254-0584(02)00153-0

Desai, S. M., & Naik, B. R. (2015). Optimum Design on Impeller of Mixed Flow Pump using CFD Simulation. *International Journal of Research in Engineering & Technology*, *4*(8), 535–537.

Dillon, T. A., Lee, R. K., & Matheson, D. (2005). Value Innovation: Passport to wealth creation. *Research Technology Management, 48*(2), 22-36.

Dixon, J. C. (2007). *The Shock Absorber Handbook* (2nd ed.). West Sussex, UK: Professional Engineering Publishing Ltd. and John Wiley and Sons Ltd. doi:10.1002/9780470516430

Doan, T. L., Brodowsky, H., & Mäder, E. (2012). Jute fibre/epoxy composites: Surface properties and interfacial adhesion. *Composites Science and Technology*, *72*(10), 1160–1166. doi:10.1016/j.compscitech.2012.03.025

Doniavi, A., Eskandarzade, M., Abdi, A., & Totonchi, A. (2008). Empirical Modelling of EDM Parameters Using Grey Relational Analysis. *Asian Journal of Scientific Research*, *1*(5), 502–509. doi:10.3923/ajsr.2008.502.509

Dornfeld, D., Min, S., & Takeuchi, Y. (2006). Recent Advances in Mechanical Micromachining. *Annals of the CIRP*, *55*(2), 1–25. doi:10.1016/j.cirp.2006.10.006

Duari, S., Barman, T. K., & Sahoo, P. (2014). Study of tribological properties of Ni-P coatings under lubrication. *IOSR Journal of Mechanical and Civil Engineering*, *6*, 22–26.

Duari, S., Mukhopadhyay, A., Barman, T. K., & Sahoo, P. (2016). Optimization of wear performance of electroless Ni-B coating under lubrication. *International Journal of Engineering and Technologies*, *7*, 94–103. doi:10.18052/www.scipress.com/IJET.7.94

Duari, S., Mukhopadhyay, A., Barman, T. K., & Sahoo, P. (2017). Study of wear and friction of chemically deposited Ni-P-W coating under dry and lubricated condition. *Surfaces and Interfaces*, *6*, 177–189. doi:10.1016/j.surfin.2017.01.009

Eilam, E. (2005). *Reversing:Secrets of Reverse Engineering*. Indianapolis, IN: John Wiley & Sons.

Elangovan, K., & Balasubramanian, V. (2008). Influences of tool pin profile and welding speed on the formation of friction stir processing zone in AA2219 aluminium alloy. *Journal of Materials Processing Technology*, *200*(1-3), 163–175. doi:10.1016/j.jmatprotec.2007.09.019

El-Sayed, A. A., El-Sherbiny, M. G., Abo-El-Ezz, A. S., & Aggag, G. A. (1995). Friction and wear properties of polymeric composite materials for bearing applications. *Wear*, *184*(1), 45–53. doi:10.1016/0043-1648(94)06546-2

El-Tayeb, N. S. M., & Mostafa, I. M. (1996). The effect of laminate orientations on friction and wear mechanisms of glass reinforced polyester composite. *Wear*, *195*(1-2), 186–191. doi:10.1016/0043-1648(95)06849-X

EngineeringToolbox.com. (n.d.). *ISO Grade Oils - Viscosities and Densities* [Table]. Retrieved August 2015, from http://www.engineeringtoolbox.com/iso-grade-oil-d_1207.html

Etcheverry, M., & Barbosa, E. (2012). Glass Fiber Reinforced Polypropylene Mechanical Properties Enhancement by Adhesion Improvement. *Materials (Basel)*, *5*(12), 1084–1113. doi:10.3390/ma5061084 PMID:28817025

Fabricio, J. P., Joao, R. F., Messias, B. S., Anderson, P. P., & Pedro, P. B. (2010). Artificial neural networks for machining processes surface roughness modeling. *International Journal of Advanced Manufacturing Technology*, *49*(9-12), 879–902. doi:10.1007/s00170-009-2456-2

Farrokhi, H., Heidarzadeh, A., & Saeid, T. (2013). Friction stir welding of copper under different welding parameters and media. *Science and Technology of Welding and Joining*, *18*(8), 697–702. doi:10.1179/1362171813Y.0000000148

Feng, A. H., & Ma, Z. Y. (2007). Enhanced mechanical properties of Mg–Al–Zn cast alloy via friction stir processing. *Scripta Materialia*, *56*(5), 397–400. doi:10.1016/j.scriptamat.2006.10.035

Feng, A. H., & Ma, Z. Y. (2007). Formation of Cu2FeAl7 phase in friction stir welded Si Cp/Al-Cu-Mg composite. *Scripta Materialia*, *57*(12), 1113–1116. doi:10.1016/j.scriptamat.2007.08.020

Ferber, R., Davis, I. M., & Williams, D. S. III. (2005). Effect of foot orthotics on rear foot and tibia joint coupling patterns and variability. *Journal of Biomechanics*, *38*(3), 477–483. doi:10.1016/j.jbiomech.2004.04.019 PMID:15652545

Fracchia, C. A., Castro, J., & Tucker, C. L. (1989). A Finite Element/Control Volume Simulation of Resin Transfer Mould Filling. *Proceedings of the American Society for Composites, Fourth Technical Conference*, 157-166.

Franco, M., Sha, W., Aldic, G., Malinov, S., & Çimenoğlu, H. (2016). Effect of reinforcement and heat treatment on elevated temperature sliding of electroless Ni–P/SiC composite coatings. *Tribology International*, *97*, 265–271. doi:10.1016/j.triboint.2016.01.047

Fujii, H., Cui, L., Maeda, M., & Nogi, K. (2006). Effect of tool shape on mechanical properties and microstructure of friction stir welded aluminum alloys'. *Materials Science and Engineering A*, *419*(1–2), 25–31. doi:10.1016/j.msea.2005.11.045

Fujii, H., Cui, L., Maeda, M., Sato, Y. S., & Nogi, K. (2006). Effect of threads on tool in friction stir welding of aluminum alloys. *Adv. Struct. Funct.Mater.Des. Proc.*, *512*, 389–394.

Fu, S. U., Feng, X. Q., Lauke, B., & Mai, Y. W. (2008). Effects of particle size, particle/matrix interface adhesion and particle loading on mechanical properties of particulate-polymer composites. *Composites. Part B, Engineering*, *39*(6), 933–961. doi:10.1016/j.compositesb.2008.01.002

Fu, Y. K. (2010). New Product success among SME's: An empirical study in Taiwan. *Journal of International Management Studies*, *5*(1), 147153.

Gaba, V. K., Tiwari, A. K., & Bhowmick, S. (2014). Thermal Performance of Functionally Graded Parabolic Annular Fins Having Constant Weight. *Journal of Mechanical Science and Technology*, *28*(10), 4309–4318. doi:10.1007/s12206-014-0945-1

Gaba, V. K., Tiwari, A. K., & Bhowmick, S. (2015). A Parametric Study of Functionally Graded Rotating Annular Fin. *Procedia Engineering*, *127*, 126–132. doi:10.1016/j.proeng.2015.11.436

Gadhari, P., & Sahoo, P. (2016c). Effect of TiO_2 particles on micro-hardness, corrosion, wear and friction of Ni–P–TiO_2 composite coatings at different annealing temperatures. *Surface Review and Letters, 23*, 1550082. DOI: 10.1142/S0218625X15500821

Gadhari, P., & Sahoo, P. (2016a). Electroless nickel-phosphorus composite coatings: A review. *International Journal of Manufacturing, Materials, and Mechanical Engineering*, *6*(1), 14–50. doi:10.4018/IJMMME.2016010102

Gadhari, P., & Sahoo, P. (2016b). *Effect of annealing temperature and alumina particles on mechanical and tribological properties of Ni-P-Al_2O_3 composite coatings.* Silicon. doi:10.1007/s12633-016-9452-6

Gahlot, V. K., & Nyiri, A. (1993). Impeller Pumps. Theory and design. M.A.C.T.

Gaidhani, Y.B. & Kalamani, V.S. (2013). Abrasive water jet review and parameter selection by AHP method. *IOSR Journal of Mechanical and Civil Engineering, 8*(5), 1-6.

Gaitonde, V. N., & Karnik, S. R. (2012). Minimizing burr size in drilling using Artificial Neural Network- Particle Swarm optimization approach (PSO). *Journal of Intelligent Manufacturing*, *23*(5), 1783–1793. doi:10.1007/s10845-010-0481-5

Gan, Y. X., Solomon, D., & Reinbolt, M. (2010). Friction Stir Processing of Particle Reinforced Composite Materials. *Materials (Basel)*, *3*(1), 329–350. doi:10.3390/ma3010329

Gene^t, F., Schnitzler, A., Mathieu, S., Autret, K., Théfenne, L., Dizien, O., & Maldjian, A. (2010). Orthotic devices and gait in polio patients. *Annals of Physical and Rehabilitation Medicine*, *53*(1), 51–59. doi:10.1016/j.rehab.2009.11.005 PMID:20022835

Gerwin, D., & Ferris, S. J. (2004). Organising New Product Development Projects in Strategic Alliances. *Organisation Science, 15*(1), 22-38.

Ghaderi, M., Rezagholizadeh, M., Vaghefi, S. M., & Heidary, A. (2016). Investigation of high temperature wear resistance of electroless nickel coating with different contents of phosphorous. *Protection of Metals and Physical Chemistry of Surfaces*, *52*(3), 538–542. doi:10.1134/S2070205116030126

Gibson, R. F. (2010). A review of recent research on mechanics of multifunctional composite materials and structures. *Composite Structures*, *92*(12), 2793–2810. doi:10.1016/j.compstruct.2010.05.003

Giovacchini, F., Vannetti, F., Fantozzi, M., & Cempinia, M. (2015). A light-weight active orthosis for hip movement assistance. *Robotics and Autonomous Systems*, *73*, 123–134. doi:10.1016/j.robot.2014.08.015

Goffin, K., & Micheli, P. (2010). Maximising the value of Industrial Design in NPD. *Research Technology Management*, *51*, 29–37.

Gorla, R. S. R., & Bakier, A. Y. (2011). Thermal analysis of natural convection and radiation in porous fins. *International Communications in Heat and Mass Transfer*, *38*(5), 638–645. doi:10.1016/j.icheatmasstransfer.2010.12.024

Goswami & Kumar. (2014). Investigation of surface integrity, material removal rate and wire wear ratio for WEDM of Nimonic 80A alloy using GRA and Taguchi method. *Engineering Science and Technology, an International Journal, 17*, 173-184.

Govardhan & Paul, D. K. (2016). Effect of Lead in Elastomeric Bearings for Structures Located in Seismic Region. *Procedia Technology, 25*, 146 – 153.

Govardhan, Paul, D.K., & Singh, Y. (2014). Non-iterative bilinear properties of structural bearings, IALCCE 2014 symposium. Tokyo, Japan: Waseda University.

Griffin, A. (1997). PDMA research on new product development practices. *Journal of Product Innovation Management*, *14*(6), 429–458. doi:10.1016/S0737-6782(97)00061-1

Griffin, D. A., & Ashwill, T. D. (2003). Alternative Composite Materials for Megawatt-Scale Wind Turbine Blades: Design Considerations and Recommended Testing. *Journal of Solar Energy Engineering*, *125*(4), 515. doi:10.1115/1.1629750

Gunawan, S. P., Tutik, S., & Muslim, M. (2012). Improvement of Machining time through workpiece vibration and graphite powder mixed in dielectric fluid. *Indian Journal of Engineering and Materials Sciences*, *19*, 375–378.

Gupta, T. V. K., Ramkumar, J., Tandon, P., & Vyas, N. S. (2012). Application of Grey Relational Analysis for Geometrical Characteristics in Abrasive Water Jet Mille Channels. *5th International & 26th All India Manufacturing Technology, Design and Research Conference, 393*, 1-6.

Gyenes, L., & Mitchell, C. G. B. (1994). *The effect of vehicle-road interaction on fuel consumption. Vehicle-Road Interaction*. ASTM International.

Han, D., & Mungal, M. G. (2001). Direct measurement of entrainment in reacting/nonreacting turbulent jets. *Combustion and Flame*, *124*(3), 370–386. doi:10.1016/S0010-2180(00)00211-X

Harborne, P., & Johne, A. (2003). Creating a project climate for successful product innovation. *European Journal of Innovation Management*, *6*(2), 118–133. doi:10.1108/14601060310475273

Hart, S. J. (1989). Design orientation and market success. In The marketing initiative – Economic and Social Research Council (ESRC) studies in British Marketing. Prentice Hall. doi:10.1016/0142-694X(89)90043-4

Hart, S. J. (1995). Where we've been and where we're going in new product development research. In Product development – Meeting the challenge of the design-marketing interface. John Wiley and Sons.

Hart, S. J., & Baker, M. J. (1996). The multiple convergent processing model of new product development. In New Product Development – A reader. The Dryden Press.

Hart, S. J., Hutlink, E., Tzokas, N., & Commandeur, H. (2003). Industrial Companies' Evaluation Criteria in new Product Development gates. *Journal of Product Innovation Management*, *20*(1), 22–36. doi:10.1111/1540-5885.201003

Hashish, M. (1984). A modeling study of metal cutting with abrasive water jets. *Journal of Engineering Materials and Technology*, *106*(1), 88–100. doi:10.1115/1.3225682

Hashish, M. (1991). Advances in composite machining with abrasive waterjets. *ASME Process Manufacturing Computer Materials*, *49*(27), 93–111.

Hatami, M., & Ganji, D. D. (2014). Thermal behavior of longitudinal convective – radiative porous fins with different section shapes and ceramic materials (SiC and Si3N4). *Ceramics International*, *40*(5), 6765–6775. doi:10.1016/j.ceramint.2013.11.140

Hattingh, D. G., Blignault, C., van Niekerk, T. I., & James, M. N. (2008). Characterization of the influences of FSW tool geometry on welding forces and weld tensile strength using an instrumented tool. *Journal of Materials Processing Technology*, *203*(1–3), 46–57. doi:10.1016/j.jmatprotec.2007.10.028

Havener, C., & Thorpe, M. (1994). Customers can tell you what they want. *Management Review*, *14*(12), 42–45.

Haykin, S. (1999). *Introduction. In Neural networks and learning machines* (3rd ed.). Pearson Prentice Hall, New Jersy.

Hebb, D. O. (1949). *The organization of behavior A Neuropsychological theory*. New York: John Wiley& Sons,Inc.

Herrera-Franco, P. J., & Valadez-Gonza'lez, A. (2004). Mechanical properties of continuous natural fibre-reinforced polymer composites. *Composites. Part A, Applied Science and Manufacturing*, *35*(3), 339–345. doi:10.1016/j.compositesa.2003.09.012

Hewidy, M. S., El-Taweel, T. A., & El-Safty, M. F. (2005). Modelling the machining parameters of wire electrical discharge machining of Inconel 601 using RSM. *Journal of Materials Processing Technology*, *169*(2), 328–336. doi:10.1016/j.jmatprotec.2005.04.078

Hill, B. J. (1972). Measurement of local entrainment rate in the initial region of axisymmetric turbulent air jets. *Journal of Fluid Mechanics*, *51*(04), 773–779. doi:10.1017/S0022112072001351

Himmelfarb, P. A. (1992). *Survival of the fittest – New product development during the 90's*. Englewood Cliffs, NJ: Prentice-Hall International.

Hodowany, J., Ravichandran, G., Rosakis, A. J., & Rosakis, P. (2000). Partition of plastic work into heat and stored energy in metals. *Experimental Mechanics*, *40*(2), 113–123. doi:10.1007/BF02325036

Ho, K. H., & Newman, S. T. (2003). State of the art electrical discharge machining (EDM). *International Journal of Machine Tools & Manufacture*, *43*(13), 1287–1300. doi:10.1016/S0890-6955(03)00162-7

Hollins, W., & Pugh, S. (1990). *Successful product design*. Butterworth.

Holmes, J. W., Brøndsted, P., Sørensen, B. F., Jiang, Z. H., Sun, Z. H., & Chen, X. H. (2009). Development of a Bamboo-Based Composite as a Sustainable Green Material for Wind Turbine Blades. *Journal of Wind Engineering*, *33*(2), 197–210. doi:10.1260/030952409789141053

Holstrom, J., Korhonen, H., Laiho, A., & Hartial, H. (2006). Managing product introductions across the supply chain: Findings from a development project. *Supply Chain Management*, *11*(2), 121–130. doi:10.1108/13598540610652519

Holt, K., Geschka, H., & Peterlongo, G. (1984). *Need assessment – A key to user orientated product innovation*. John Wiley and Sons Ltd.

Hong, Y., Chunmei, W., Wujun, F., & Xuyang, C. (2017). Analysis of mechanism based on two types of pulse generators in micro-EDM using single pulse discharge. *International Journal of Advanced Manufacturing Technology*, *89*(9), 3217–3230.

Horlock, J. H. (1958). *Axial Flow Compressor –Fluid mechanics and thermodynamics*. London: Butterworth Scientific Publications.

Houdaille, M. (1908). *Improvements in or Relating To Shock-Absorbing or Dash-Pot Apparatus for Motor Vehicles and the Like*. France Patent number FR394081-A.

Houdaille, M. (1909). *Shock-absorbing apparatus*. United States patent number US933076 A.

Houman, L. (1983). Total EDM. In E. C. Jameson (Ed.), *Electrical Discharge Machining: Tooling, Methods and Applications* (pp. 5–19). Dearborn, MI: Society of Manufacturing Engineers.

Hsu, Y. (2006). Comparative study of product design strategy and related design issues. *Journal of Engineering Design*, *17*(4), 357–370. doi:10.1080/09544820600664655

Huang, C. H., & Chung, Y. L. (2016). A nonlinear fin design problem in determining the optimum shapes of fully wet annular fins. *Applied Thermal Engineering*, *103*, 195–204. doi:10.1016/j.applthermaleng.2016.03.066

Hu, C., & Baker, T. N. (1997). A new aluminium silicon carbide formed in laser processing. *Journal of Materials Science*, *32*, 504–764.

Hultink, E. J., & Hart, S. (1998). The world's path to the better mousetrap: Myth or reality? An empirical investigation into the launch strategies of high and low advantage new products. *European Journal of Innovation Management*, *1*(3), 106–121. doi:10.1108/14601069810230199

Huo, L. A., Lan, J., & Wang, Z. (2011). New parametric prioritization methods for an analytical hierarchy process based on a pairwise comparison matrix. *Mathematical and Computer Modelling*, *54*(11-12), 2736–2749. doi:10.1016/j.mcm.2011.06.062

Hur, K. H., Jeong, J. H., & Lee, D. N. (1990). Microstructures and crystallization of electroless Ni-P deposits. *Journal of Materials Science*, *25*(5), 2573–2584. doi:10.1007/BF00638061

Hutyrová, Z., Ščučka, J., Hlaváček, H. P., & Zeleňák, M. (2015). Turning of wood plastic composites by water jet and abrasive water jet. *International Journal of Advanced Manufacturing Technology*. doi:10.1007/s00170-015-7831-6

Ishizaki, M. H., Visconte, L. L., Furtado, C. R., Oliveira, M. G., & Leblanc, J. L. (2008). Characterization of Green Coconut Fiber Composites. *International Journal of Polymer Analysis and Characterization*, *13*(5), 319–330. doi:10.1080/10236660802351409

Islam, A. B. M. S., Jumaat, M. Z., Hussain, R., & Alam, M. A. (2013). Incorporation of rubber-steel bearing isolation in multi-storey building. *Journal of Civil Engineering and Management*, *19*(sup1), 33–49. doi:10.3846/13923730.2013.801904

Jafarzadeh, B., Hajari, A., Alishahi, M. M., & Akbari, M. H. (2011). The flow simulation of a low-specific-speed high-speed centrifugal pump. *Applied Mathematical Modelling*, *35*(1), 242–249. doi:10.1016/j.apm.2010.05.021

Jagadish. Bhowmik, S., & Ray, A. (2015). Prediction of surface roughness quality of green abrasive water jet machining: A soft computing approach. *Journal of Intelligent Manufacturing*. doi:10.1007/s10845-015-1169-7

Jagadish, B. S., & Ray, A. (2015). Prediction and optimization of process parameters of green composites in AWJM process using response surface methodology. *International Journal of Advanced Manufacturing Technology*. doi:10.1007/s00170-015-8281-x

Jagannatha & Harish. (2015). Mechanical properties of carbon/glass fiber reinforced epoxy hybrid polymer composites. *International Journal of Mechanical Engineering and Robotics Research*, *4*(2), 131 – 137.

Jahan, M. P. (2013). Micro-Electrical Discharge Machining. In J.P. Davim (Ed.), Non traditional Machining Processes. Springer-Verlag.

Jahan, M. P., Kakavand, P., Kwang, E. L. M., Rahman, M., & Wong, Y. S. (2015). An experimental investigation into the micro-electro-discharge machining behavior of aluminium alloy (AA 2024). *International Journal of Advanced Manufacturing Technology*, *78*(5-8), 1127–1139. doi:10.1007/s00170-014-6712-8

Jameson, E. C. (2001). Description and development of electrical discharge machining (EDM). In *Electrical Discharge Machining* (Vol. 12). Dearborn, MI: Society of Manufacturing Engineers.

Jana, S., Mishra, R. S., Baumann, J. B., & Grant, G. (2009). Effect of stress ratio on the fatigue behavior of a friction stir processed cast Al-Si-Mg alloy. *Scripta Materialia*, *61*(10), 992–995. doi:10.1016/j.scriptamat.2009.08.011

Jang, S., Yoon, Y., Lee, I., & Kim, J. (2009). Design Orientated NPD. *Research Technology Management*, 36–46.

Jenkins, S., Forbes, S., Durrani, T. S., & Banerjee, S. K. (1997). Managing the product development process [Part I: An assessment]. *International Journal of Technology Management*, *13*(4), 359–378. doi:10.1504/IJTM.1997.001670

Jie, L. (2011). *Environmental Effects of Vehicle Exhausts, Global and Local Effects – A Comparison between Gasoline and Diesel* (Published Master thesis of Applied Environmental Science). Halmstad University.

Jobber, D. (2009). *Principles and Practice of Marketing* (6th ed.). McGraw-Hill Education.

John, A. (2009). Fast numerical method for flow analysis and blade design in centrifugal pump impellers. *Computers & Fluids*, *32*, 284–289.

Johnsen, T. E. (2011). Supply network delegation and intervention strategies during supplier involvement in new product development. *International Journal of Operations & Production Management*, *31*(6), 686–708. doi:10.1108/01443571111131999

Jonker, J. B., & Essen, V. G. T. (1997). A finite element perturbation method for computing fluid induced forces on a centrifugal impeller, rotating and whirling in a volute casing. *International Journal for Numerical Methods in Engineering*, *40*(2), 269–294. doi:10.1002/(SICI)1097-0207(19970130)40:2<269::AID-NME63>3.0.CO;2-G

Jorgensen, B., & Messner, M. (2009). Management control in NPD: The dynamics of managing flexibility and efficiency. *Journal of Management Accounting Research*, *21*, 99–124. doi:10.2308/jmar.2009.21.1.99

Joseph, S., Joseph, K., & Thomas, S. (2006). Green Composites from Natural Rubber and Oil Palm Fiber: Physical and Mechanical Properties. *International Journal of Polymeric Materials*, *55*(11), 925–945. doi:10.1080/00914030600550505

Joshi, S. N., & Pande, S. S. (2009). Development of an intelligent process model for Electro-Discharge Machining. *International Journal of Advanced Manufacturing Technology*, *45*, 300–317. doi:10.1007/s00170-009-1972-4

Kahn, K. B., Barczak, G., & Moss, R. (2006). Establishing an NPD best practice framework. *Journal of Product Innovation Management*, *23*(2), 106–116. doi:10.1111/j.1540-5885.2006.00186.x

Kaipia, R., & Holmstrom, J. (2007). Selecting the right planning approach for a product. *Supply Chain Management*, *12*(1), 3–13. doi:10.1108/13598540710724347

Kalyanaram, G., & Krishnan, V. (1997). Deliberate product definition: Customising the product definition process. *JMR, Journal of Marketing Research*, *34*(2), 276–286. doi:10.2307/3151864

Kaneko, S., & Hayashi, I. (2014). Pressure pulsations in piping system excited by a centrifugal turbo machinery taking the damping characteristics into consideration. *Journal of Fluids and Structures*, *45*, 216–234. doi:10.1016/j.jfluidstructs.2013.11.012

Kao, J. Y., Tsao, C. C., Wang, S. S., & Hsu, C. Y. (2010). Optimization of the EDM parameters on machining Ti–6Al–4V with multiple quality characteristics. *International Journal of Advanced Manufacturing Technology*, *47*(1-4), 395–402. doi:10.1007/s00170-009-2208-3

Kapoor, R., Kumar, N., Mishra, R. S., Huskamp, C. S., & Sankaran, K. K. (2010). Influence of fraction of high angle boundaries on the mechanical behavior of an ultra fine grained Al-Mg alloy. *Materials Science and Engineering A*, *527*(20), 5246–5254. doi:10.1016/j.msea.2010.04.086

Kapoor, R., & Nemat-Nasser, S. (1998). Determination of temperature rise during high strain-rate deformation. *Mechanics of Materials*, *27*(1), 1–12. doi:10.1016/S0167-6636(97)00036-7

Karande, P., & Chakraborty, S. (2012). Application of multi-objective optimization on the basis of ratio analysis (MOORA) method for materials selection. *Materials & Design*, *37*(1), 317–324. doi:10.1016/j.matdes.2012.01.013

Karthikeyam, L., & Senthilkumar, V. S. (2011). Relationship between process parameters and mechanical properties of friction stir processed AA 6063-T3 aluminium alloy. *Materials & Design*, *32*(5), 3085–3091. doi:10.1016/j.matdes.2010.12.049

Kastner, J., Plank, B., Salaberger, D., & Sekelja, J. (2010). Defect and Porosity Determination of Fibre Reinforced Polymers by X-ray Computed Tomography. *2nd International Symposium on NDT in Aerospace 2010*, 1 – 12.

Kawanami, S. (2010). Analysis of effect of tool geometry on plastic flow during friction stir spot welding using particle method. *Journal of Materials Processing Technology*, *210*(11), 1455–1463. doi:10.1016/j.jmatprotec.2010.04.003

Keith Hargrove, S., & Ding, D. (2007). Determining cutting parameters in wire EDM based on workpiece surface temperature distribution. *International Journal of Advanced Manufacturing Technology*, *34*(3-4), 295–299. doi:10.1007/s00170-006-0609-0

Kelly, J. M. (1997). *Earthquake-Resistant Design with Rubber* (2nd ed.). London: Springer-Verlag. doi:10.1007/978-1-4471-0971-6

Kikuyama, K., Hasegawa, H., & Maeda, T. (1992). Unsteady pressure change in centrifugal pump impeller passages due to inlet swirl. *Journal of Fluids and Structures*, *6*(3), 337–351. doi:10.1016/0889-9746(92)90013-S

Kinsella. (2000). Innovative Automotive Applications Based on ZenTron® High Strength Glass Fiber. *Proc.* 45th*Science of Advanced Materials and Process Engineering Series, 45*, 1979 – 1994.

Kleinschmidt, E. J., de Brentani, U., & Salomo, S. (2007). Performance of global new product development programs: A resource based view. *Journal of Product Innovation Management*, *24*(5), 419–441. doi:10.1111/j.1540-5885.2007.00261.x

Kou, H., Lee, J., & Lai, C. (2005). Thermal analysis of a longitudinal fin with variable thermal properties by recursive formulation. *International Journal of Heat and Mass Transfer*, *48*(11), 2266–2277. doi:10.1016/j.ijheatmasstransfer.2004.12.038

Krasteva, N., Fotty, V., & Armyanov, S. (1994). Thermal stability of Ni-P and Ni-Cu-P amorphous alloys. *Journal of the Electrochemical Society*, *141*(10), 2864–2867. doi:10.1149/1.2059246

Kröling, H., Fleckenstein, J., Nubbo, N., Endres, A., Miletzky, F., & Schabel, S. (2014). Non-Woven and Paper Based Epoxy Composites. *Science and Technology DAS PAPIER*, *6*(7), 2 – 5.

Ku, H., Wang, H., Pattarachaiyakoop, N., & Trada, M. (2011). A review on the tensile properties of natural fiber reinforced polymer composites. *Composites. Part B, Engineering*, *42*(4), 856–873. doi:10.1016/j.compositesb.2011.01.010

Kumar, D., & Bhingole, P. P. (2015). CFD Based Analysis of Combined Effect of Cavitation and Silt Erosion on Kaplan Turbine. *Materials Today: Proceedings*, *2*(4-5), 2314–2322. doi:10.1016/j.matpr.2015.07.276

Kumar, E. A., Balakrishna, M. V., & Chandra, M. R. (2015). Effect of Shim Percentage on Non-Linear Static Behavior of Flex Seal of Rocket Nozzle. *Materials Today: Proceedings*, *2*(4-5), 1427–1434. doi:10.1016/j.matpr.2015.07.063

Kumar, K., & Agarwal, S. (2012). Multi-objective parametric optimization on machining with wire electric discharge machining. *International Journal of Advanced Manufacturing Technology*, *62*(5-8), 617–633. doi:10.1007/s00170-011-3833-1

Kumar, K., Kailas, S. V., & Srivatsan, T. S. (2008). Influence of tool geometry in friction stir welding. *Materials and Manufacturing Processes*, *23*(2), 189–195. doi:10.1080/10426910701774734

Kumar, N., Mishra, R. S., Huskamp, C. S., & Sankaran, K. K. (2011). Microstructure and mechanical properties of friction stir processed ultra fine grained Al-Mg-Sc alloy. *Materials Science and Engineering A*, *528*(18), 5883–5887. doi:10.1016/j.msea.2011.03.109

Kumar, R., Bhowmik, S., & Kumar, K. (2017). Establishment and Effect of Constraint on Different Mechanical Properties of Bamboo Filler Reinforced Epoxy Composite. *International Polymer Processing*, *32*(3), 308–315. doi:10.3139/217.3311

Kumar, R., Kumar, K., Sahoo, P., & Bhowmik, S. (2014). Study of Mechanical Properties of Wood Dust Reinforced Epoxy Composite, *Procedia. Materials Science*, *6*, 551–556.

Kumbhalkar Manoj, A., Bhope Dr, D. V., & Vanalkar Dr, A. V. (2015). Material and Stress Analysis of Railroad Vehicle Suspension: A Failure Investigation. *Procedia Materials Science*, *10*, 331–343. doi:10.1016/j.mspro.2015.06.064

Kundu, S., Das, S. K., & Sahoo, P. (2016). *Tribological behaviour of electroless Ni-P deposits under elevated temperature*. Silicon. doi:10.1007/s12633-016-9450-8

Kung, K. Y., Horng, J. T., & Chiang, K. T. (2009). Material removal rate and electrode wear ratio study on the powder mixed electrical discharge machining of cobalt-bonded tungsten carbide. *International Journal of Advanced Manufacturing Technology*, *40*(1-2), 95–104. doi:10.1007/s00170-007-1307-2

Kuriachen, B., & Mathew, J. (2015). Effect of Powder Mixed Dielectric on Material Removal and Surface Modification in Micro Electric Discharge Machining of Ti-6Al-4V. *Materials and Manufacturing Processes*, 1–30.

La Mantia, F. P., & Morreale, M. (2011). Green composites: A brief review. *Composites. Part A, Applied Science and Manufacturing*, *42*(6), 579–588. doi:10.1016/j.compositesa.2011.01.017

Lai, C.-Y., Kou, H.-S., & Lee, J.-J. (2009). Recursive formulation on thermal analysis of an annular fin with variable thermal properties. *Applied Thermal Engineering*, *29*(4), 779–786. doi:10.1016/j.applthermaleng.2008.04.007

Lang, H. H. (1997). *A Study of the Characteristics of Automotive Hydraulic Dampers at High Stroking Frequencies. Ph.D.* Ann Arbor, MI: University of Michigan.

Lee, C. J., Huang, J. C., & Hsieh, P. J. (2006). Mg based nano-composites fabricated by friction stir processing. *Scripta Materialia*, *54*(7), 1415–1420. doi:10.1016/j.scriptamat.2005.11.056

Lee, C. J., Huang, J. C., & Hsieh, P. L. (2006). Mg based nano-composites fabricated by friction stir processing. *Compos. Mater.*, *313*, 69.

Lee, H., Chou, H., & Yang, Y.-C. (2004). The function estimation in predicting heat flux of pin fins with variable heat transfer coefficients. *Energy Conversion and Management*, *45*(11-12), 1749–1758. doi:10.1016/j.enconman.2003.09.017

Lemeš, S., & Zaimović-Uzunović, N. (2002). Mode shapes of centrifugal pump impeller. *Trends in the Development of Machinery and Associated Technology B&H*, 18-22.

Lienert, T. J., Stellwag, W. L. Jr., Grimmett, B. B., &Warke, R. W. (2003). Friction Stir Welding Studies on Mild Steel. *Welding Journal*, 48-59.

Liew, K. M., Lei, Z. X., & Zhang, L. W. (2015). Mechanical analysis of functionally graded carbon nanotube reinforced composites: A review. *Composite Structures*, *120*, 90–97. doi:10.1016/j.compstruct.2014.09.041

Lim, S. T., & Lee, W. I. (2000). An Analysis of the Three-Dimensional Resin Transfer Mould Filling Process. *Composites Science and Technology*, *60*(7), 961–975. doi:10.1016/S0266-3538(99)00160-8

Li, N., Zhou, Q., Chen, X., Xu, T., Hui, S., & Zhang, D. (2008). Liquid drop impact on solid surface with application to water drop erosion on turbine blades, Part I: Nonlinear wave model and solution of one-dimensional impact. *International Journal of Mechanical Sciences*, *50*(10-11), 1526–1542. doi:10.1016/j.ijmecsci.2008.08.001

Lin, J. L., & Lin, C. L. (2002). The use of the orthogonal array with gray relational analysis to optimize the electrical discharge machining process with multiple performance characteristics. *International Journal of Machine Tools & Manufacture*, *42*(2), 237–244. doi:10.1016/S0890-6955(01)00107-9

Lin, M., & Hahn, T. (1996). A Robust and Efficient Approach for RTM Simulation. *Advanced Materials*, *74*, 75–76.

Lin, Y., & Lee, H. (2009). Optimization of machining parameters using magnetic-force-assisted EDM based on gray relational Analysis. *International Journal of Advanced Manufacturing Technology*, *42*(11-12), 1052–1064. doi:10.1007/s00170-008-1662-7

Li, T. (1999). The impact of the marketing-R&D interface on new product export performance: A contingency analysis. *Journal of International Marketing*, *7*(1), 10–34.

Liu, D., Tang, Y., & Cong, W. L. (2012). A review of mechanical drilling for composite laminates. *Composite Structures*, *94*(4), 1265–1279. doi:10.1016/j.compstruct.2011.11.024

Liu, F. C., & Ma, Z. Y. (2008, October). Achieving exceptionally high superplasticity at high strain rates in a micro grained Al–Mg–Sc alloy produced by friction stir processing. *Scripta Materialia*, *59*(8), 882–885. doi:10.1016/j.scriptamat.2008.06.035

Liu, F., Zhang, H., & Cheng, X. H. (1999). A decision-making framework model for green manufacturing and the case study. *Journal of Mechanical Engineering*, *35*(5), 11–15.

Liu, H., Lv, Y. Y., Liu, Z., & Thompson, G. E. (2016). Dry sliding wear behaviour and structural characteristics of laser-annealed electroless Ni–P/Ni–Mo–P duplex coatings. *Tribology International*, *103*, 343–351. doi:10.1016/j.triboint.2016.07.014

Liu, J., Wang, X., Tian, Z., Yuan, M., & Ma, X. (2015). Effect of copper content on the properties of electroless Ni-Cu-P coatings prepared on magnesium alloys. *Applied Surface Science*, *356*, 289–293. doi:10.1016/j.apsusc.2015.08.072

Liu, Y., & Zhao, Q. (2004). Study of electroless Ni–Cu–P coatings and their anti-corrosion properties. *Applied Surface Science*, *228*(1), 57–62. doi:10.1016/j.apsusc.2003.12.031

Li, Z., Wang, J., Lu, J., & Meng, J. (2013). Tribological characteristics of electroless Ni–P–MoS_2 composite coatings at elevated temperatures. *Applied Surface Science*, *264*, 516–521. doi:10.1016/j.apsusc.2012.10.055

Li, Z., Zeng, Y., Liu, X., & Wang, H. (2012). Optimum design on impeller blade of mixed-flow pump based on CFD. *Procedia Engineering*, *31*, 187–195. doi:10.1016/j.proeng.2012.01.1011

Lorrain, O., Favier, V., Zahrouni, H., & Lawrjaniec, D. (2010). Understanding the material flow path of friction stir welding process using unthreaded tools. *Journal of Materials Processing Technology*, *210*(4), 603–609. doi:10.1016/j.jmatprotec.2009.11.005

Loto, C. A. (2016). Electroless nickel plating–A review. *Silicon*, *8*(2), 177–186. doi:10.1007/s12633-015-9367-7

MacCormack, A., Crandall, W., Henderson, P., & Toft, P. (2012). Do you need a NPD strategy? Aligning process with context. *Research Technology Management*, (Jan-Feb): 34–43. doi:10.5437/08956308X5501014

Machining Data Center. (1980). Cincinnati, OH: Machining Data Handbook, Inc.

Mahendra, S., Devarajan, R., Nagarajan, T., & Majdi, A. (2010). A review of Micro-EDM. *Proceedings of International Multiconference of Engineers & computer scientists*, 2.

Maher, I., Sarhan, A. A. D., & Hamdi, M. (2015). Review of improvements in wire electrode properties for longer working time and utilization in wire EDM machining. *International Journal of Advanced Manufacturing Technology*, *76*(1-4), 329–351. doi:10.1007/s00170-014-6243-3

Mahoney, M. W., Binget, W. H., & Mishra, R. S. (2003). Microstructural Modification and Resultant Properties of Friction Stir Processed Cast NiAl Bronze. *Materials Science Forum*, *426*(432), 2843–2853. doi:10.4028/www.scientific.net/MSF.426-432.2843

Majumder, A. (2014). Comparative study of three evolutionary algorithms coupled with neural network model for optimization of electric discharge machining process parameters. *Proceedings of the Institution of Mechanical Engineers, Part B: Journal of Engineering Manufacture*, 1-13.

Mandal, D., Pal, S. K., & Saha, M. (2007). Modelling of EDM process using back propagation neural network and multi-objective optimization using non-dominating sorting genetic algorithm-II. *Journal of Materials Processing Technology*, *186*, 154–162. doi:10.1016/j.jmatprotec.2006.12.030

Manna, A., & Bhattacharyya, B. (2006). Taguchi and Gauss elimination method: A dual response approach for parametric optimization of CNC wire cut EDM of PRAlSiCMMC. *International Journal of Advanced Manufacturing Technology*, *28*(1-2), 67–75. doi:10.1007/s00170-004-2331-0

Markarian, J. (2002). Additive developments aid growth in wood-plastic composites. *Plastics Additives & Compounding*, *4*(11), 18–21. doi:10.1016/S1464-391X(02)80174-X

Markatos, N. C. (1986). The mathematical modelling of turbulent flows. *Applied Mathematical Modelling*, *10*(3), 190–220. doi:10.1016/0307-904X(86)90045-4

Martande, S., Jangale, Y. N., & Motgi, N. S. (2013). Design and Analysis of Shock Absorber. *International Journal of Application or Innovation in Engineering & Management*, *2*(3), 195–199.

Masoumi, F., Ghasemi, H. R., Ziaei, A. A., & Shahriari, D. (2012). Tribological characterization of electroless Ni–10% P coatings at elevated test temperature under dry conditions. *International Journal of Advanced Manufacturing Technology*, *62*(9-12), 1063–1070. doi:10.1007/s00170-011-3832-2

Masuzawa, T. (2000). State of the art micromachining. *Annual CIRP*, *49*(2), 473–488. doi:10.1016/S0007-8506(07)63451-9

Ma, Z. Y., Mishra, R. S., & Mahoney, M. W. (2002). Super Plastic Deformation Behavior of Friction Stir processed 7075 Aluminium alloy. *Acta Materialia*, *50*(17), 4419–4430. doi:10.1016/S1359-6454(02)00278-1

McCulloch, W. S., & Pitts, W. (1943). A logical calculus of the ideas immanent in nervous activity. *Bulletin of Mathematical Biology*, *52*(1/2), 99–115. PMID:2185863

McGeough, J. A. (1988). *Electro Discharge Machining. In Advanced Methods Machine* (p. 130). London: Chapman & Hall.

McGrath, M. E., Anthony, M. T., & Shipiro, A. R. (1992). Product development: Success through cycle time excellence. In *PRTM*. Butterworth-Heinemann Publishing.

Mehul, A. R., & Chirag, P. P. (2013). Parametric Optimization of Magnetic Abrasive Water Jet Machining of AISI 52100 Steel using Grey Relational Analysis. *International Journal of Engineering Research and Applications*, *3*(4), 527–530.

Melngk, S. A., & Smith, R. T. (1996). *Green Manufacturing*. Dearborn, MI: Society of Manufacturing Engineers.

Meredith, J., Bilson, E., & Powe, R. (2015). A performance versus cost analysis of prepreg carbon fibre epoxy energy absorption structures. Composite Structures, 124, 206–213.

Mhamunkar, M. M., & Rau, N. (2016). Optimization of Process Parameter of CNC Abrasive Water Jet Machine for Titanium Ti 6Al 4V material. *International Journal of Advanced Research in Science, Engineering, and Technology.*, *3*(3), 1640–1646.

Michalewicz, Z. (1999). *Genetic algorithms + data structures = evolution programs*. Berlin: Springer-Verlag.

Mingxiong, O., Weidong, S., Weidong, J., & Qiang, F. (2015). Numerical simulation and experimental validation on hydrodynamic radial force of mixed-flow pump impeller. *Nongye Gongcheng Xuebao (Beijing)*, *9*, 71–76.

Miranda, R. M., Gandra, J., & Vilaça, P. (2013). *Surface Modification by Friction Based Processes*. Intech Open.

Mishra, R. S., & Mahoney, M. W. (2001). Friction stir processing: A new grain refinement technique to achieve high strain rate superplasticity in commercial alloys. In Materials Science Forum. Trans Tech Publ. doi:10.4028/www.scientific.net/MSF.357-359.507

Mishra, D. P., & Dash, S. K. (2010a). Numerical investigation of air suction through louvers of a funnel due to high velocity air jet. *Computers & Fluids*, *39*(9), 1597–1608. doi:10.1016/j.compfluid.2010.05.012

Mishra, D. P., & Dash, S. K. (2010b). Prediction of entrance length and mass suction rate for cylindrical sucking funnel. *International Journal for Numerical Methods in Fluids*, *63*, 681–700.

Mishra, D. P., & Dash, S. K. (2011). Maximum air suction into a louvered funnel through optimum design. *J Ship Res*, *55*, 1–11.

Mishra, D. P., Dash, S. K., & Kishan, P. A. (2010c). Isothermal jet suction through the lateral openings of cylindrical funnel. *J. Ship Res.*, *54*(4), 268–280.

Mishra, D. P., Samantaray, M. K., & Dash, S. K. (2014). Maximum air entrainment into a mixing pipe through optimum design. *Ships and Offshore Structures*, *9*(6), 605–618. doi:10.1080/17445302.2014.881246

Mishra, P., & Mozaffari, F. E. (2017). Using computational fluid dynamics to analyze the performance of the Maxblend impeller in solid-liquid mixing operations. *International Journal of Multiphase Flow*, *91*, 194–207. doi:10.1016/j.ijmultiphaseflow.2017.01.009

Mishra, R. S., & Kumar, N. (2014). *Friction Stir Welding and Processing Science and Engineering*. Springer.

Mishra, R. S., & Mahoney, M. W. (Eds.). (2007). *Friction stir welding and processing*. Materials Park, OH: ASM International.

Mishra, R. S., Mahoney, M., McFadden, S., Mara, N., & Mukherjee, A. (1999). High strain rate superplasticity in a friction stir processed 7075 Al alloy. *Scripta Materialia*, *42*(2), 163–168. doi:10.1016/S1359-6462(99)00329-2

Mishra, R. S., & Ma, Z. Y. (2005). Friction stir welding and processing. *Materials Science and Engineering A*, *50*(1–2), 1–78.

Mishra, R. S., & Ma, Z. Y. (2005). Friction stir welding and processing. *Materials Science and Engineering R Reports*, *50*, 1–78.

Mishra, R. S., Ma, Z. Y., & Charit, I. (2003). Friction stir processing: A novel technique for fabrication of surface composite. *Materials Science and Engineering A*, *341*(1), 307–310. doi:10.1016/S0921-5093(02)00199-5

Mishra, S., Misra, M., Tripathy, S. S., Nayak, S. K., & Mohanty, A. K. (2002). The influence of chemical surface modification on the performance of sisal-polyester biocomposites. *Polymer Composites*, *23*(2), 164–170. doi:10.1002/pc.10422

Mitra, B. C. (2014, May). Environment Friendly Composite Materials: Biocomposites and Green Composites. *Defence Science Journal*, *64*(3), 244–261. doi:10.14429/dsj.64.7323

Mittal, V., Saini, R., & Sinha, S. (2016). Natural fiber-mediated epoxy composites: A review. *Compos Part B*, *99*, 425–435. doi:10.1016/j.compositesb.2016.06.051

Mkrtychev, O. V., Dzhinchvelashvili, G. A., & Bunov, A. A. (2014). Study of Lead Rubber Bearings Operation with Varying Height Buildings at Earthquake. *Procedia Engineering*, *91*, 48–53. doi:10.1016/j.proeng.2014.12.010

Modica, F., Morrocco, V., & Fassi, I. (2017). *Micro-Electrodischarge Machining. In Micro-Manufacturing Technologies & their Applications*. Springer International Publishing.

Moreno, J. C., Rocon de Lima, E., & Andr'es, F. (2006). Design and implementation of an inertial measurement unit for control of artificial limbs: Application on leg orthoses. *Sensors and Actuators. B, Chemical*, *118*(1-2), 333–337. doi:10.1016/j.snb.2006.04.039

Morisada, Y., Fujii, H., Nagaoka, T., & Fukusumi, M. (2006). MWCNTs/AZ31 surface composites fabricated by friction stir processing. *Materials Science and Engineering A*, *419*(1-2), 344–348. doi:10.1016/j.msea.2006.01.016

Motavalli, S., & Shamsaasef, R. (1996). Object-Oriented Modeling of a Feature-Based Reverse Engineering System. *International Journal of Computer Integrated Manufacturing*, *9*(5), 354–368. doi:10.1080/095119296131463

Moultrie, J., Clarkson, P. J., & Probert, D. (2006). A tool to evaluate design performance in SME's. *International Journal of Productivity and Performance Management*, *55*(3 / 4), 184–216. doi:10.1108/17410400610653192

Moultrie, J., Clarkson, P. J., & Probert, D. (2007). Development of a design audit tool for SME's. *Journal of Product Innovation Management*, *24*(4), 335–368. doi:10.1111/j.1540-5885.2007.00255.x

Mukhopadhyay, A., Duari, S., Barman, T. K., & Sahoo, P. (2016b). Tribological performance optimization of electroless Ni–B coating under lubricated condition using hybrid grey fuzzy logic. *Journal of the Institution of Engineers (India): Series D*, *97*(2), 215-231.

Mukhopadhyay, A., Duari, S., Barman, T. K., & Sahoo, P. (2016d). Optimization of friction and wear properties of electroless Ni–P coatings under lubrication using grey fuzzy logic. *Journal of the Institution of Engineers (India): Series D*. DOI: 10.1007/s40033-016-0133-9

Mukhopadhyay, A., Duari, S., Barman, T. K., & Sahoo, P. (2016f). Wear behavior of electroless Ni-P-W coating under lubricated condition-a Taguchi based approach. IOP Conference Series: Materials Science and Engineering, 149, 012004.

Mukhopadhyay, A., Barman, T. K., & Sahoo, P. (2016c). Effect of operating temperature on tribological behavior of as-plated Ni-B coating deposited by electroless method. *Tribology Transactions*. Doi:10.1080/10402004.2016.1271929

Mukhopadhyay, A., Barman, T. K., & Sahoo, P. (2017b). Tribological behavior of sodium borohydride reduced electroless nickel alloy coatings at room and elevated temperatures. *Surface and Coatings Technology*, *321*, 464–476. doi:10.1016/j.surfcoat.2017.05.015

Mukhopadhyay, A., Barman, T. K., & Sahoo, P. (2017c). Effects of heat treatment on tribological behavior of electroless Ni–B coating at elevated temperatures. *Surface Review and Letters*, 1850014. doi:10.1142/S0218625X18500142

Mukhopadhyay, A., Duari, S., Barman, T. K., & Sahoo, P. (2015). Application of grey fuzzy logic for the optimization of tribological performance of electroless Ni-B coating. *Journal of Manufacturing Technology Research*, *7*(1/2), 1–24.

Mukhopadhyay, A., Duari, S., Barman, T. K., & Sahoo, P. (2016a). Wear analysis of electroless Ni-P coating under lubricated condition using fuzzy logic. *Portugaliae Electrochimica Acta*, *34*(1), 63–83. doi:10.4152/pea.201601063

Mukhopadhyay, A., Duari, S., Barman, T. K., & Sahoo, P. (2016e). Investigation of wear behavior of electroless Ni-P-W Coating under dry and lubricated conditions using RSM and fuzzy logic. *Portugaliae Electrochimica Acta*, *34*(4), 231–255. doi:10.4152/pea.201604231

Mukhopadhyay, A., Duari, S., Barman, T. K., & Sahoo, P. (2016g). Optimization of wear behavior of electroless Ni-P-W coating under dry and lubricated conditions using genetic algorithm (GA). *Jurnal Tribologi*, *11*, 24–48.

Mukhopadhyay, A., Duari, S., Barman, T. K., & Sahoo, P. (2017a). Tribology of electroless Ni-P coating under lubricated condition: An RSM and GA approach. *International Journal of Surface Engineering and Interdisciplinary Materials Science*, *5*(1), 37–57. doi:10.4018/IJSEIMS.2017010103

Myles, D. J. (1965). *A design method for mixed flow fans and pumps*. Report No. 117 National Engineering Laboratory.

Mylsamy, K., & Rajendran, I. (2011). The mechanical properties, deformation and thermo mechanical properties of alkali treated and untreated Agave continuous fibre reinforced epoxy composites. *Materials & Design*, *32*(5), 3076–3084. doi:10.1016/j.matdes.2010.12.051

Naeim, F., & Kelly, J. M. (1999). *Design of seismic isolated structures: From theory to practice*. New York: John Willey and Sons Inc. doi:10.1002/9780470172742

Nagarajaiah, S., Reinhorn, A. M., & Constantinou, M.C. (1991). *3D-BASIS: Nonlinear dynamic analysis of three dimensional base isolated structures: Part II*. Technical Report NCEER-91-0005. National Center for Earthquake Engineering Research, State University of New York at Buffalo.

Nagarajan, V., Mohanty, A. K., & Misra, M. (2016). Perspective on Polylactic Acid (PLA) based Sustainable Materials for Durable Applications: Focus on Toughness and Heat Resistance. *ACS Sustainable Chemistry & Engineering*, *4*(6), 2899–2916. doi:10.1021/acssuschemeng.6b00321

Nain, S. S., Garg, D., & Kumar, S. (2017). Prediction of the Performance Characteristics of WEDM on Udimet-L605 Using Different Modelling Techniques. *Materials Today: Proceedings*, *4*(2), 546–556. doi:10.1016/j.matpr.2017.01.056

Najafi, M., Nasiri, A. M., & Kokabi, A. H. (2008). Microstructure and hardness of friction stir processed AZ31 with Si Cp. *International Journal of Modern Physics B*, *22B*(18–19), 2879–2885. doi:10.1142/S0217979208047717

Nandan, R., Deb Roy, T., & Bhadeshia, H. (2008). Recent advances in friction-stir welding – process, weldment structure and properties. *Progress in Materials Science*, *53*(6), 980–1023. doi:10.1016/j.pmatsci.2008.05.001

Naphon, P. (2006). Study on the heat transfer characteristics of the annular fin under dry-surface, partially wet-surface, and fully wet-surface conditions. *International Communications in Heat and Mass Transfer*, *33*(1), 112–121. doi:10.1016/j.icheatmasstransfer.2005.08.009

Nayak & Mahapatra. (2016). Optimization of WEDM process parameters using deep cryo-treated Inconel 718 as work material. *Engineering Science and Technology, an International Journal*, *19*, 161–170.

Netravali, A. N., & Chabba, S. (2003). Composites get greener. *Materials Today: Proceedings*, *6*(4), 22–29. doi:10.1016/S1369-7021(03)00427-9

Neumann, B. (1991). *The interaction Between Geometry and Performance of a centrifugal Pump*. London: Mechanical Engineering Publications.

Newey, L., & Verreynne, M.-L. (2011). Multilevel absorptive capacity and interorganisational new product development: A process study. *Journal of Management & Organization*, *17*(01), 39–55. doi:10.1017/S183336720000170X

Newkirk, J.W., Mishra, R.S., & Ply, J. (2005). Advances in Powder Metallurgy and Particulate Materials. *MPIF*, *6*, 62.

Newton, Melkote, Watkins, Trejo, & Reister. (2009). Investigation of the effect of process parameters on the formation and characteristics of recast layer in wire-EDM of Inconel 718. *Materials Science and Engineering A*, (513–514), 208–215.

Nicholas, J., Ledwith, A., & Perks, H. (2011). NPD best practice in SME and large organisations: Theory versus practice. *European Journal of Innovation Management*, *14*(2), 227–251. doi:10.1108/14601061111124902

Nielsen, L. E., & Landel, R. F. (1994). *Mechanical Properties of Polymers and Composites* (2nd ed.). New York: Marcel Deckker.

Noke, H., & Hughes, M. (2010). Climbing the value chain: Strategies to create NPD capability in mature SME's. *International Journal of Operations & Production Management, 30*(2), 132–154. doi:10.1108/01443571011018680

Noke, H., & Radnor, Z. J. (2004). Navigating innovation: A diagnostic tool supporting the process. *Journal of Manufacturing Technology Management, 15*(2), 172–185. doi:10.1108/09576060410513751

Nopparut, A., & Amornsakchai, T. (2016). Influence of pineapple leaf fiber and it's surface treatment on molecular orientation in, and mechanical properties of, injection molded nylon composites. *Polymer Testing, 52*, 141–149. doi:10.1016/j.polymertesting.2016.04.012

Norman, P. A., & Peterson, T. (1999). Quality improvement through innovation. TQM and Innovation. In *Proceedings of the fourth international conference on ISO 9000 and TQM*. HKBU Press.

O'Dwyer, M., & Ledwith, A. (2009). Determinant of New Product performance in Small Firms. *International Journal of Entrepreneurial Behaviour & Research, 15*(2), 124–136. doi:10.1108/13552550910944548

Oakley, M. (1990). Design and design management. In Design management- A handbook of issues and methods. Basil Blackwell Ltd.

Ogama, S., & Pillar, F. (2006). Reducing the risk of NPD. *MIT Sloan Management Review, 47*(2), 65–71.

Osterlund, J. (1997). Competence management by informatics in R&D: The corporate level. *IEEE Transactions on Engineering Management, 44*(2), 135–145. doi:10.1109/17.584922

Otheguy, M. E., Gibson, A. G., Robinson, M., Findon, E., Cripps, B., & Ochoa Mendoz, A. (2009). Recycling of end of life thermoplastic composite boats.*Plastics. Rubber and Composites:Macromolecular Engineering, 38*(9-10), 406–411. doi:10.1179/146580109X12540995045642

Owens, J. D. (2009). Modelling the NPD process: The value of a Product Development Process Model Approach, as a means for Business Survival in the 21st Century. In D. Jemielniak & J. Kociatkiewics (Eds.), Handbook of Research on Knowledge Intensive Organisations (pp. 208–227). Academic Press. doi:10.4018/978-1-60566-176-6.ch013

Owens, J. D. (2007). Why do some UK SME's still find the implementation of a New Product Development process problematical? An Exploratory Investigation. *Management Decision Journal, 45*(2), 235–251. doi:10.1108/00251740710727269

Owens, J. D., & Atherton, A. (2015). Can the development of a conceptual model for UK engineering companies help understand what the customer wants from new products? *Journal of Multidisciplinary Engineering Science and Technology, 2*(11), 3155–3165.

Owens, J. D., & Davies, J. (2000). The Importance of a New Product Development (NPD) Process: Getting Started. *First European Conference on Knowledge Management*. Bled School of Management.

Palaniappa, M., & Seshadri, S. K. (2008). Friction and wear behavior of electroless Ni–P and Ni–W–P alloy coatings. *Wear, 265*(5), 735–740. doi:10.1016/j.wear.2008.01.002

Panda, D. K., & Bhoi, R. K. (2005). Artificial Neural Network Prediction of Material Removal Rate in Electro Discharge Machining. *Materials and Manufacturing Processes, 20*(4), 645–672. doi:10.1081/AMP-200055033

Panja, B., Das, S. K., & Sahoo, P. (2016). Tribological behavior of electroless Ni–P coatings in various corrosive environments. *Surface Review and Letters, 23*, 1650040. DOI: 10.1142/S0218625X16500402

Panja, B., & Sahoo, P. (2014). Taguchi-Grey based optimization of friction and wear of electroless Ni-P coatings in acidic environment. *International Journal of Surface Engineering and Interdisciplinary Materials Science*, *2*(2), 53–69. doi:10.4018/IJSEIMS.2014070104

Panja, B., & Sahoo, P. (2015). Tribological behavior of electroless Ni-P coatings in alkaline environment and optimization of coating parameters using Taguchi based grey relational analysis. *Indian Journal of Engineering and Materials Sciences*, *22*, 503–512.

Park, H. (2015). A study on structural design and analysis of small wind turbine blade with natural fibre (flax) composite. *Advanced Composite Materials*, *25*(2), 125–142. doi:10.1080/09243046.2015.1052186

Park, S., Sato, Y., Kokawa, H., Okamoto, K., Hirano, S., & Inagaki, M. (2009). Boride formation induced by pcBN tool wear in friction-stir-welded stainless steels. *Metallurgical and Materials Transactions. A, Physical Metallurgy and Materials Science*, *40A*(3), 625–636. doi:10.1007/s11661-008-9709-9

Park, Y. J., Went, Y. K., & Ang, A. H. S. (1986). Random vibration of hysteretic systems under bi directional ground motions. *Earthquake Engineering & Structural Dynamics*, *14*(4), 543–557. doi:10.1002/eqe.4290140405

PartInfo. (n.d.). *Boge Monotube Shock* [Electronic Image]. Retrieved 2 August 2015 from http://www.partinfo.co.uk/articles/74

Pashah, S., Moinuddin, A., & Zubair, S. M. (2016). Thermal Performance and Optimization of hyperbolic annular fins under dehumidifying operating conditions-analytical and numerical solutions. *International Journal of Refrigeration*, *65*, 42–54. doi:10.1016/j.ijrefrig.2016.01.006

Patel, K. M., Pandey, P. M., & Rao, P. V. (2010). Optimization of process parameters for multi-performance characteristics in EDM of Al2O3 ceramic composite. *International Journal of Advanced Manufacturing Technology*, *47*(9-12), 1137–1147. doi:10.1007/s00170-009-2249-7

Patel, K. N., Gelot, B. N., & Gupta, A. H. (2014). A Review Study – Design and Analysis of Suspension System. *International Journal for Scientific Research & Development*, *2*(3).

Patel, S. R., & Shaikh, A. A. (2013). A Review on Machining of Fiber Reinforced Plastic using Abrasive Waterjet. *International Journal of Innovative Technology & Adaptive Management*, *1*(3), 10–18.

Pattanaik, L. N. (2017). Artificial Neural Network(MATLAB). In Analytical Tools in Research (pp. 250-261). Dwarka, India: Educreation Publishing.

Pavani, P. N. L., Prafulla, B. K., Rao, R. P., & Srikiran, S. (2014). Design, Modeling and Structural Analysis of Wave Springs. *Procedia Materials Science*, *6*, 988–995. doi:10.1016/j.mspro.2014.07.169

Pei, H., Dohmen, H. J., Yuan, S. Q., & Benra, F. K. (2012). Investigation of unsteady flow-induced impeller oscillations of a single-blade pump under off-design conditions. *Journal of Fluids and Structures*, *35*, 89–104. doi:10.1016/j.jfluidstructs.2012.08.005

Pei, J., Yuan, S., Yuan, J., & Wang, W. (2013). The influence of the flow rate on periodic flow unsteadiness behaviors in asewage centrifugal pump. *Journal of Hydrodynamics*, *25*(5), 702–709. doi:10.1016/S1001-6058(13)60415-1

Peng, G., Cao, S., Ishizuka, M., & Hayama, S. (2002). Design optimization of axial flow hydraulic turbine runner: Part I – an improved Q3D inverse method. *International Journal for Numerical Methods in Fluids*, *39*(6), 517–531. doi:10.1002/fld.342

Perks, H., Cooper, R., & Jones, C. (2005). Characterising the role of design in NPD: An empirically derived taxonomy. *Journal of Product Innovation Management*, *22*(2), 111–127. doi:10.1111/j.0737-6782.2005.00109.x

Petrie, A. (2008). Developing products with a holistic process. *Design Management Review*, *19*(3), 68–73. doi:10.1111/j.1948-7169.2008.tb00131.x

Phelan, R. F. (1997). Simulation of the Injection Process in Resin Transfer Moulding. *Polymer Composites*, *18*(4), 460–476. doi:10.1002/pc.10298

Pirondi, A., Collini, L., & Fersini, D. (2008). Fracture and fatigue crack growth behavior of PMMC friction stir welded butt joints. *Engineering Fracture Mechanics*, *75*(15), 4333–4342. doi:10.1016/j.engfracmech.2008.05.001

Popoola, A. P. I., Loto, C. A., Osifuye, C. O., Aigbodion, V. S., & Popoola, O. M. (2016). Corrosion and wear properties of Ni-Sn-P ternary deposits on mild steel via electroless method. *Alexandria Engineering Journal*, *55*(3), 2901–2908. doi:10.1016/j.aej.2016.06.018

Porwal, R.K., & Yadava, V. (2012). ANN Modelling For The Prediction Of Material Removal Rate And Machined Hole Overcut In Hole Drilling Electro Discharge Micro Machining. *International Journal of Mechanical Engineering and Robotics Research, 1*(2), 1-16.

Powell, R. M. (1994). Utilization of Natural Fibers in Plastic Composites, *Lignocellulosic. Plastic Composites*, *4*(27), 23–51.

Prabhakaran, D., & Thirumalai, R. (2014). A critical review of future materials for wind turbine blades, *Int. J. Materials Engineering Innovation*, *5*(2), 2014.

Prabhu, S., & Vinayagam, B. K. (2015). Adaptive neuro fuzzy inference system modelling of multi-objective optimisation of electrical discharge machining process using single-wall carbon nanotubes. *Australian Journal of Mechanical Engineering*, *13*(2), 97–117. doi:10.7158/M13-074.2015.13.2

Pradhan, M. K., & Chandan, B. K. (2010). Neuro- fuzzy and Neural Network Based prediction of various responses in EDM of AISI D2 steel. *International Journal of Advanced Manufacturing Technology*, *50*, 591–610. doi:10.1007/s00170-010-2531-8

Pragya Shandilya, P. K. (2012). Parametric optimization during wire electrical discharge machining using response surface methodology. *Procedia Engineering*, *38*, 2371–2377. doi:10.1016/j.proeng.2012.06.283

Prater, T., Strauss Alvin, M., Cook George, E., Gibson Brian, T., & Cox Chase, D. (2013). A;Comparative evaluation of the wear resistance of various tool materials in friction stirs welding of metal matrix composites. *Journal of Materials Engineering and Performance*, *22*(6), 1807–1813. doi:10.1007/s11665-012-0468-9

Prihandana, G. S., Sriani, T., & Mahardika, M. (2012). Improvement of machining time in micro-EDM with workpiece vibration and graphite powder mixed in dielectric fluid. *Indian Journal of Engineering and Materials Sciences*, *19*, 375–378.

Pritchard, G. (2004). Two technologies merge wood-plastic composites. *Plastics Additives & Compounding*, *6*(4), 18–21. doi:10.1016/S1464-391X(04)00234-X

Pritchard, R., Guy, J. J., & Conner, N. E. (1977). *Industrial Gas Utilization*. New Providence, NJ: Bowker.

Pugh, S. (1991). *Total design – Integrated methods for successful product engineering*. Addison-Wesley Publishing.

Pugh, S., & Moreley, I. E. (1988). *Total Design: Towards a theory of total design and Total Design: Some questions and some answers. Design division*. University of Strathclyde.

Puglia, G. (2013). *Life cycle cost analysis on wind turbines* (Master's thesis). Department of Energy and Environment, Division of Electric Power Engineering, Chalmers University of Technology, Gothenburg, Sweden.

Puri, A. B., & Bhattacharyya, B. (2005). Modeling and analysis of white layer depth in a wire-cut EDM process through response surface methodology. *International Journal of Advanced Manufacturing Technology*, *25*(3-4), 301–307. doi:10.1007/s00170-003-2045-8

Rahman, M. M. (2012). Modeling of Machining Parameters of Ti-6Al-4V for EDM: A neural network approach. *Scientific Research and Essays*, *7*(8), 881–890. doi:10.5897/SRE10.1116

Rai, R., De, A., Bhadeshia, H., & Deb Roy, T. (2011). Review: Friction stir welding tools. *Science and Technology of Welding and Joining*, *16*(4), 325–342. doi:10.1179/1362171811Y.0000000023

Raja, V. (2008). Introduction to Reverse Engineering. In V. Raja & K. J. Fernandes (Eds.), *Reverse Engineering: An Industrial Perspective* (pp. 1–9). London: Springer Verlag. doi:10.1007/978-1-84628-856-2_1

Rajyalakshmi, G., & Venkata Ramaiah, P. (2013). Multiple process parameter optimization of wire electrical discharge machining on Inconel 825 using Taguchi grey relational analysis. *International Journal of Advanced Manufacturing Technology*, *69*(5-8), 1249–1262. doi:10.1007/s00170-013-5081-z

Rakib, M. I., Choudhury, I. A., Hussain, S., & Osman, N. A. A. (2015). Design and biomechanical performance analysis of a user-friendly orthotic device. *Materials & Design*, *65*, 716–725. doi:10.1016/j.matdes.2014.09.075

Ramachandran, K. K., Murugan, N., & Shashi Kumar, S. (2014). An assessment on friction stir welding of high melting temperature materials. *AMM*, *592*, 43–47. doi:10.4028/www.scientific.net/AMM.592-594.43

Ramakrishnan, R., & Karunamoorthy, L. (2006). Multi response optimization of wire EDM operations using robust design of experiments. *International Journal of Advanced Manufacturing Technology*, *29*(1-2), 105–112. doi:10.1007/s00170-004-2496-6

Ramakrishnan, R., & Karunamoorthy, L. (2008). Modeling and multi-response optimization of Inconel 718 on machining of CNC WEDM process. *Journal of Materials Processing Technology*, *207*(1-3), 343–349. doi:10.1016/j.jmatprotec.2008.06.040

Ramalho, A., & Miranda, J. C. (2007). Tribological characterization of electroless NiP coatings lubricated with biolubricants. *Wear*, *263*(1), 592–597. doi:10.1016/j.wear.2006.10.027

Ramamurti, V., & Balasubramanian, P. (1987). Steady state stress analysis of centrifugal fan impellers. *Computers & Structures*, *25*(1), 129–135. doi:10.1016/0045-7949(87)90224-0

Rao, P. N. (2009). Introduction. In Manufacturing Technology (vol. 2, pp. 1-5). New Delhi: Tata McGraw-Hill.

Rao, G. K. M., Janardhana, G. R., Rao, D. H., & Rao, M. S. (2008). Development of hybrid model & optimization of Material Removal Rate in EDM using Artificial neural Network and Genetic Algorithm. *Journal of Engineering and Applied Sciences (Asian Research Publishing Network)*, *3*(1), 19–30.

Rao, R. R., & Kalyankar, V. D. (2014). Optimization of Modern Machining Processes Using Advanced Optimization Techniques: A Review. *International Journal of Advanced Manufacturing Technology*, *73*(5-8), 1159–1188. doi:10.1007/s00170-014-5894-4

Ras, A., & Boumechra, N. (2017Dissipation's Capacity Study of Lead–Rubber Bearing System in Seismic Steel Structures Design. *Arabian Journal for Science and Engineering*. doi:10.1007/s13369-017-2489-z

Ravikumar, K., Kiran, K., & Sreebalaji, V. S. (2017). Characterization of mechanical properties of aluminium/tungsten carbide composites. *Measurement*, *102*, 142–149. doi:10.1016/j.measurement.2017.01.045

Reddy, V. V. N., Ramamoorthy, B., & Nair, P. K. (2000). A study on the wear resistance of electroless Ni–P/Diamond composite coatings. *Wear*, *239*(1), 111–116. doi:10.1016/S0043-1648(00)00330-6

Reid, S. E., & de Brentani, U. (2004). The fuzzy front end of the new product development for discontinuous innovations: A theoretical model. *Journal of Product Innovation Management*, *21*(3), 170–198. doi:10.1111/j.0737-6782.2004.00068.x

Reshad Seighalani, K., Besharati Givi, M. K., Nasiri, A. M., & Bahemmat, P. (2010). Investigations on the effects of the tool material, geometry and tilt angle on friction stir welding of pure titanium. *Journal of Materials Engineering and Performance*, *19*(7), 955–962. doi:10.1007/s11665-009-9582-8

Rhodes, C. G., Mahoney, M. W., Bingal, W., Spurling, R. A., & Bamption, C. C. (1997). Effects of friction stir welding on microstructure of 7075 aluminum. *Scripta Materialia*, *36*(1), 69–75. doi:10.1016/S1359-6462(96)00344-2

Ricou, F. P., & Spalding, D. B. (1961). Measurements of entrainment by axisymmetrical turbulent jets. *Journal of Fluid Mechanics*, *11*(01), 21–32. doi:10.1017/S0022112061000834

Rocha, I. B. C. M., Raijmaekers, S., Nijssen, R. P. L., van der Meer, F. P., & Sluys, L. J. (2017). Hygrothermal ageing behaviour of a glass/epoxy composite used in wind turbine blades. *Composite Structures*, *174*, 110–122. doi:10.1016/j.compstruct.2017.04.028

Rodriguez, N. G., Perez, M. J. S., & Gutierrez, J A T. (2008). Can good organisational climate compensate for a lack of top management commitment to NPD? *Journal of Business Research*, *61*(2), 118–131. doi:10.1016/j.jbusres.2007.06.011

Roy, S., & Sahoo, P. (2014). Parametric optimization of corrosion and wear of electroless Ni-P-Cu coating using grey relational coefficient coupled with weighted principal component analysis. International Journal of Mechanical and Materials Engineering, 9(1).

Roy, B. (2005). Paradigms and challenges. In J. Figueira, S. Greco, & M. Ehrgott (Eds.), *multiple criteria decision analysis. State of the art surveys*. New York, NY: Springer. doi:10.1007/0-387-23081-5_1

Roy, S., Davim, J. P., Bhowmik, S., & Kumar, K. (2016). Estimation of mechanical and tribological properties of epoxy based green composites. In D. Verma, S. Jain, X. Zhang, & P. C. Gope (Eds.), *Green Approaches to Biocomposite Materials Science and Engineering* (pp. 96–124). Hershey, PA: Engineering Science Reference. doi:10.4018/978-1-5225-0424-5.ch005

Roy, S., & Sahoo, P. (2012). Corrosion study of electroless Ni-P-W coatings using electrochemical impedance spectroscopy. *Portugaliae Electrochimica Acta*, *30*(3), 203–220. doi:10.4152/pea.201203203

Rozman, H. D., Lai, C. Y., Ismail, H., & Mohd-Ishak, Z. A. (2000). The effect of coupling agents on the mechanical and physical properties of oil palm empty fruit bunch polypropylen composites. *Polymer International*, *49*(11), 1273–1278. doi:10.1002/1097-0126(200011)49:11<1273::AID-PI469>3.0.CO;2-U

Rumelhart, D. E., Hinton, G. E., & Williams, R. J. (1986). Learning representations by back-propagating errors. *Nature*, *323*(6088), 533–536. doi:10.1038/323533a0

Saaty, T. L. (1989). Decision making with the analytic hierarchy process. *Journal de Chirurgie*, *1*(1), 83–98.

Sabbaghian, M., Shamanian, M., Akramifard, H. R., & Esmailzadeh, M. (2014). Effect of friction stir processing on the microstructure and mechanical properties of Cu–TiC composite. *Ceramics International*, *40*(8), 12969–12976. doi:10.1016/j.ceramint.2014.04.158

Sadeghzadeh-Attar, A., AyubiKia, G., & Ehteshamzadeh, M. (2016). Improvement in tribological behavior of novel sol-enhanced electroless Ni-P-SiO_2 nanocomposite coatings. *Surface and Coatings Technology*, *307*, 837–848. doi:10.1016/j.surfcoat.2016.10.026

Sahoo, P. (2009). Wear behaviour of electroless Ni–P coatings and optimization of process parameters using Taguchi method. *Materials & Design*, *30*(4), 1341–1349. doi:10.1016/j.matdes.2008.06.031

Sahoo, P., & Das, S. K. (2011). Tribology of electroless nickel coatings – A review. *Materials & Design*, *32*(4), 1760–1775. doi:10.1016/j.matdes.2010.11.013

Sahoo, P., & Das, S. K. (2015). Wear performance optimization for electroless Ni-P Coating. *International Journal of Surface Engineering and Interdisciplinary Materials Science*, *3*(2), 1–17. doi:10.4018/IJSEIMS.2015070101

Sahoo, P., & Pal, S. K. (2007). Tribological performance optimization of electroless Ni–P coatings using the Taguchi method and grey relational analysis. *Tribology Letters*, *28*(2), 191–201. doi:10.1007/s11249-007-9264-3

Sahoo, P., & Roy, S. (2017). Tribological behavior of electroless Ni-P, Ni-P-W and Ni-P-Cu coatings: A comparison. *International Journal of Surface Engineering and Interdisciplinary Materials Science*, *5*(1), 1–15. doi:10.4018/IJSEIMS.2017010101

Sahoo, P., Roy, S., & Davim, J. P. (2014). Design and selection of chemically deposited Ni-P-W coatings for optimum tribological behavior. In L. Santo & J. P. Davim (Eds.), *Surface Engineering Techniques and Applications: Research Advancements* (pp. 45–72). Engineering Science Reference (an imprint of IGI Global). doi:10.4018/978-1-4666-5141-8.ch002

Saito, N., Shigematsu, I., Komaya, T., Tamaki, T., Yamauchi, G., & Nakamura, M. (2001). Grain refinement of 1050 aluminum alloy by friction stir processing. *Journal of Materials Science Letters*, *20*(20), 1913–1918. doi:10.1023/A:1012882413401

Şakar, G., & Sabuncu, M. (2008). Dynamic stability analysis of pretwisted aerofoil cross-section blade packets under rotating conditions. *International Journal of Mechanical Sciences*, *50*(1), 1–13. doi:10.1016/j.ijmecsci.2007.06.001

Sankara Narayanan, T. S. N., Selvakumar, S., & Stephen, A. (2003). Electroless Ni–Co–P ternary alloy deposits: Preparation and characteristics. *Surface and Coatings Technology*, *172*(2), 298–307. doi:10.1016/S0257-8972(03)00315-3

Santella, M. L., Engstrom, T., Storjohann, D., & Pan, T. Y. (2005). Effect of friction stir processing on mechanical properties of the cast aluminium alloys A319 and A356. *Scripta Materialia*, *53*(2), 201–206. doi:10.1016/j.scriptamat.2005.03.040

Sapuan, S. M., & Maleque, M. A. (2005). Design and fabrication of natural woven fabric reinforced epoxy composite for household telephone stand. *Materials & Design*, *26*(1), 65–71. doi:10.1016/j.matdes.2004.03.015

Saren, M A (1984). A classification and review of models of the intra-firm innovation process. *R&D Magazine, 14*(1), 11-24.

Sarkar, A., Panja, S. C., Das, D., & Sarkar, B. (2015). Developing an efficient decision support system for non-traditional machine selection: An application of MOORA and MOOSRA. *Journal of Production & Manufacturing Research*, *3*(1), 324–342. doi:10.1080/21693277.2014.895688

Sathiskumar, R., Murugan, N., Dinaharan, I., & Vijay, S. J. (2013). Role of friction stir processing parameters on microstructure and micro hardness of boron carbide particulate reinforced copper surface composites. *Sadhana*, *38*(6), 1433–1450. doi:10.1007/s12046-013-0184-7

Saxsena, K. K., Srivastava, A. S., & Agarwal, S. (2016). Experimental Investigation into micro- EDM characteristics of conductive SiC. *Ceramics International*, *42*(1), 1597–1610. doi:10.1016/j.ceramint.2015.09.111

Schmidt, J., Sarangee, K., & Montoya, M. (2009). Exploring NPD project review practices. *Journal of Product Innovation Management*, *26*(5), 520–535. doi:10.1111/j.1540-5885.2009.00678.x

Sethi, P. K. (1990). Designing Aids for Physically Handicapped in Developing Countries. *Indian Journal of Physical Medicine and Rehabilitation*, *3*, 1–10.

Sethi, P. K., Udawat, M. P., Kasliwal, S. C., & Chandra, R. (1978). Vulcanized rubber foot for lower limb amputees. *Prosthetics and Orthotics International*, *2*, 125–136. PMID:733459

Sethi, R. (2000). New product Quality and Product Development Teams. *Journal of Marketing*, *64*(2), 1–14. doi:10.1509/jmkg.64.2.1.17999

Shafiei-Zarghani, A., Kashani-Bozorg, S. F., & Zarei-Hanzaki, A. (2009). Microstructures and mechanical properties of Al/Al2 O 3 surface nano-composite layer produced by friction stir processing. A. *Materials Science and Engineering A*, *500*(1-2), 84–91. doi:10.1016/j.msea.2008.09.064

Shah, D. U., Schubel, P. J., & Clifford, M. J. (2013). Can flax replace E-glass in structural composites? A small wind turbine blade case study. *Composites. Part B, Engineering*, *52*, 172–181. doi:10.1016/j.compositesb.2013.04.027

Sharma, N., Khanna, R., Gupta, R. D., & Sharma, R. (2013). Modeling and multi response optimization on WEDM for HSLA by RSM. *International Journal of Advanced Manufacturing Technology*, *67*(9-12), 2269–2281. doi:10.1007/s00170-012-4648-4

Sheng, P., & Srinivasan, M. (1995). Multi-objective process planning in environmentally conscious manufacturing: A feature-based approach. *CIRP Annals - Manufacturing Technology*, *44*(1), 433-437.

Shenhar, A. J. (2001). One size does not fit all projects: Exploring classical contingency domains. *Management Science*, *47*(3), 394–414. doi:10.1287/mnsc.47.3.394.9772

Shivakumar, S., & Guggari, G. S. (2011). Literature review of fiber reinforced polymer composites. *International Journal of Advances in Engineering and Technology*, *1*(5), 218–226.

Shojaeia, A., & Ghaffariana, S. R. (2003). Modeling and Simulation Approaches in the Resin Transfer Moulding Process: A Review. *Polymer Composites*, *24*(4), 525–544. doi:10.1002/pc.10050

Shokrani, A., Dhokia, V., & Newman, S. T. (2012). Environmentally conscious machining of difficult-to-machine materials with regard to cutting fluids. *International Journal of Machine Tools & Manufacture*, *57*, 83–101. doi:10.1016/j.ijmachtools.2012.02.002

Shukla, A., & Harsha, S. P. (2015). An Experimental and FEM Modal Analysis of Cracked and Normal Steam Turbine Blade. *Materials Today: Proceedings*, *2*(4-5), 2056–2063. doi:10.1016/j.matpr.2015.07.191

Siddiquee, A. N., Khan, Z. A., & Mallick, Z. (2010). Grey relational analysis coupled with principal component analysis for optimization design of the process parameters in in-feed centreless cylindrical grinding. *International Journal of Advanced Manufacturing Technology*, *46*(9-12), 983–992. doi:10.1007/s00170-009-2159-8

Singaravel, B., Selvaraj, T., & Vinod, S. (2016). Multi-objective optimization of turning parameters using the combined MOORA and Entropy method. *Transactions of the Canadian Society for Mechanical Engineering*, *40*(1), 101–111.

Singh, G. S., Sundarajan, T., & Bhaskaran, K. A. (2003). Mixing and entrainment characteristics of circular and non-circular confined jets. *Journal of Fluids Engineering*, *125*(5), 835–842. doi:10.1115/1.1595676

Singh, G. S., Sundarajan, T., & Shet, U. S. P. (1999). Entrainment and mixing studies for a variable density confined jet. *Numer. Heat Transfer. Part A*, *35*, 205–223.

Singh, H., & Garg, R. (2009). Effects of process parameters on material removal rate in WEDM. *Journal of Achievements in Materials and Manufacturing Engineering of Achievements in Materials and Manufacturing Engineering*, *32*, 70–74.

Singh, S. (2012). Optimization of machining characteristics in electric discharge machining of 6061Al/Al2O3p/20P composites by grey relational analysis. *International Journal of Advanced Manufacturing Technology*, *63*(9-12), 1191–1202. doi:10.1007/s00170-012-3984-8

Singh, S., & Verma, M. (2012). Parametric Optimization of Electric Discharge Drill Machine Using Taguchi Approach. *Journal of Engineering and Applied Sciences (Asian Research Publishing Network)*, *1*(3), 39–47.

Sivandipoor, I., & Ashrafizadeh, F. (2012). Synthesis and tribological behaviour of electroless Ni-P-WS_2 composite coatings. *Applied Surface Science*, *263*, 314–319. doi:10.1016/j.apsusc.2012.09.051

Sivapirakasam, S. P., Mathew, J., & Surianarayanan, M. (2011). Multi-attribute decision making for green electrical discharge machining. *Expert Systems with Applications*, *38*(7), 8370–8374. doi:10.1016/j.eswa.2011.01.026

Skinner, R. I., Robinson, W. H., & McVerry, G. H. (1999). *An Introduction to Seismic Isolation*. New York: John Wiley & Sons.

Smith, A. D. (2011). Competitive approaches to NPD: A comparison of successful organisations in an unstable economic environment. *Team Performance Management*, *17*(3/4), 124–145. doi:10.1108/13527591111143682

Soares, M. E., Soares, P., Souza, P. R., Souza, R. M., & Torres, R. D. (2017). The effect of nitriding on adhesion and mechanical properties of electroless Ni–P coating on AISI 4140 steel. *Surface Engineering*, *33*(2), 116–121. doi:10.1080/02670844.2016.1148831

Somashekhar, K. P., Ramchandran, N., & Mathew, J. (2010). Optimization of Material Removal Rate in Micro-EDM Using Artificial Neural Network and Genetic Algorithms. *Materials and Manufacturing Processes*, *25*(6), 467–475. doi:10.1080/10426910903365760

Song, G. S., Sun, S., Wang, Z. C., Luo, C. Z., & Pan, C. X. (2017). Synthesis and characterization of electroless Ni–P/Ni–Mo–P duplex coating with different thickness combinations. *Acta Metallurgica Sinica. English Letters*. doi:10.1007/s40195-017-0603-6

Song, M., & Benedetto, A. D. (2008). Supplier's involvement and success of radical NPD in new ventures. *Journal of Operations Management*, *26*(1), 1–22. doi:10.1016/j.jom.2007.06.001

Sorensen, C. D., & Nelson, T. W. (2007). Friction stir welding of ferrous and nickel alloys in friction stir welding and processing. ASM International, 111–121.

Sorensen, C. D., & Nielsen, B. (2009). Exploring geometry effects for convex scrolled shoulder, step spiral probe FSW tools. *Proc. TMS Annual Meet.*, 85–92.

Sorensen, C. D., & Stahl, A. L. (2007). Experimental measurements of load distributions on friction stir weld pin tools. *Metallurgical and Materials Transactions. B, Process Metallurgy and Materials Processing Science*, *38B*(3), 451–459. doi:10.1007/s11663-007-9041-6

Sorenson, C. D., Nelson, T. W., Packer, S. M., & Steel, R. J. (2004). Innovative technology applications in FSW of high softening temperature materials. *Proc. 5th Int. Conf. on 'Friction stir welding'*.

Spowart, J. E., Ma, Z. Y., & Mishra, R. S. (2003). The effect of friction stir processing (FSP) on the spatial heterogeneity of discontinuously-reinforced aluminum (DRA) microstructures. Friction Stir Welding and Processing II, 243.

Srivastava, S., Roy, A. K., & Kumar, K. (2014). Design of a mixed flow pump impeller blade and its validation using stress analysis. *Procedia Materials Science*, *6*, 417–424. doi:10.1016/j.mspro.2014.07.053

Staia, M. H., Castillo, E. J., Puchi, E. S., Lewis, B., & Hintermann, H. E. (1996). Wear performance and mechanism of electroless Ni-P coating. *Surface and Coatings Technology*, *86*, 598–602. doi:10.1016/S0257-8972(96)03086-1

Stanton, J., & Roeder, C. (1991). Advantages and limitations of seismic isolation. *Earthquake Spectra*, *2*(2), 301–322. doi:10.1193/1.1585630

Stepanoff, A. J. (1957). *Centrifugal and Axial flow pumps*. New York: John Wiley and sons.

Stodola, A. (1945). *Steam and gas turbine*. New York: Peter Smith.

Subash, T., & Pillai, S. N. (2015). Bast fibers reinforced green composites for aircraft indoor structures applications [review]. *Journal of Chemical and Pharmaceutical Sciences*, *7*, 305–307.

Sudagar, J., Lian, J., & Sha, W. (2013). Electroless nickel, alloy, composite and nano coatings–A critical review. *Journal of Alloys and Compounds*, *571*, 183–204. doi:10.1016/j.jallcom.2013.03.107

Suganthi, X. H., Natarajan, U., Sathiyamurthy, S., & Chidambaram, K. (2013). Prediction of quality responses in micro-EDM process using adaptive neuro- fuzzy inference system (ANFIS) model. *International Journal of Advanced Manufacturing Technology*, *68*(1-4), 339–347. doi:10.1007/s00170-013-4731-5

Suh, J. (2014). *Redesign and Stress Analysis of Simplified Landing Gear* (BEng thesis). University of Exeter.

Sun, N., Yin, Y. H., Gerlich, A. P., & North, T. H. (2009). Tool design and stir zone grain size in AZ31 friction stir spot welds. *Science and Technology of Welding and Joining*, *14*(8), 747–752. doi:10.1179/136217109X12518083193559

Suomala, P., & Jokionen, I. (2003). The pattern of success in product development: A case study. *European Journal of Innovation Management*, *6*(4), 213–227. doi:10.1108/14601060310500931

Suresh, K. P., Niranjan, H. B., Martin Jebaraj, P., & Chowdiah, M. P. (2003). Tensile and wear properties of aluminum composites. *Wear*, *255*(1-6), 638–642. doi:10.1016/S0043-1648(03)00292-8

Talbott, M. S. (2002). *An Experimentally Validated Physical Model of a High Performance Automotive Damper. M.S.* Purdue University.

Talla, G., Gangopadhyay, S., & Biswas, C. K. (2016). Influence of Graphite Powder Mixed EDM on the Surface Integrity Characteristics of Inconel 625. *Particulate Science and Technology*, 1–26.

Tang, L., & Du, Y. T. (2014). Experimental study on green electrical discharge machining in the tap water of Ti–6Al–4V and parameters optimization. *International Journal of Advanced Manufacturing Technology*, *70*(1-4), 469–475. doi:10.1007/s00170-013-5274-5

Taylor, A. (2010). The next generation: Technology adoption and integration through internal competition in NPD. *Organization Science*, *21*(1), 23–41. doi:10.1287/orsc.1080.0399

Tewari, A., Spowart, J.E., Gokhale, A.M., Mishra, R.S., & Miracle, D.B. (n.d.). Characterization of the effects of friction stir processing on microstructural changes in DRA composites. *Mater. Sci. Eng. A: Struct. Mater. Prop. Micro struct. Process.*, *428*, 80.

Thakur, V. K., Singha, A. S., & Thakur, M. K. (2012). Green Composites from Natural Fibers: Mechanical and chemical aging properties, *International Journal of Polymer Analysis and. Characterization*, *17*, 401–407.

Thomas, W. M., Nicholas, E. D., Needham, J. C., & Church, M. G. (1991). *International Patent Application No. PCT/GB92/02203 and GB Patent Application No. 9125978.8*. TWI.

Tien, S. K., & Duh, J. G. (2004). Thermal reliability of electroless Ni–P–W coating during the aging treatment. *Thin Solid Films*, *469*, 268–273. doi:10.1016/j.tsf.2004.08.179

Toshimitsu, R., Okada, A., Kitada, R., & Okamoto, Y. (2016). Improvement in Surface Characteristics by EDM with Chromium Powder Mixed Fluid. *Procedia CIRP*, *42*, 231–235. doi:10.1016/j.procir.2016.02.277

Tosun, N., & Cogun, C. (2013). An investigation on wire wear in WEDM. *Journal of Materials Processing Technology*, *134*(3), 273–278. doi:10.1016/S0924-0136(02)01045-2

Tosun, N., Cogun, C., & Pihtili, H. (2003). The Effect of Cutting Parameters on Wire Crater Sizes in Wire EDM. *International Journal of Advanced Manufacturing Technology*, *21*(10-11), 857–865. doi:10.1007/s00170-002-1404-1

Tripathy, S., & Tripathy, D. K. (2017). Surface Characterization and Multi-response optimization of EDM process parameters using powder mixed dielectric. *Materials Today: Proceedings*, *4*(2), 2058–2067. doi:10.1016/j.matpr.2017.02.051

Troy, L. C., Hirunyawipada, T., & Paswan, A. K. (2008). Cross-functional and new product success: An empirical investigation of the findings. *Journal of Marketing*, *72*(Nov), 132–146. doi:10.1509/jmkg.72.6.132

Tsai, K. M., & Wang, P. J. (2001). Comparisons of neural network models on Material Removal Rate in EDM. *Journal of Materials Processing Technology*, *117*, 111–124. doi:10.1016/S0924-0136(01)01146-3

Tsai, K. M., & Wang, P. J. (2001). Predictions on surface finish in EDM based upon Neural Network Models. *International Journal of Machine Tools & Manufacture*, *41*, 1385–1403. doi:10.1016/S0890-6955(01)00028-1

Turkyilmazoglu, M. (2014). Efficiency of heat and mass transfer in fully wet porous fins: Exponential fins versus straight fins. *International Journal of Refrigeration*, *46*(1970), 158–164.

Turkyilmazoglu, M. (2012). International Journal of Thermal Sciences Exact solutions to heat transfer in straight fins of varying exponential shape having temperature dependent properties. *International Journal of Thermal Sciences*, *55*, 69–75. doi:10.1016/j.ijthermalsci.2011.12.019

Tzeng, G. H., & Huang, J. J. (2011). *Multiple attribute decision making: methods and applications*. CRC Press.

Tzeng, Y. F., & Lee, C. Y. (2001). Effects of Powder Characteristics on EDMing Efficiency. *International Journal of Advanced Manufacturing Technology*, *17*, 586–592. doi:10.1007/s001700170142

United States Environmental Protection Agency. (1980). *Passenger Car Fuel Economy: EPA and Road*: a *Report to the Congress in Response to the National Energy Conservation Policy Act of 1978, Public Law 95-619, Title IV, Part 1, Section 404*. Author

Unune & Mali. (2017). Experimental investigation on low-frequency vibration assisted micro-WEDM of Inconel 718. *Engineering Science and Technology, an International Journal*, *20*, 222–231.

Upadhyay, A., Prakash, V., Pandey, V., & Sharma, V. (2016, November). *Effect of different types of Additives Added in Dielectric Fluid on Micro-EDM of Brass sheet*. Paper presented at International Conference on Evolutions in Manufacturing: Technologies and Business strategies for Global Competitiveness, Mesra, India.

Van Echtelt, F. E. A., Wynstra, F., Van Weele, A. J., & Duyesters, G. (2008). Managing supplier Involvement in NPD: A multiple case study. *Journal of Product Innovation Management*, *25*, 180–201. doi:10.1111/j.1540-5885.2008.00293.x

Varchola, M., & Hlbocan, P. (2012). Geometory design of a mixed flow pump using experimental results of an internal impeller flow. *Procedia Engineering*, *39*, 168–174. doi:10.1016/j.proeng.2012.07.021

Varma, R., & Advani, S. G. (1994). Three-dimensional Simulations of Filling in Resin Transfer Moulding, Advances in Finite Element Analysis in Fluid Dynamics (ASME). *FED*, *200*, 21–27.

Vedansh, C. (2015). Multi-Response Optimization of Process Parameters of Abrasive Water Jet Machining for Stainless Steel AISI 304 Using VIKOR Approach Coupled with Signal to Noise Ratio Methodology. *Journal of Advanced Manufacturing Systems, 14*(2), 107–121. doi:10.1142/S0219686715500080

Veeraraghavan, B., Kim, H., & Popov, B. (2004). Optimization of electroless Ni–Zn–P deposition process: Experimental study and mathematical modeling. *Electrochimica Acta, 49*(19), 3143–3154. doi:10.1016/j.electacta.2004.01.035

Verrey, J., Wakeman, M. D., Michaud, V., & Manson, A. E. (2006). Manufacturing Cost Comparison of Thermoplastic and Thermoset RTM for an Automotive Floor Pan. *Composites. Part A, Applied Science and Manufacturing, 37*(1), 9–22. doi:10.1016/j.compositesa.2005.05.048

Vijay, S. J., & Murugan, N. (2010). Influence of tool pin profile on the metallurgical and mechanical properties of friction stir welded Al– 10 wt.% TiB2 metal matrix composite. *Materials & Design, 31*(7), 3585–3589. doi:10.1016/j.matdes.2010.01.018

Vitry, V., & Bonin, L. (2017). Formation and characterization of multilayers borohydride and hypophosphite reduced electroless nickel deposits. *Electrochimica Acta, 243*, 7–17. doi:10.1016/j.electacta.2017.04.152

Wagner, S. M., & Hoegl, M. (2006). Involving suppliers in product development: Insights from R&D directors and project managers. *Industrial Marketing Management, 35*, 936-943.

Wang, H. P., & Lee, H. S. (1989). Fundamentals of Computer Modeling for Polymer Processing. Hanser Publishers.

Wang, T., Wang, C., Kong, F., Gou, Q., & Yang, S. (2017). Theoretical, experimental, and numerical study of special impeller used in turbine mode of centrifugal pump as turbine. *Energy*.

Wang, Y. S., Zhang, X. Y., Zeng, G. T., & Li, F. C. (2001). In situ production of Fe–VC and Fe–TiC surface composites by cast-sintering. *Composites. Part A, Applied Science and Manufacturing, 32*, 281.

Watson, J. C. & Serrano, J. C. (2011). Composite Materials for Wind Blades. *Wind Systems Magazine*, 46-51.

Webzell, S. (2002). That first step into EDM. Machinery, (4040), 41.

Weller, E.J. (1984). *Non-traditional machining processes*. SME Dearborn.

Wind Energy Technology. (2009). *50th AIAA Structures*. Structural Dynamics, & Materials Conference Palm Springs.

Wislicenus, G. F. (1947). *Fluid mechanics of turbomachinery*. New York: McGraw-Hill.

Wright, I. C., & Swain, E. (1995). New product engineering: The quality paradox. *Journal of Engineering Design, 6*(1), 49–56. doi:10.1080/09544829508907902

Wright, T., & Gerhart, P. (2009). *Fluid Machinery application, selection and design*. Boca Raton: CRC press. doi:10.1201/b15836

Wu, F. B., Tien, S. K., Chen, W. Y., & Duh, J. G. (2004). Microstructure evaluation and strengthening mechanism of Ni-P-W alloy coatings. *Surface and Coatings Technology, 177-178*, 312–316. doi:10.1016/j.surfcoat.2003.09.010

Xu, J. Z., & Gu, C. W. (1992). A numerical procedure of three-dimensional design problem in turbo machinery. *ASME Trans J Turbomach, 114*(3), 548–582. doi:10.1115/1.2929178

Xu, Y., Zheng, X., Hu, X., Yin, Y., & Lei, T. (2014). Preparation of the electroless Ni–P and Ni–Cu–P coatings on engine cylinder and their tribological behaviors under bio-oil lubricated conditions. *Surface and Coatings Technology, 258*, 790–796. doi:10.1016/j.surfcoat.2014.07.079

Yadav, S. K., & Patel, S. K. (2013). *Optimization of green electro-discharge machining using VIKOR* (Thesis). Dept of Mech Eng., NIT, Rourkela, India.

Yadav, S., Mathur, M. K., Dr, S. C., & Kasliwa, S. C. (1993). Saltiel's Design Modification: Short leg brace for knee extensor weakness. *Indian Journal of Physical Medicine and Rehabilitation.*, *6*, 24–29.

Yang, Y. B., Chang, K. C., & Yau, J. D. (2003). *Base Isolation Earthquake Engineering Handbook*. Washington, DC: CRC Press.

Yan, M.-T., Fang, G.-R., & Liu, Y.-T. (2013). An experimental study on micro wire-EDM of polycrystalline diamond using a novel pulse generator. *International Journal of Advanced Manufacturing Technology*, *66*(9-12), 1633–1640. doi:10.1007/s00170-012-4446-z

Yeo, S. H., Neo, K. G., & Tan, H. C. (1998). Assessment of health hazard in the production of printed paper packages. *International Journal of Advanced Manufacturing Technology*, *14*(5), 376–384. doi:10.1007/BF01178918

Yeo, S. H., & New, A. K. (1999). A method for green process planning in EDM. *International Journal of Advanced Manufacturing Technology*, *15*(4), 287–291. doi:10.1007/s001700050068

Yuan, J., Wang, K., Yu, T., & Fang, M. (2008). Reliable multi-objective optimization of high-speed WEDM process based on Gaussian process regression. *International Journal of Machine Tools & Manufacture*, *48*(1), 47–60. doi:10.1016/j.ijmachtools.2007.07.011

Yu, G.-C., Wu, L., Feng, L., & Yang, W. (2016). Thermal and mechanical properties of carbon fiber polymer-matrix composites with a 3D thermal conductive pathway. *Composite Structures*, *149*, 213–219. doi:10.1016/j.compstruct.2016.04.010

Yu, H. S., Luo, S. F., & Wang, Y. R. (2001). A comparative study on the crystallization behavior of electroless Ni-P and Ni-Cu-P deposits. *Surface and Coatings Technology*, *148*(2-3), 143–148. doi:10.1016/S0257-8972(01)01345-7

Zain, Z. M., Ndaliman, M. B., Khan, A. A., & Ali, M. Y. (2014). Improving micro-hardness of stainless steel through powder-mixed electrical discharge machining. *Proceedings of the Institution of Mechanical Engineers, Part C: Journal of Mechanical Engineering Science*, 1-8. doi:10.1177/0954406214530872

Zangeneh, M., Goto, A., & Takemura, T. (1996). Suppression of secondary flows in a mixed flow pump impeller by application of three-dimensional inverse design method: Part 1 – design and numerical validation. *ASME Trans J Turbomach*, *118*(3), 536–543. doi:10.1115/1.2836700

Zavadskas, E. K., & Turskis, Z. (2010). A new additive ratio assessment (ARAS) method in multicriteria decision-making. *Technological and Economic Development of Economy*, *16*(2), 159–172. doi:10.3846/tede.2010.10

Zavadskas, E. K., Turskis, Z., Antucheviciene, J., & Zakarevicius, A. (2012). Optimization of weighted aggregated sum product assessment. *Elektronika ir Elektrotechnika*, *122*(6), 3–6. doi:10.5755/j01.eee.122.6.1810

Zavadskas, E. K., Turskis, Z., & Kildiene, S. (2014). State of art surveys of overviews on MCDM/MADM methods. *Technological and Economic Development of Economy*, *20*(1), 165–179. doi:10.3846/20294913.2014.892037

Zhang, Y., Sato, Y. S., Kokawa, H., Park, S. H. C., & Hirano, S. (2008). Stir zone microstructure of commercial purity titanium friction stir welded using pcBN tool. *Materials Science and Engineering A*, *488*(1-2), 25–30. doi:10.1016/j.msea.2007.10.062

Zhao, Z., Yap, C. M., & Ang, B. W. (1999). A framework for facilitating quality tools' applications in new product development. *TQM and Innovation proceedings of the fourth international conference on ISO 9000 and TQM*. HKBU Press.

Zheng, J., Mishra, R. S., Berbon, P. B., & Mahoney, M. W. (2001). *Microstructure and mechanical behavior of friction stir processed Al-Ti-Cu alloy. Friction Stir Welding and Processing.*

Zollo, M., Reuer, J. J., & Singh, H. (2002). Interorganisational routines and performance in strategic alliances. *Organization Science*, *13*(6), 701–713. doi:10.1287/orsc.13.6.701.503

Zordan, T., Liu, T., Briseghella, B., & Zhang, Q. (2014). Improved equivalent viscous damping model for base-isolated structures with lead rubber bearings. *Engineering Structures*, *75*, 340–352. doi:10.1016/j.engstruct.2014.05.044

Zuleta, A. A., Correa, E., Castaño, J. G., Echeverría, F., Baron-Wiecheć, A., Skeldon, P., & Thompson, G. E. (2017). Study of the formation of alkaline electroless Ni-P coating on magnesium and AZ31B magnesium alloy. *Surface and Coatings Technology*, *321*, 309–320. doi:10.1016/j.surfcoat.2017.04.059

About the Contributors

Kaushik Kumar, B.Tech (Mechanical Engineering, REC (Now NIT), Warangal), MBA (Marketing, IGNOU) and Ph.D (Engineering, Jadavpur University), is presently an Associate Professor in the Department of Mechanical Engineering, Birla Institute of Technology, Mesra, Ranchi, India. He has 14 years of Teaching & Research and over 11 years of industrial experience in a manufacturing unit of Global repute. His areas of teaching and research interest are Quality Management Systems, Optimization, Non-conventional machining, CAD / CAM, Rapid Prototyping and Composites. He has 9 Patents,12 Book, 3 Edited Book 22 Book Chapters, 113 international Journal publications, 18 International and 8 National Conference publications to his credit. He is on the editorial board and review panel of 7 International and 1 National Journals of repute. He has been felicitated with many awards and honours.

J. Paulo Davim received his Ph.D. in Mechanical Engineering in 1997, M.Sc. degree in Mechanical Engineering (materials and manufacturing processes) in 1991, Licentiate degree (5 years) in Mechanical Engineering in 1986, from the University of Porto (FEUP), the Aggregate title from the University of Coimbra in 2005 and a D.Sc. from London Metropolitan University in 2013. He is Eur Ing and Senior Chartered Engineer by the Portuguese Institution of Engineers with a MBA and Specialist title in Engineering and Industrial Management. Currently, he is Professor at the Department of Mechanical Engineering of the University of Aveiro. He has more than 30 years of teaching and research experience in Manufacturing, Materials and Mechanical Engineering with special emphasis in Machining & Tribology. Recently, he has also interest in Management/Industrial Engineering and Higher Education for Sustainability/Engineering Education. He has received several scientific awards. He has worked as evaluator of projects for international research agencies as well as examiner of Ph.D. thesis for many universities. He is the Editor in Chief of several international journals, Guest Editor of journals, books Editor, book Series Editor and Scientific Advisory for many international journals and conferences. Presently, he is an Editorial Board member of 30 international journals and acts as reviewer for more than 80 prestigious Web of Science journals. In addition, he has also published as editor (and co-editor) more than 80 books and as author (and co-author) more than 10 books, 60 book chapters and 350 articles in journals and conferences (more than 200 articles in journals indexed in Web of Science/h-index 35+ and SCOPUS/h-index 44+).

***

Andrew Atherton is a Professor of Enterprise and Deputy Vice-Chancellor at Lancaster University and an expert in SME policy and China. Professor Atherton joined Lancaster University in January 2013 from Lincoln University where he was Senior Deputy Vice-Chancellor and led on major university initiatives on employer engagement and enterprise. This included establishing the first new School of Engineering in the UK for more than 20 years, in partnership with Siemens, and developing new approaches and strategies to enterprise and employability that have seen increases in partnerships with employers, improved graduate employability and higher levels of external research and related grant income. Andrew worked at Durham University from 1995 to 2002, latterly as Head of Department and before that as Director of the Policy Research Unit within the Small Business Centre, Durham University Business School. He has advised national and regional government as well as major bilateral and multilateral agencies on policy development and implementation, as well as leading major projects to establish enterprise development centres in Central and Eastern Europe and China. Professor Atherton has degrees from the School of Oriental and African Studies, University of London and Yale and publishes on enterprise, SME development and related areas of economic development and regeneration in leading international journals and policy publications. His current research interests and areas of activity include: enterprise and SME policy; innovation and entrepreneurship; local and regional development; social dynamics and aspects of entrepreneurship; and business start-up/new venture creation.

Tapan Kumar Barman received his BME, MME and PhD from Jadavpur University, Kolkata, India. He is presently working as Assistant Professor in Mechanical Engineering at Jadavpur University, Kolkata. His areas of interests are characterisation of machined surface topography and coatings tribology. He has 60 journals and 40 conference publications to his credit.

Tarakant Bhanu Currently pursing M.tech in Thermal Engineering from NIT Raipur. Obtained B.Tech in Mechanical Engg. from Orissa Engineering College, Bhubneswar in 2014.My research area includes heat transfer and fluid mechanics.

Govardhan Bhatt obtained Ph.D in the area of Seismic Base Isolation of Structures from IITROORKEE, India. He is engaged in teaching and research activities since the last 10 years.He is presently working as Assistant Professor in NIT Raipur. His area of expertise includes: Structural Dynamics, Geotechnical Engineering, Earthquake Engineering, Modeling, Vedic Studies, etc.. Published many peer reviewed national and international journal and conference papers. He is also actively engaged in ndustrial consultancy projects.

Shubhankar Bhowmick, Assistant Professor, obtained Ph.D. (Engg.) in the area of Computational Mechanics from Jadavpur University, Kolkata, in 2011. He is engaged in teaching and research activities since the last 15 years. He was awarded with CSIR (SRF) Fellowship in the year 2007. His research area includes Investigation of static and dynamic behavior of axis-symmetric structures in elastic and post-elastic regime. His present areas of interest also include fluid-structure interaction and coupled field analysis. He is also actively engaged in FEA based industrial consultancy projects.

Sumit Bhowmik is an Assistant Professor in Department of Mechanical Engineering, National Institute of Technology Silchar, Assam. He received his B. E. in Mechanical Engineering from Tripura Engineering College (Tripura University), India, Master in Mechanical Engineering and PhD from Jadavpur University, Kolkata. He has over 15 years of teaching and research experience. He has more than 20 international journals and international Conference publications to his credit. His areas of interests are Fracture Mechanics, Materials properties, Composites, and Optimization etc.

Dheeman Bhuyan, an alumnus of Birla Institute of Technology, Dheeman Bhuyan has varied experience spread over more than 2 years including engineering education, research and industry. He has published works in 4 international journals and has 5 patent applications under review to his credit. He currently holds the position of Assistant Professor in the Department of Mechanical Engineering, Girijananda Chowdhury Institute of Management and Technology, Guwahati, India.

Gurmeet Singh is currently working as Associate Professor in the Department of Mechanical Engineering at Bhai Gurdas Institute of Engineering and Technology, Sangrur, Punjab, India. He obtained his Ph.D degree in Mechanical Engg. from Sant Longowal Institute of Engineering and Technology, Longowal, Sangrur, Punjab, India. He has published twenty-five research papers in national and international conferences and journals. His research interests are in broad areas of friction stir welding and production engineering.

Santanu Duari is pursuing a PhD from Jadavpur University in the area of Tribology. He passed Masters in Engineering in the year of 2013 from the same university in the field of Mechanical Engineering (Machine Design). He has 3 years in teaching and 8 years in industrial experience.

Vivek Kumar Gaba obtained his ME (Heat Power) in 2003 from BIT Mesra (Ranchi), India and he did is PhD on "Non Linear Analysis of Variable Geometry Annular Fins of Varying Thermal Properties" in 2016 from NIT Raipur (CG), India. He worked in BIT Mesra as Sr. Lecturer. during 2003- 2009, after that he is with NIT Raipur as Assistant Professor in Mechanical Engineering Department. He has research and academic experience of about thirteen years. His research area involves Heat Transfer, Alternative fuels and Energy. Apart from academic duties, he has successfully completed a research project of Government of India. He has, to his name, research papers in International Journals and has also presented his work in International Conferences. He also supervised more than 20 M-Tech thesis.

Jagadish, is an Assistant Professor in Department of Mechanical Engineering, National Institute of Technology Silchar, Assam. He received his Diploma (Mechanical) from GPT Bidar (Karnataka); B.E. (Mechanical) from REC Bhalki (Karnataka); M.Tech (Product Design and Development) from NIT Warangal (Andhra Pradesh), Ph.D (Engineering) from NIT Silchar. He has over 3 years of industrial experience in the field of Design and Analysis and 4 years of teaching and research experience. He received Institutional Award (Gold Medal) by institution of engineers india and Best Innovative Award by springer for his outstanding research contribution and published more than 12 research papers. His areas of interests are Modeling of advanced manufacturing process, Green manufacturing, Applied soft computing techniques and optimization etc.

Pradeep Kumar Karsh is Ph.D. scholar at mechanical engineering department, National institute of technology, Silchar, Assam. He received bachelor degree in industrial and production engineering from Guru Ghasidas Central University, Bilaspur, Chhattisgarh and post graduate in industrial and production engineering from National institute of technology, Kurukshetra, Haryana. He has published 4 research papers in journals and has 4 international conferences to his credit.

Rahul Kumar is currently a Ph.D. candidate in the Department of Mechanical Engineering, National Institute of Technology, Silchar, India. Mr. Kumar graduated with a Master's degree from the National Institute of Technology, Silchar in 2014. He is a recipient of GATE scholarship from the Ministry of Human Resource Development, Government of India during his Master's degree. He holds a Bachelor's degree in mechanical engineering from Orissa Engineering College, Bhubaneswar, India. The primary focus of his research is development and characterization of plant based bio-fibres and its polymer composites for commercial applications. He has published 4 peer reviewed journal articles and 1 book chapter.

Nisha Kumari, (BE, Mechanical Engineering, Institute of Technical Education and Research, Bhubaneswar), presently pursuing M.E. (Design of Mechanical Equipment, BIT Mesra). She has over 1 year of Industrial experience. Her areas of interests are Bio-mechanics, Product and Process Design, CAD/CAM/CAE and Rapid Prototyping. She has 1 Book, 3 Book Chapter, 4 International journal publications to her credit.

Raghu Raja Pandiyan earned his Bachelor of Technology in Chemical Engineering at Bharathiar University, India (2003) and obtained his Master of Technology in Chemical Engineering at Anna University, India (2005). Then, he carried out his doctoral program at Composite Applications Laboratory, Department of Chemical Engineering, IIT Kharagpur, India, where he received his Ph. D. in composite product development. Dr. Raghu Raja Pandiyan won prestigious MARIE CURIE - AMAROUT EUROPE PROGRAMME FELLOWSHIP and joined the Materials Department of Madrid's Institute for Advanced Studies (IMDEA-Materials) as Research Associate in 2011. In IMDEA, his Post Doctoral work was on the title "Development of Integrated Composites for Applications in Aerospace Industries", Sponsored by Airbus Operations S.L., Getafe, Madrid, Spain. Dr. Raghu Raja is currently working as Assistant Professor, Department of Chemical Engineering, Birla Institute of Technology (BIT), Mesra: Ranchi and actively involved in the research and development of new innovative composite material and technology for high performance applications.

Dipti Mishra BE Mechanical Engineering, U.C.E., Burla, ME Mechanical Engineering (Heat Power Engineering), U.C.E., Burla, Ph.D. Mechanical Engineering (CFD), IIT Kharagpur.

Arkadeb Mukhopadhyay received Bachelor of Technology in Mechanical Engineering from West Bengal University of Technology and Master of Mechanical Engineering from Jadavpur University. Currently he is pursuing his doctoral work at Jadavpur University under DST PURSE II scheme of Govt of India. His area of research interest is tribology of coatings.

Jonathan D Owens PhD is an Associate Professor in Operations Management at the University of Salford, United Kingdom. His research interests are in New Product Development and the appreciation of having a structured technology road map incorporated into the process, the learning issues to be understood in the design process, in particular the NPD process, innovation and design and knowledge transfer through distance learning. He is a Chartered Engineer with over twenty five years' work experience and fifteen of those in academia. He has extensive experience of working at undergraduate, postgraduate taught and postgraduate research where the Operations Management discipline is involved. In addition, he has travelled widely and subsequently has experience of teaching in Europe, United States, Mexico and China. He currently lectures and researches in the Operations Management discipline within Salford Business School.

D K Paul PhD from Univ. of Wales, Swansea, U.K. in 1982 and PDF from Univ. of Wales, Swansea, U.K. in 1986. Distinguished Professor in the domain of Earthquake Engineering. Forty years Teaching and Research experience at IIT Roorkee; Research Interests: Application of Finite Element Method in coupled Field Problems such as Fluid-Structure Interaction and Soil-Pore fluid-Structure Interaction; Nonlinear Structural Dynamics; Earthquake Engineering, Earthquake Disaster Mitigation. Executed more than 150 consultancy projects.

Ved Prakash is currently working as Scientist In CSIR-Central Mechanical Engineering Research Institute, (CMERI) Durgapur. He has done hid B.tech from NIFFT Ranchi, M.E from BIT Mesra In Automated manufacturing System (AMS) and Pursuing Ph.D from IIT(ISM) Dhanbad(Part Time). The Author's Research area is non-Conventional machining and casting Technology. He has several SCI journals and international Conference publication in micro-EDM field.

Apurba Kumar Roy, B. E. (Mechanical Engineering, REC (Now NIT), Jaipur), M.E. (Mechanical Engineering, Jadavpur University, Kolkata) and PhD (Engineering IIT Kharagpur), is an Associate Professor at the Department of Mechanical Engineering, Birla Institute of Technology, Mesra, Ranchi, India. He has over 27 years of teaching, research and industrial experience. His areas of interests are Fluid Dynamics, Turbo machines, Multiphase Flow, CFD, Optimization and Non Conventional Energy, Direct Energy Conversion. He has 1 Book, 2 Book Chapter, 30 international Journal, 5 International and 15 National Conference publications to his credit. He is on the editorial board and review panel of 1 International Journal of repute.

Prasanta Sahoo is a Professor at the Department of Mechanical Engineering, Jadavpur University, Kolkata, India. He received Bachelor of Mechanical Engineering from Jadavpur University and MTech and PhD in Mechanical Engineering from Indian Institute of Technology, Kharagpur, India. His research area is tribology and structural mechanics. He is the supervisor of 22 PhD theses and 25 Master's theses. He has authored a textbook on Engineering Tribology (Prentice Hall of India, 2005) and book-chapters in Tribology Research Advances (Nova Science, 2009) and Tribology for Engineers – a practical guide (Woodhead Publishing, 2010). He has coauthored more than 400 technical papers in international journals and conference proceedings. He is Associate Editor (computational mechanics and tribology) of International Journal of Engineering, Science and Technology (Multicraft). He is on the editorial board of Journal of Tribology Research (Serials), Journal of Tribology and Surface Engineering (Nova Science), International Journal of Manufacturing, Materials and Mechanical Engineering (IGI Global), Materials and Design (Elsevier).

Sameer Sahu B.Tech. Mechanical Engineering, IACR Engineering College, Rayagada, ME Mechanical Engineering (Heat Power Engineering), BIT Mesra.

Vinay Sharma has obtained his master degree from Sambalpur University, Odisha and has completed his Doctoral research in "Water Jet Cutting", from Birla Institute of Technology, Mesra Ranchi. He has joined DRDO as Senior Research Fellow and successfully completed a number of projects. For his excellent work in DRDO he got DRDO award. He joined the Department of Production Engineering, Birla Institute of Technology Mesra Ranchi India in February 1999 as Lecturer. He became Reader in the same Department in 2005 and Associate Professor in 2007. In year 2011, he became Professor. He served the Institute for various administrative position such as Dean Student Welfare and Head of the Department. He has published about 50 research papers in Journals and 55 papers in proceedings of national/international seminar/ conferences and has been awarded certificate of merit for some of his research papers. He has guided more than 25 students in Master levels and 03 in Doctoral research level. 06 students are pursuing their doctoral research work under his guidance. He is a six-sigma black belt holder given by NCQM. He is the Principal Investigator of UGC Major Research project "Microwave Welding", Coordinator of Nationally Coordinated Project (NCP) of AICTE, "Deformation and Fracturing Behavior of Metallic Foams" and have obtained FIST-II project from DST. He was the coordinator of the project for "Time and Motion Study for Payment of Waged to Workers" in the State under National Rural Employment Guarantee Act 2005(NREGA), Rural Development Department Government of Jharkhand February 2008, He has organized more than 25 conferences of national and international level. He has delivered a number of professional lecturers in the field of Micromanufacturing, Precision Manufacturing and Quality Engineering in various workshops and summer school. He is in the editorial board of a number of Journals. He is reviewer of journals published by Springer, Elsevier, Emerald and Taylor Francis. He has published books entitled "Water Jet System for Mining under Indian Environment", Vdm Verlag Publishing House Germany, ISBN 978-3-8443-9506-8, "IT Based Manufacturing", Narosa Publishing House, New Delhi, ISBN: 81-7139-509-9 & "Intelligent Materials and Manufacturing", Satya Prakashan, New Delhi ISBN: 81-7684-325-3. He is member of "American Society of Mechanical Engineering (ASME)" (since 2006), "Fellow" of Institution of Engineers (India) (IEI)", Life Member of "Indian Society for Technical Education (ISTE)", Life Member of "Indian Science Congress", Life Member of Powder Metallurgy Association of India (PMAI), Life Member of Additive Manufacturing Society of India (AMSI), Member of "International Production and Operation Management Society" His field of interest is Advance manufacturing and Quality engineering. At present, he is working in the field of Microwave welding, Additive mixed µEDM, µforming, µMachining, Rapid Prototyping, waste free Manufacturing and cloud manufacturing.

Sanjay Sharma is a scholar with I K Gujral Punjab Technical University and serving as Associate Professor in Department of Mechanical Engineering, Graphic Era Hill University Dehradun. He is having 25 years of rich experience of around 20 years of teaching and 5 years in industries. He is teaching manufacturing science and engineering and Industrial Automation, which are mechanical engineering specializations. Presently working on Friction Stir processing technology for metals and polymers.

Hari Singh is currently Professor in Mechanical Engineering and Coordinator, School of Renewable Energy and Efficiency, at NIT Kurukshetra. He completed his Bachelor in Mechanical Engineering with Honours in 1987, Master in Mechanical Engg in 1994 and PhD in 2001 from Regional Engineering College, Kurukshetra (now, NIT Kurukshetra). He has 30 years of teaching and research experience in the present institute. He has published 130 research papers in various journals of repute and national/ international confernece proceedings. He has attended a number of conferences abroad in Brazil, Australia, Hongkong, Singapore, France,and Italy. He has supervised eight PhDs and an equal number of PhDs are in progress. He has also supervised 30 MTech students for their dissertations. He has two patents to his credit.

Sahib Sartaj Singh is now working as Workshop Superintendent Punjabi University Patiala, Former Principal of CGC Landran and RIMT Mandi Gobindgarh India. He obtained his bachelor's degree in Mechanical Engineering from G.N.E.C. Ludhiana, Panjab University Chandigarh, Master engineering from Thapar Institute of engg and tech Patiala, He received his PhD degree for his work on Entrepreneurship (TBIs) from N.I.T. Kurukeshtra in 2010. He has 17 years of teaching and research experience. He has over 50 publications in international journals/conferences. He has presented papers at international conferences in India, Singapore, US, Australia, Dubai,Canada and U.K. His areas of research interests include, welding, entrepreneurship, operations management, energy management and Six Sigma applications.

Anil Kumar Tiwari, Professor, Mechanical Engineering, obtained his Ph.D. (Engg.) in the area of Heat and MassTransfer in Solar Distillation from CES,IIT Delhi, India. His present areas of interest include Refrigeration and Airconditioning,Solar Thermal Applications, Heat and Mass Transfer, and Alternative Fuel in Internal Combustion Engines. He is credited with scientific publications in International journals of repute and also has co-authored a book on Solar Distillation Practice.He has been contributing as reviewer for many International Journals.

Ananya Upadhyay completed her masters in Production Engineering with specialization in Automated Manufacturing System from Birla Institute of Technology, Mesra Ranchi. She completed her thesis " Experimental Investigation and Artificial Neural Network Modelling for Blind Hole Drilling by Micro-EDM " under the supervision of Dr. Vinay Sharma, Professor, Department of Production Engineering, Birla Institute of Technology, Mesra Ranchi. She has participated and presented a project in LnT Techgium 2016 on " Detection & Stoppage of leakage using pressure sensors", at L&T Technology Services at Chennai. At present, she is working as a fellow at Teach for India, Chennai.

Deepak Verma is Assistant Professor in Department of Mechanical Engineering, Graphic Era ill University, Dehradun. He is having around 9 years of experience of Teaching(Both UG and PG)..He is having around 14 national and international papers and around 27 book chapters published by the reputed publishers such as Springer, IGI, Elsevier. He also edited two books, one with Lap Lambert and other with IGI global. Now he is in process of editing another book on biocomposites and biofuels with reputed publisher WOODHEAD (ELSEVIER).

Divya Zindani, (BE, Mechanical Engineering, Rajasthan Technical University, Kota), M.E. (Design of Mechanical Equipment, BIT Mesra), presently pursuing PhD (National Institute of Technology, Silchar). He has over 2 years of Industrial experience. His areas of interests are Optimization, Product and Process Design, CAD/CAM/CAE, Rapid prototyping and Material Selection. He has 1 Patent, 2 Books, 1 Book Chapter, 1 Journal (Web of Sc.), 5 Scopus Indexed international journal and 4 International Conference publications to his credit.

Index

www.ingramcontent.com/pod-product-compliance
Lightning Source LLC
LaVergne TN
LVHW062135300926
849161LV00010B/2540